Into the Arctic Ice

The Largest Polar Expedition of All Time

Esther Horvath

Sebastian Grote
Katharina Weiss-Tuider

With a foreword
by Markus Rex

With additional images by
UFA Show & Factual

PRESTEL
Munich · London · New York

Into the Arctic Ice

The Largest Polar Expedition of All Time

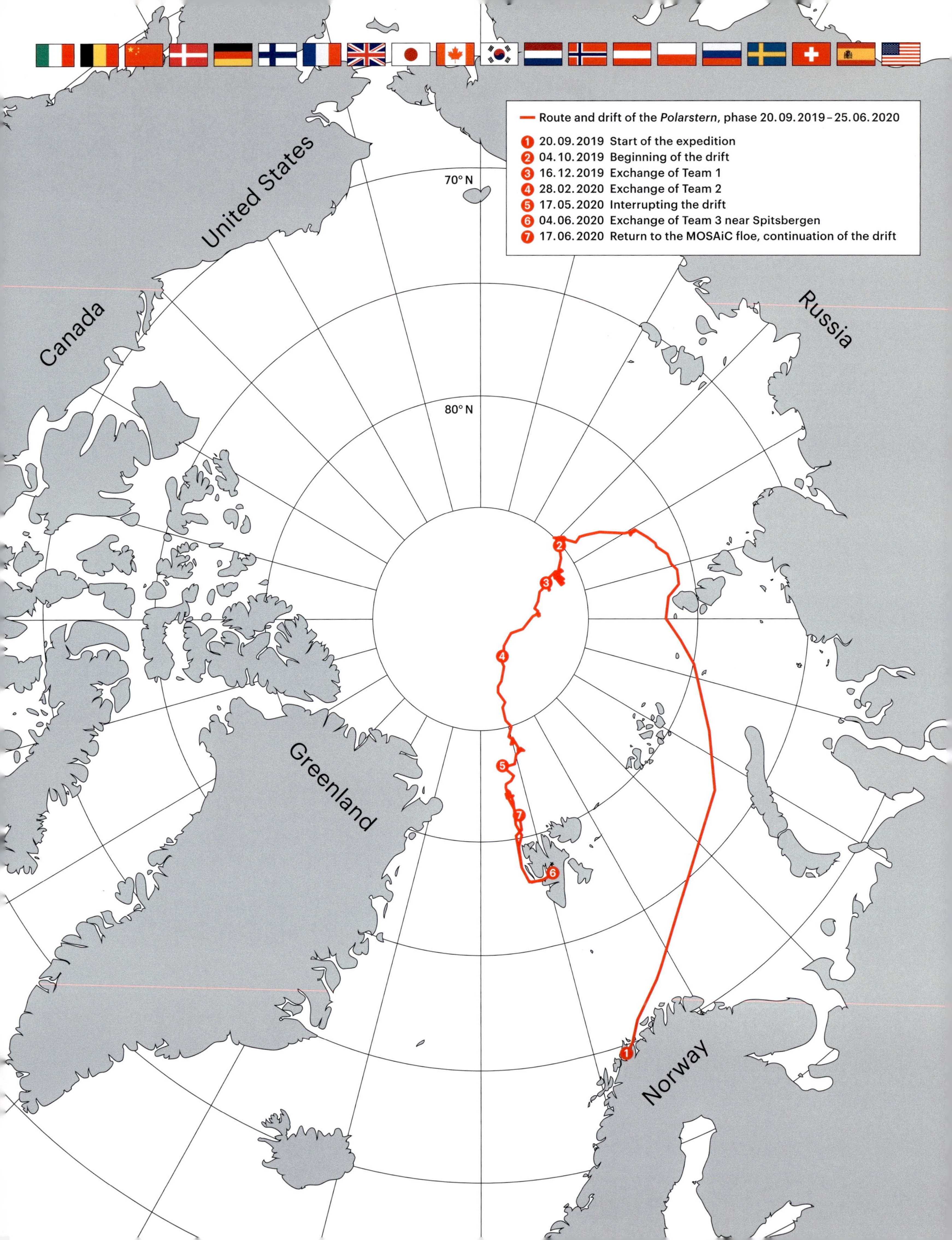

Route and drift of the *Polarstern*, phase 20. 09. 2019 – 25. 06. 2020
1 20. 09. 2019 Start of the expedition
2 04. 10. 2019 Beginning of the drift
3 16. 12. 2019 Exchange of Team 1
4 28. 02. 2020 Exchange of Team 2
5 17. 05. 2020 Interrupting the drift
6 04. 06. 2020 Exchange of Team 3 near Spitsbergen
7 17. 06. 2020 Return to the MOSAiC floe, continuation of the drift
70° N
80° N
United States
Canada
Russia
Greenland
Norway

Contents

T

During the MOSAiC expedition, scientists from twenty countries are investigating the Arctic over the course of a year. From autumn 2019 until autumn 2020 the German icebreaker *Polarstern*, frozen into the ice, is drifting through the northern polar seas. MOSAiC is being carried out under the direction of the Alfred Wegener Institute, Helmholtz Centre for Polar and Marine Research (AWI). To ensure the success of this unique project and to collect as much valuable data as possible, more than eighty institutions are working together in a research consortium. The expedition has a budget of more than 140 million euros.

This book of photographs documents the MOSAiC expedition, the largest Arctic expedition ever undertaken. Following in the footsteps of the Norwegian explorer Fridtjof Nansen, a total of about 500 expedition members set out in its various phases to understand the climate system of the central Arctic. They have set new milestones in polar research: no ship has ever ventured so far north into the central Arctic in winter as the *Polarstern* during the MOSAiC expedition. Even for the best research icebreaker this is not possible under the ship's own power: the ice is too thick in winter. For this reason the expedition locked itself into the ice in late summer on the Siberian side of the Arctic and allowed itself to be carried by the natural drift of the ice, almost across the North Pole to the sea ice edge in the Atlantic – completely at the mercy of wind and currents and the forces of nature.

While storms raged through the polar night, the tremendous pressure of the ice piled up floes into metres-high mounds around the ship, and while the world on the ice shrank to a small bubble of light from headlamps in the absolute blackness of polar night, the expedition team had unique experiences. The appearance of the first light after a months-long winter night, at first hesitant and fleeting on the horizon, and finally the first sunrise, the summer thaw, azure melt pools on the ice and the sun circling permanently around the horizon – these were indeed impressive moments. And then there were the numerous visits from polar bears, beautiful, awe-inspiring animals that roam nimbly across their icy habitat, making us humans appear small.

But there was also normal life on board, everyday things like baking bread rolls, cutting hair, constantly putting on and taking off expedition equipment in order to go out into the icy cold and the storms on the ice. Esther Horvath captured all of this, uniquely documenting everything for us with her fascinating photographs. To browse through this book is to take a trip in time through the various phases of the expedition. As I look at the pictures I too am transported back to this remarkable landscape and can once again feel the atmosphere and moods of the expedition.

But why did we carry out this expedition? Why, in an unprecedented logistical achievement, did we deploy a total of five icebreakers, helicopters and aircraft to supply the main research vessel, the *Polarstern*, and to make possible this once-in-a-century expedition? Why did hundreds of expedition members take great hardships to operate their instruments under the most forbidding conditions on the ice of the central Arctic?

The Arctic is the epicentre of climate change. No other part of our planet is warming as quickly as the Arctic. Here, the rate of the warming is at least twice as fast as on the rest of the planet – and this is even more pronounced in the winter months. I have been travelling to the Arctic since the early 1990s, and it is already a different world to the one that I knew back then. When I arrived at our research station on Svalbard in winter in the 1990s, there was only snow and ice. It was a hard-frozen landscape of glittering white snow crystals and deep-blue blocks of ice. The station lies on the shore of a fjord, but it was frozen in winter and barely discernible. I crossed it countless times on skis and snowmobiles.

When I come to the station today, in the middle of winter, liquid water splashes at my feet. The fjord has rarely frozen over for more than a decade, and waves play merrily in the wind where once there was solid ice. We travel by boat to places where I used to make trips on skis. The station data reveal warming in the winter months of approximately 3°C (5.4°F) per decade – more than 6°C (10.8°F) since the mid-1990s. Nowhere is climate change more apparent. Here there is no need for highly precise measuring instruments or

> "We need solid scientific foundations in order to be able to accurately shape the upcoming political decisions on climate protection based on solid evidence.... Only in this way can our societies take decisions based on well-founded knowledge."
>
> **Markus Rex**

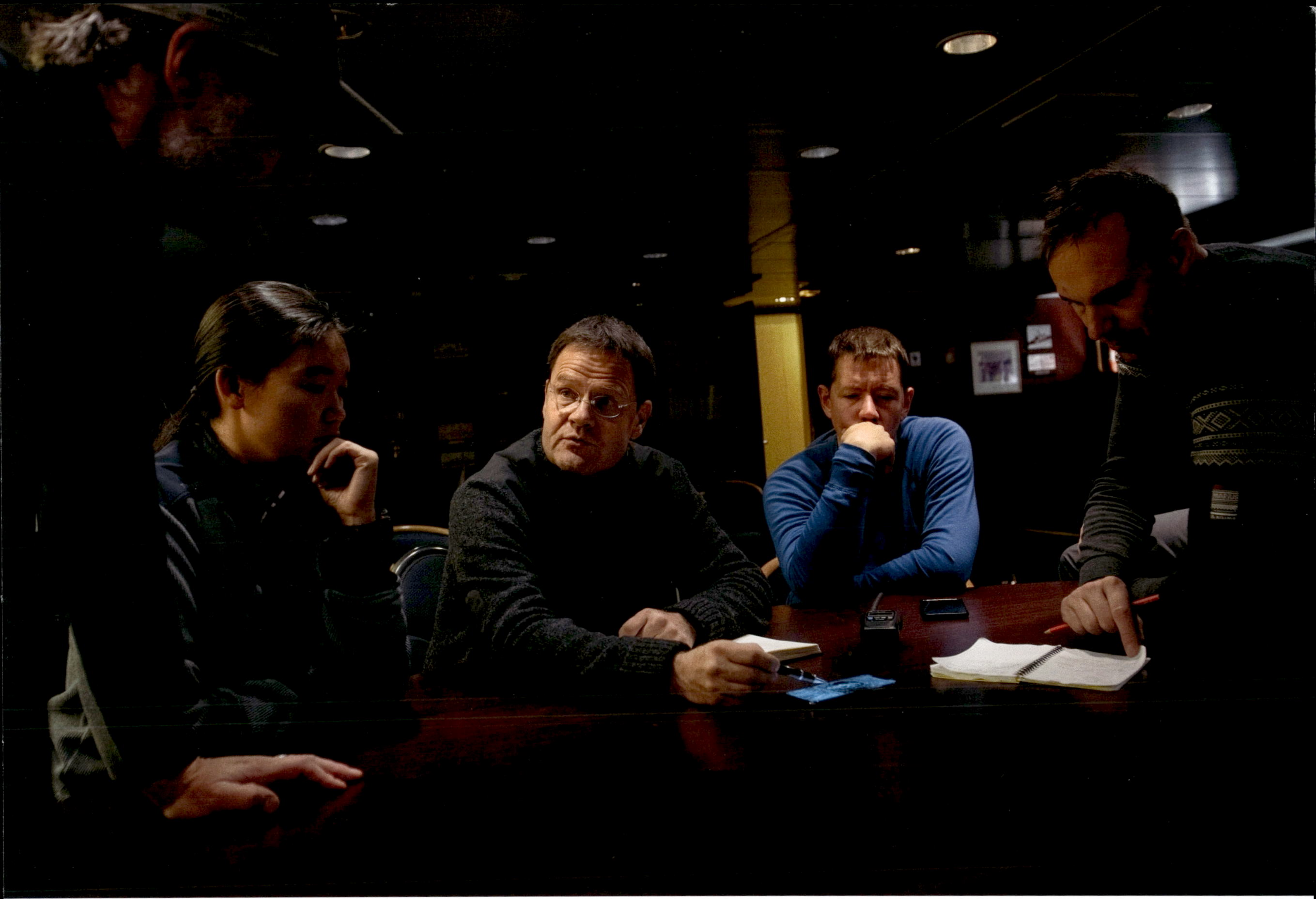

sophisticated statistics to demonstrate climate change: you need only open your eyes. But to understand this change and to be able to predict the future, we need precise observations of dozens of highly complex climatic processes in the central Arctic.

Unfortunately, the Arctic is the region of our planet whose climate system we at present least understand. For the same CO_2 emission scenario, the predictions of different climate models for the warming of the Arctic diverge by a factor of three. For a pessimistic emission scenario, the forecasts for the warming of the Arctic vary between 5 and 15°C (9 and 27°F). And the conditions in the Arctic heavily affect the weather in North America, Europe and northern Asia. The temperature contrast between the cold Arctic and the warmer mid-latitudes drives the main wind system of the Northern Hemisphere, the polar jet stream. As a result of the stronger warming of the Arctic, this wind system is changing, with direct and far-reaching consequences for the weather and climate in Europe, North America and Asia.

We need solid scientific foundations in order to be able to accurately shape the upcoming political decisions on climate protection based on solid evidence. We need reliable climate models so that we can inform people of the respective consequences of the different pathways for climate protection measures that are under discussion. Only in this way can our societies take decisions based on well-founded knowledge. To this end we must better understand the climate processes of the central Arctic and better represent them in our climate models. This is why we set out on our unique, year-long mission to the central Arctic. Thanks to the photographs in this book, you can come with us.

Markus Rex
Expedition leader, MOSAiC
(Multidisciplinary drifting Observatory
for the Study of Arctic Climate)

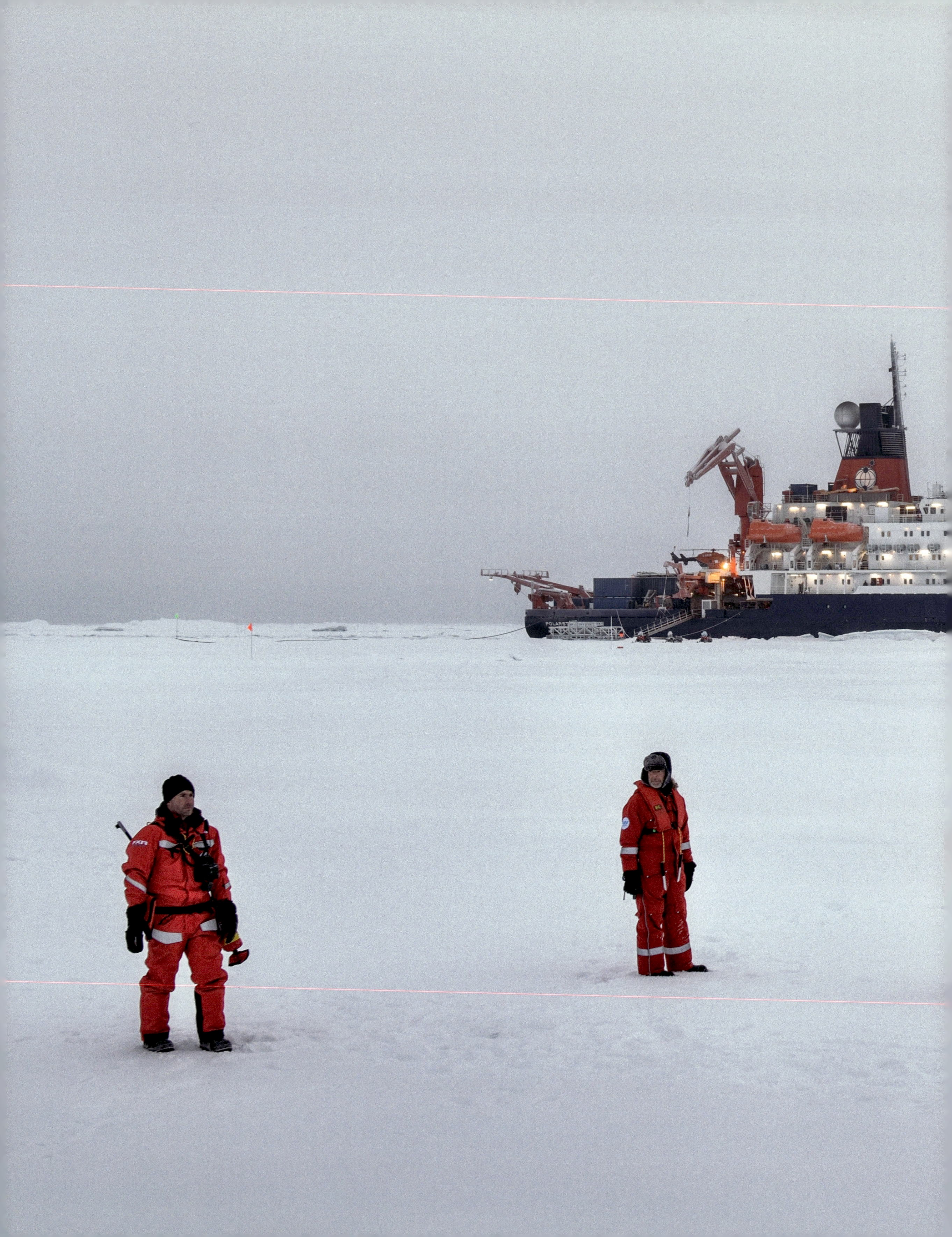

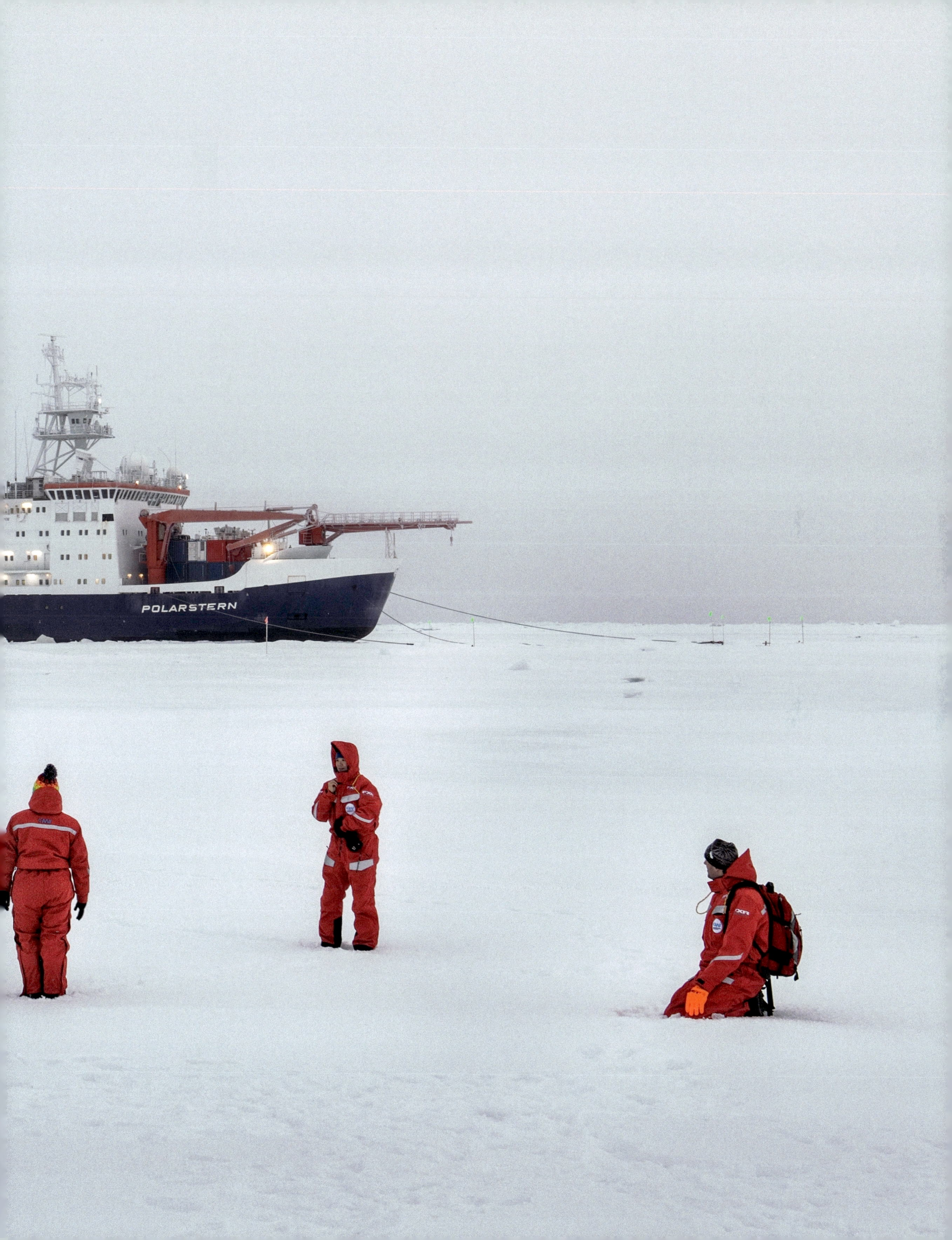
POLARSTERN

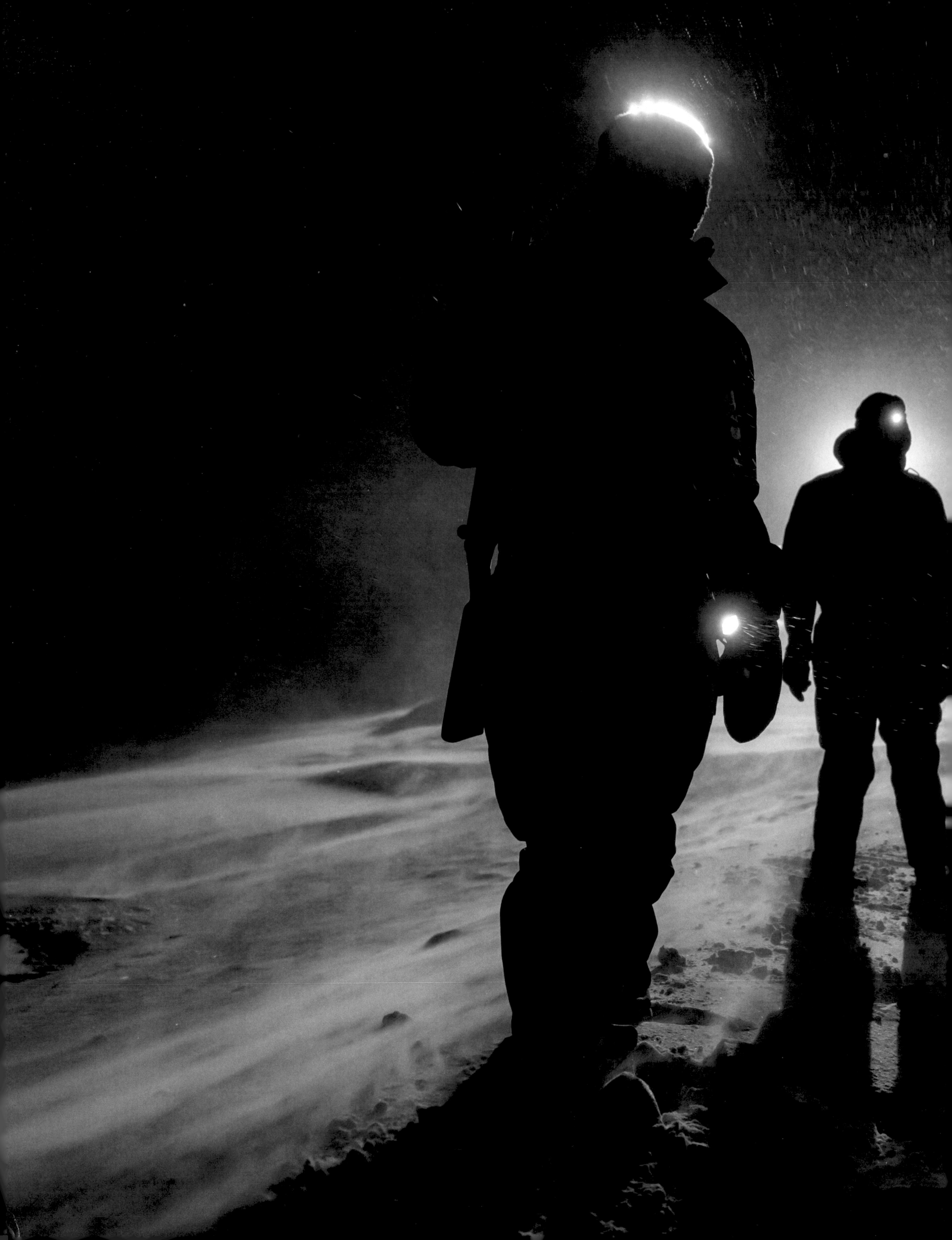

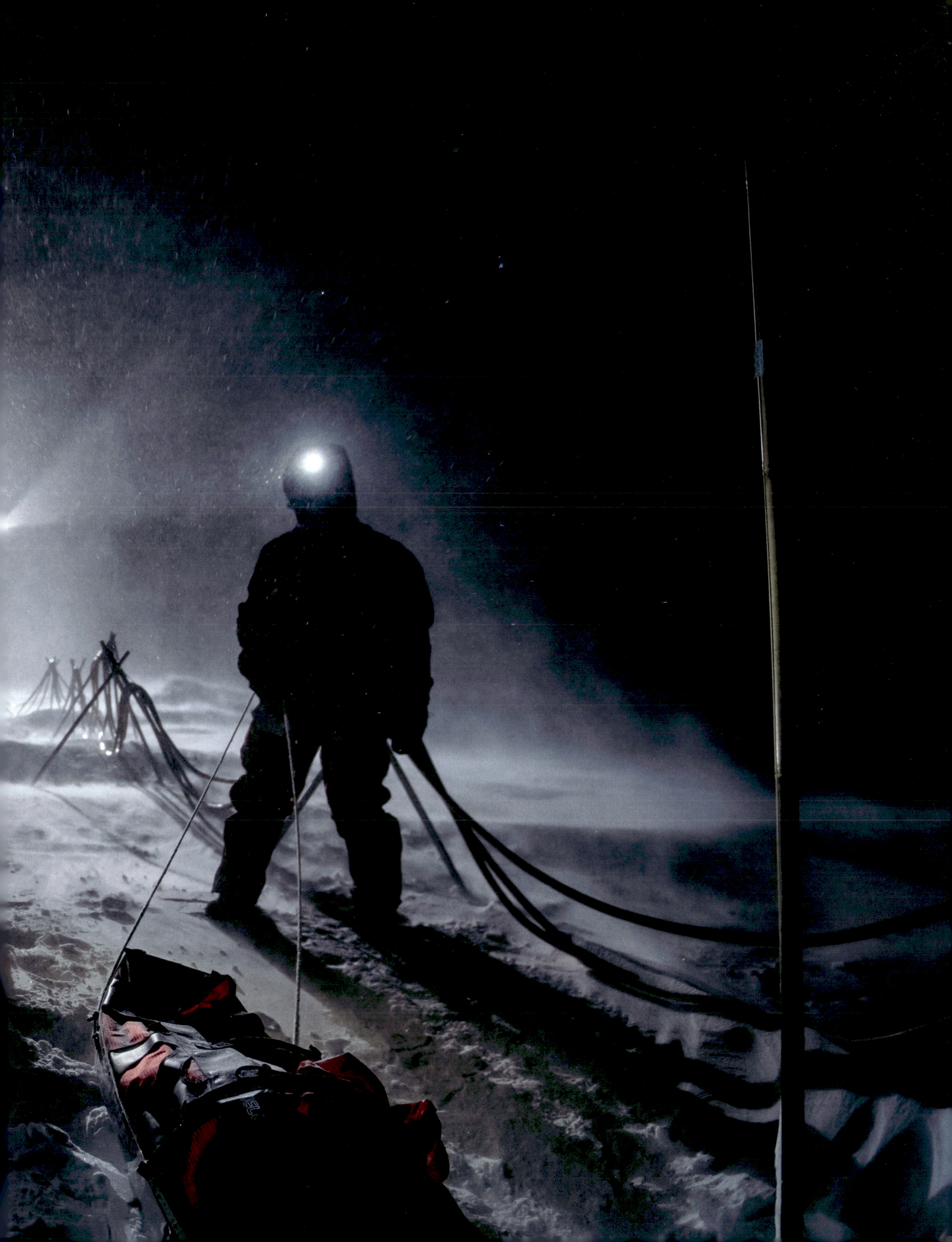

Measuring a vanishing world

Hardly anyone has ever made it this far: hundreds of kilometres from the world's northernmost coastline, in the middle of the Arctic Ocean, surrounded by pack ice, cold and darkness. Storms rage on the surface, keeping the ice in motion. Creaking and groaning, it pushes up to form metres-high towers. During the polar night, only the moonlight provides an idea of the bizarre silhouettes of the ice formations.

It is as if the wind and sea want to reconceive this unique landscape each and every day. At any moment a wide crack could open in the sea ice and reveal the ocean beneath. For a human being, this ocean with its thousands of metres of darkness is a mortal danger. Yet for the life of our planet it plays a crucial role, as does its icy cover – and the entirety of the supposedly distant Arctic.

The Arctic is the epicentre of climate change. Hardly any region on earth has warmed up so quickly in recent decades. In contrast to the temperate latitudes, here the changes are already clearly visible. And the dramatic transformation is progressing faster than expected. It has long been clear that we must say farewell to the Arctic as we know it today. The ice will retreat further in increasingly warmer summers. Massive ice floes that were once metres thick will grow thinner, and those that last for several years will become ever less frequent. The unique setting of the Arctic will disintegrate – with consequences far beyond the seas around the North Pole.

No one can yet say what the loss of the "old" Arctic means for our future. There are no research stations at the North Pole and thus hardly any measurement data – especially not for the polar night, when the central Arctic is shrouded in darkness for the entire winter half-year. The region around the North Pole is largely a blind spot in climate research. It would be an understatement to say that this is the only missing piece of the puzzle. In fact, the framework of the entire region remains unknown. But without understanding the Arctic, it is impossible to understand climate change.

For a decade, scientists have therefore made plans to close this gap in our knowledge. The result is a logistical choreography that has never taken place before in the central Artic. It envisages the German research icebreaker *Polarstern* drifting for a year through the northern polar seas – frozen into the ice – with one hundred expedition members from more than twenty countries on board for each of five legs. They are not the first to venture into the harsh, remote far north to carry out measurements, but, if all goes according to plan, they will be the first to come so close to the North Pole on a modern research icebreaker in the depths of winter; the first to spend the winter in a region almost inaccessible in the darkness and cold of the polar night, on a route determined solely by the natural force of the ice drift.

For participants from all over the world, this is the expedition of their lives. They have prepared for it with great care, learning how to survive at −40 °C (−40 °F) and to work in the dark, how to handle the sensitive measuring instruments and to recognize dangers on the ice before they pose a threat. They know that the opportunity for such a mission does not come a second time.

On 20 September 2019 the MOSAiC expedition of the *Polarstern* finally begins. The first hundred participants set out to sea from Tromsø, Norway, accompanied in the first weeks by additional team members on a second research ship, the Russian vessel *Akademik Fedorov*. Behind them is one of the warmest Artic summers on record, while a months-long journey into the unknown lies ahead. After days on the open sea they break into the ice, and a race against time begins. They set up their research camp on an ice floe and link it to a widespread network of measuring stations. This piece of ice is then simultaneously their home, place of work and research subject. Day after day the sun begins to retreat, until it disappears completely behind the horizon. The polar night begins and covers the ice for 150 days. In the midst of this barren landscape now shrouded in darkness, the *Polarstern* offers protection from icy storms, life-threatening temperatures and hungry polar bears.

The ship is the researchers' observatory, an eye in this vastly unknown region. Here they measure the atmosphere and the ocean with its living creatures. They investigate the ice and the snow that

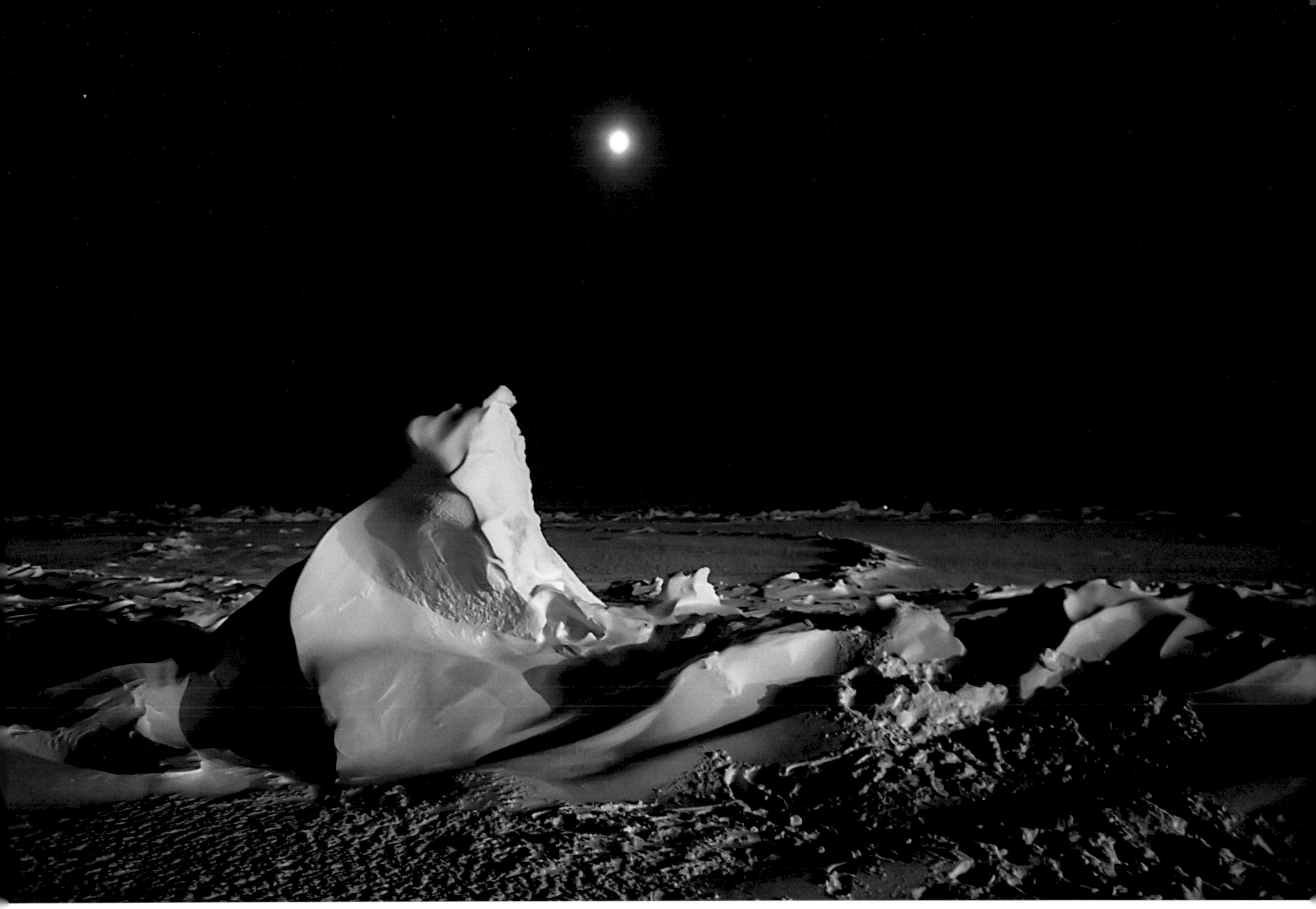

lies upon it. They are concerned with contexts both large and small. Their questions are closely linked: Exactly how is sea ice formed? What happens when the ice layer breaks open and relatively warm ocean water comes into contact with extremely cold air? What happens to the ecosystem under the extreme conditions of the polar night before explosive new life emerges under the ice in spring with the increasing sunlight? And of course the scientists on this expedition want to find answers to one of the most important questions of our day: To what extent is the Arctic affected by and, at the same time, a driver of climate change?

Their work will be a milestone in climate research. Yet the mission also entails unprecedented challenges: scientific, logistical but also interpersonal. People, machines and the sensitive technology of measuring instruments have to operate in the extreme conditions of the Arctic. The calculations must be right and the numbers must square out, because if food, fuel or spare parts run short, for example, the mission can be at risk. And in spite of the most modern technology, success ultimately depends on team members – however diverse they may be in profession, place of origin and age – being able rely on each other during this extreme experience.

Today we still have an idea of what the polar regions of our planet looked like around a hundred years ago. Explorers such as Roald Amundsen, Robert Scott and Fridtjof Nansen recorded their expeditions and races to the poles with their cameras: grainy grey-white and sepia images, taken in what was thought to be the eternal ice of days past. The images recorded during the *Polarstern* MOSAiC expedition are high-definition. They show in great detail how humans set out to record a disappearing world. And they are an invitation to take part – in the largest Arctic expedition of our time.

Esther Horvath, Sebastian Grote and Katharina Weiss-Tuider, expedition participants. Central Arctic, October 2019

Baptism of Fire

Ready to work at the limit

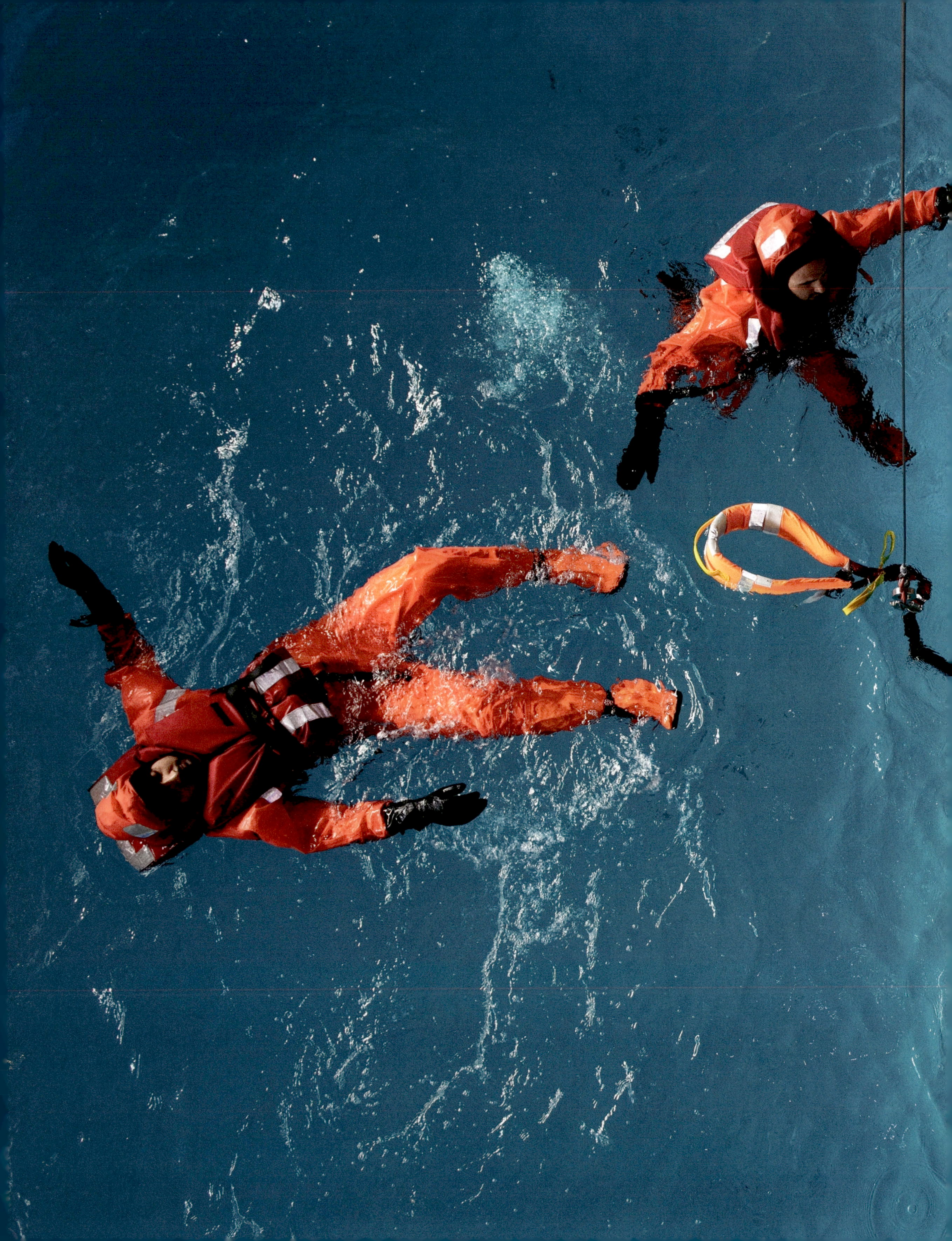

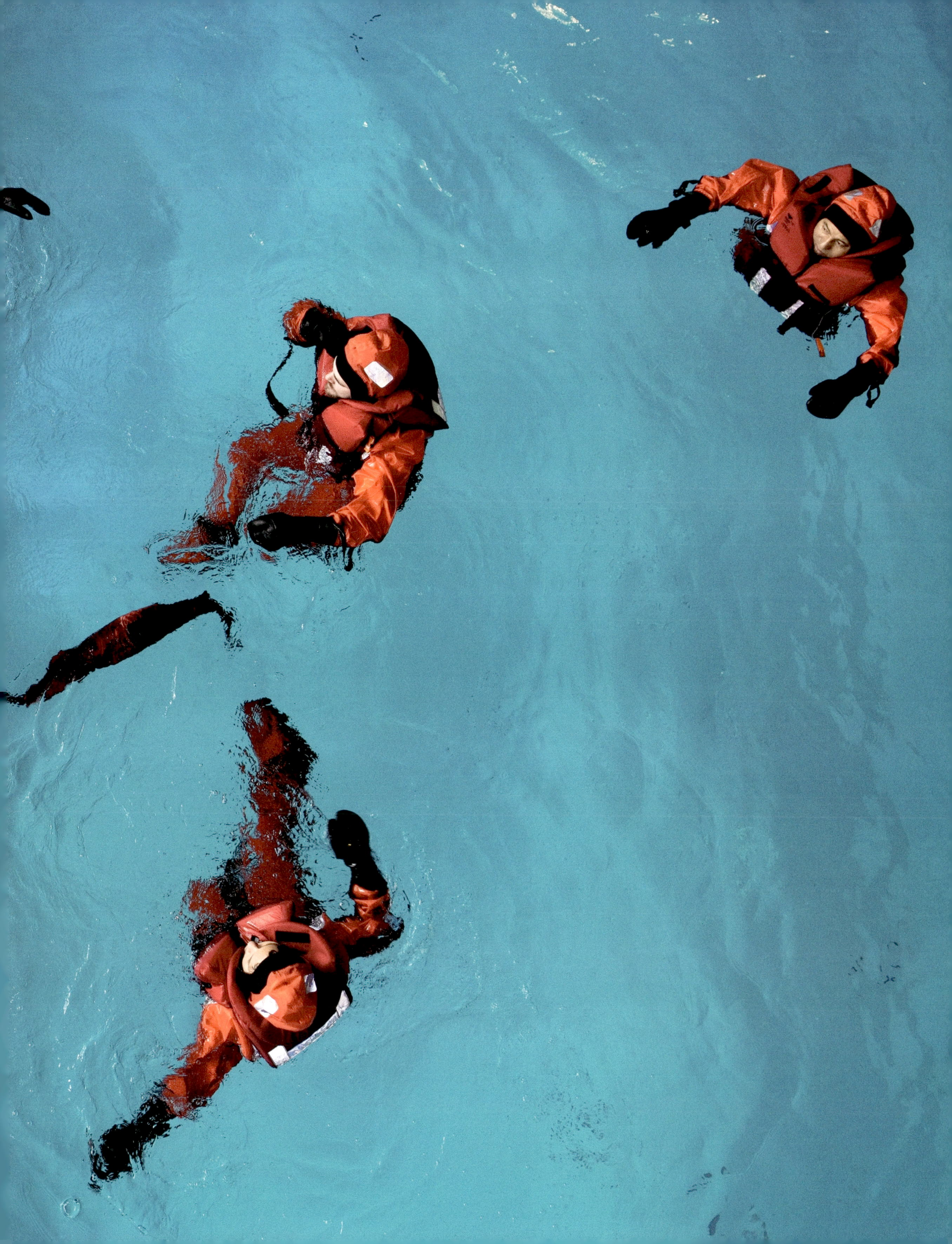

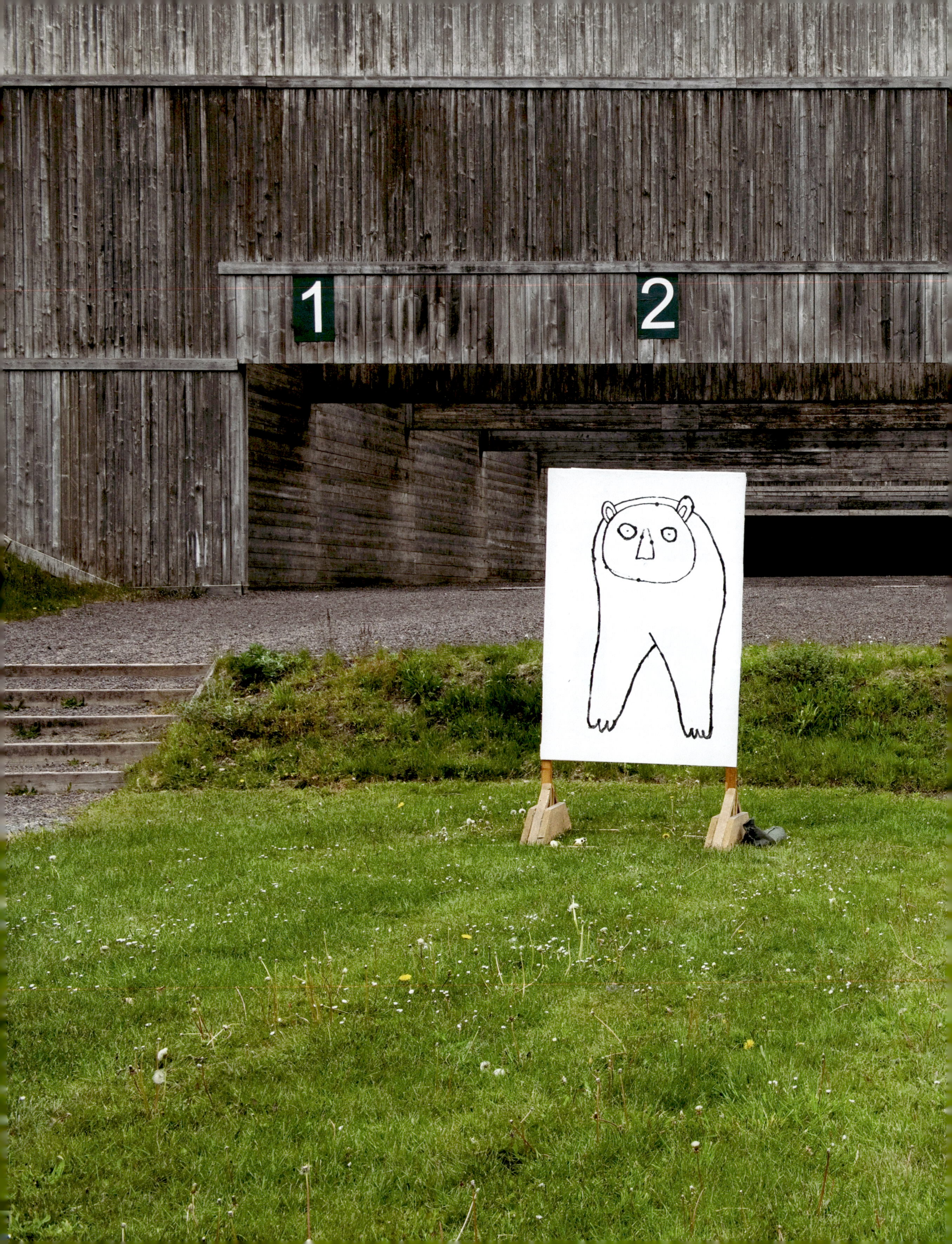
1
2

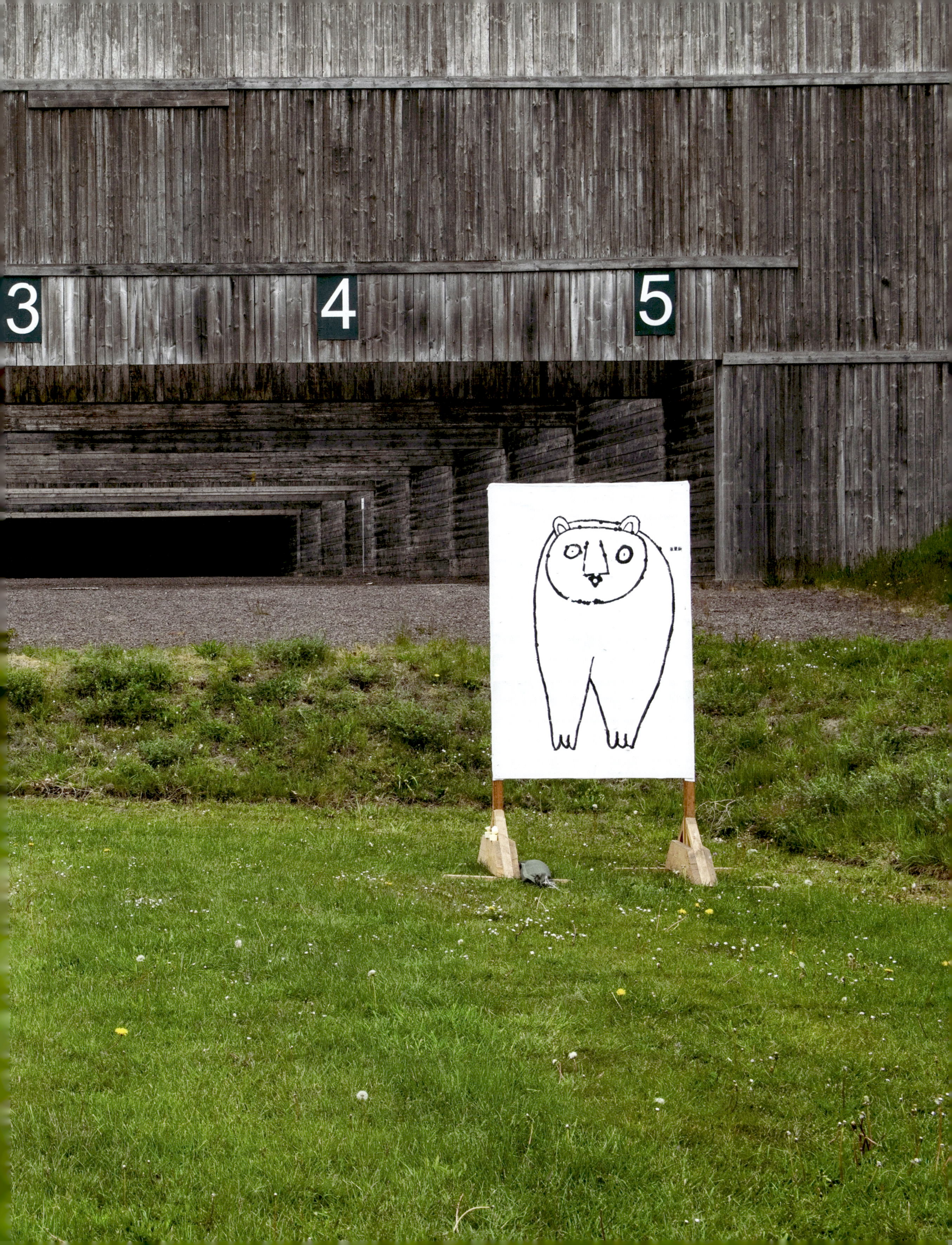
3
4
5

In the summer before the start of the expedition, the *Polarstern* lies in the Lloyd shipyard at Bremerhaven. The dimensions of the icebreaker can be appreciated more clearly from the bottom of the dry dock than from anywhere else. The research ship is 118 metres long and has an enormous draught of 11.2 metres. She is the symbol of German polar research. This icebreaker has chalked up more than 120 expeditions to the Arctic and Antarctic, sailing 1.7 million nautical miles. However, no expedition is comparable to what now lies ahead for the *Polarstern*. The hull that the workers are preparing today will shortly be surrounded by ice for a period of months.

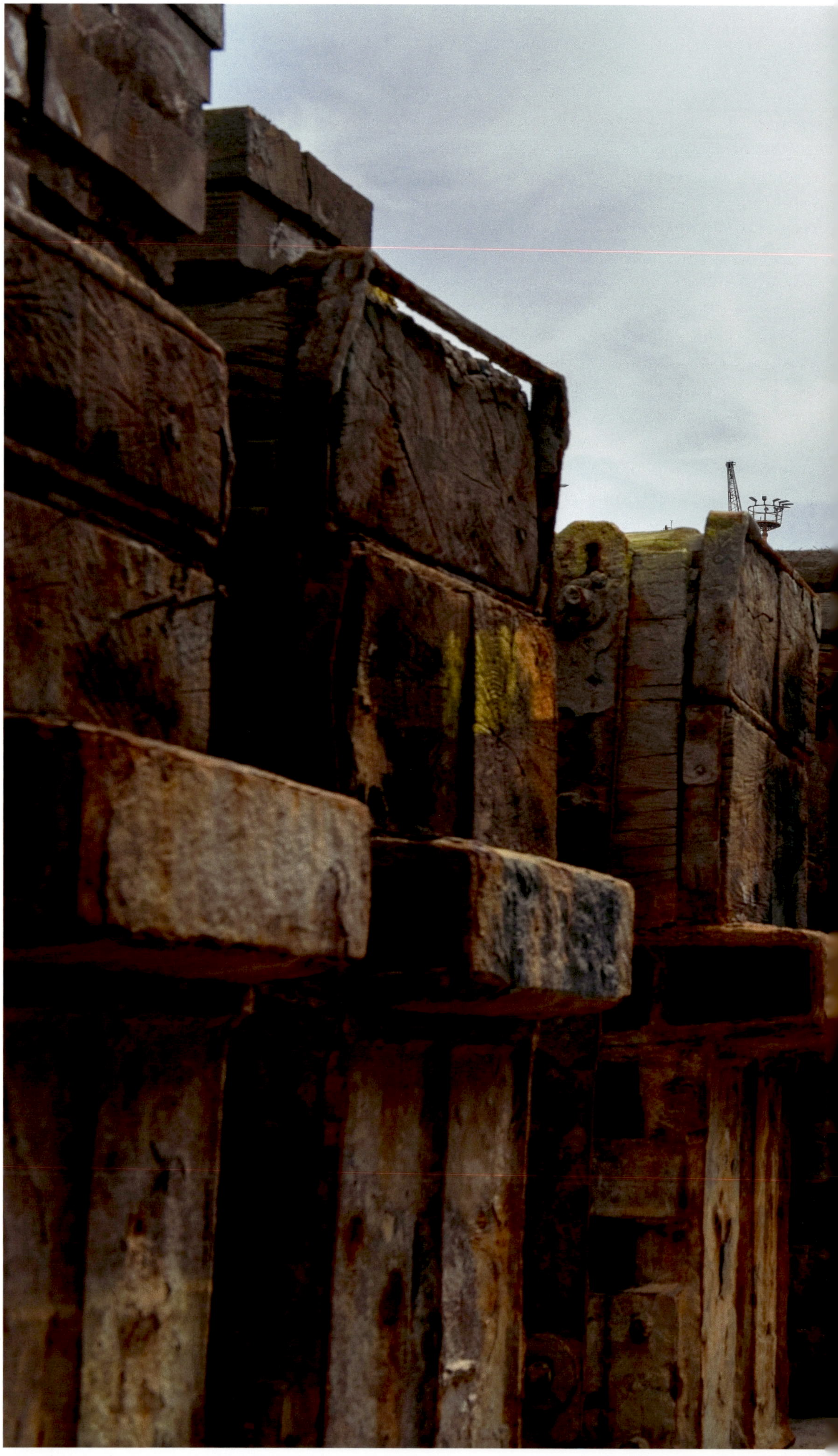

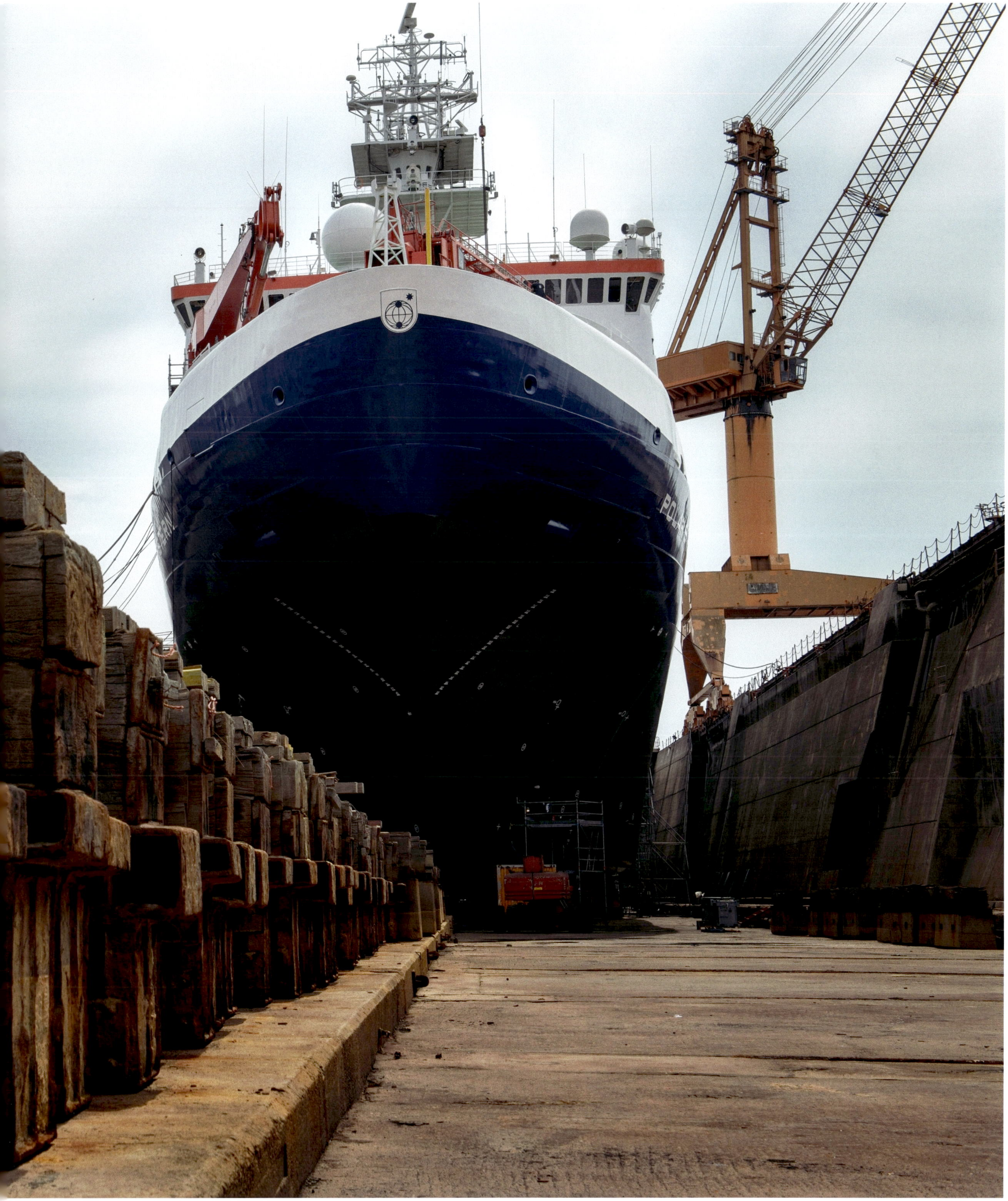

I

It is a bitterly cold March day in Ny-Ålesund, Norway, one of the northernmost villages in the world, as a few people float motionless in the icy water of the fjord. They remain on the surface, as if borne by invisible hands. A few minutes pass before they laboriously paddle with their arms back to the coast and pull themselves ashore.

They are now convinced that they have found the right clothing for their mission: a buoyant survival suit that provides the necessary protection but still allows them to work with relative ease. They will need the best equipment for what they plan to do. The international research settlement at Ny-Ålesund is situated on the Arctic archipelago of Svalbard and is thus an ideal testing ground for the MOSAiC expedition. Yet they know that where they're going the temperatures will be even lower and the winds much stronger.

Back on land, more exercises await. The programme includes getting around on snowshoes and skis. They also set up a large tent to house a tethered balloon for atmospheric research. Now they are taking their time, time that later on in the polar night they will not have. They will practise putting up the tent until it becomes so routine that they can do it with their eyes closed.

A few weeks earlier, in northern Finland: the temperatures have fallen below −20 °C (−4 °F). This is a popular tourist region, but the start of the season is still months away. The Arctic Lighthouse Hotel on the beach of the island of Hailuoto is opening especially for this group of about sixty scientists. Every morning after breakfast they leave the beach popular with tourists and go to the nearby sea ice. As if in extended circuit training, the participants pass through various stations: driving a snowmobile, measuring the thickness of ice, sawing holes in the ice and taking out ice cores. One working group has also developed a special navigation system which they are now testing under real conditions for the first time. Because a floe moves during the ice drift, a normal GPS device is useless there: every measuring point has moved somewhere else after only a few minutes. This is why a few sea ice physicists have developed the so-called Iceflow navigator. The tablet app displays a coordinate system that moves with the ice floe. The scientists can therefore always see where they took a particular sample, where dangerous regions lie, or how to find their way back to the *Polarstern* in dense fog.

The working conditions on the MOSAiC expedition will be extremely demanding. For most of the participants, working on an ice floe in the depths of winter close to the North Pole will be a new challenge. They have therefore all undergone training for several weeks in places north of the Arctic Circle under conditions that are at least partially comparable. They have gained familiarity with the contents of emergency packs and practised handling their measuring instruments in the cold and dark. They have rowed to the point of exhaustion wearing life jackets in a swimming pool, and have extinguished fires while blindfolded. Because their expedition is taking them to the realm of polar bears, they have learned how to act if they encounter these animals. Ultimately, this could mean shooting one with a rifle while under great stress. Finally, a small group of captains and officers has trained for a situation of absolute emergency: How can the team survive if they have to evacuate the ship in the middle of the Arctic sea ice – more than a thousand kilometres from the nearest lifeboat station?

While they persevere in the cold and wait to be rescued, shipyard workers in Bremerhaven are preparing the *Polarstern* for its year-long mission. In the dry dock they are welding, hammering and painting. This time the maintenance work goes beyond what is typically required. As a protective measure for the approaching Arctic winter, they are installing, among other things, a new tank

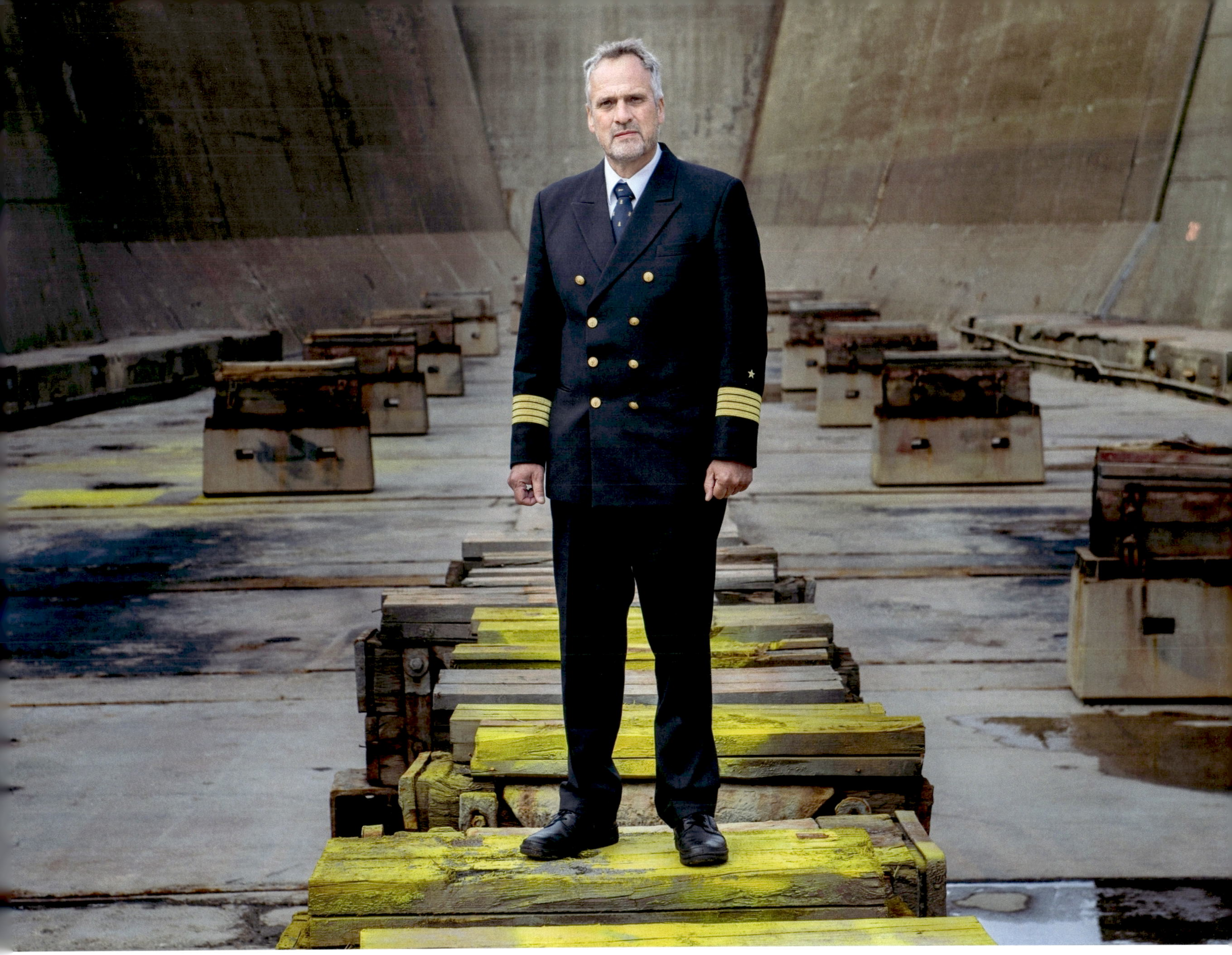

"The *Polarstern* is the heart of MOSAiC – an anchor in heavy seas, as it were. If anything goes wrong, the *Polarstern* will still be there. I have to ensure that. That is how I see my task." Captain Stefan Schwarze

heater. They are mounting an additional crane in the bows for placing heavy measuring equipment on the floe. The ship is also being fitted with a new winch for the so-called moon pool, a shaft that establishes a direct connection to the water in the ice. The port storage of the Alfred Wegener Institute is only a few hundred metres away. Here the shelves are filling up. The expedition will take a total of 500 tonnes of cargo – from heavy machinery such as snowcats to a great number of screws. Anything that they fail to load now might be sorely missed later when setting up the research camp, as the nearest hardware store is inaccessibly far away.

In the fjord of Ny-Ålesund, on the shooting range and in simulated emergency situations, expedition members have not only learned a great deal about the work and the dangers on the ice, for many of them the expedition has truly already started. Time will tell whether these intense preparations will prove their worth in the relentless reality of the Arctic. A plan means almost nothing in the polar region, yet continuous planning means everything. Above all, they have learned to bring along a great deal of flexibility and creativity, because one thing is certain: unforeseen things will undoubtedly occur on expeditions of this kind, and especially on this one.

Putting out fires blindfolded and shooting live ammunition in the dark – the preparations are a stress test, an examination for the members of the expedition. Many of them have never before held a fire hose or a loaded rifle in their hands. But this is the only way to prepare them for a real emergency.

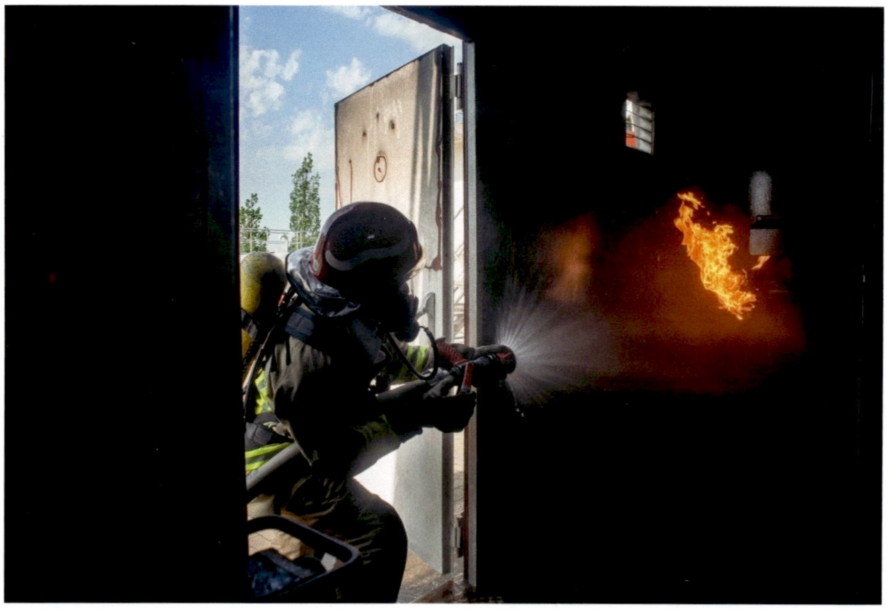

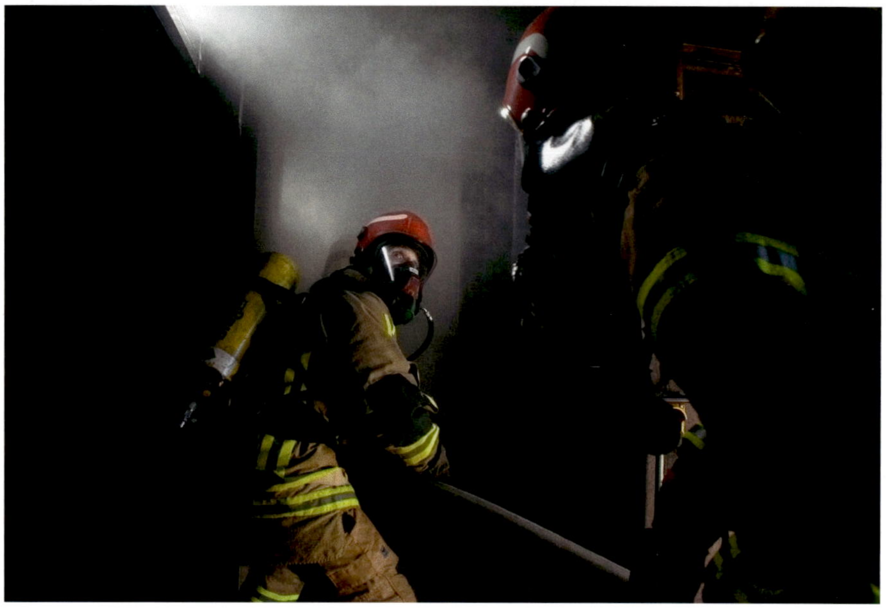

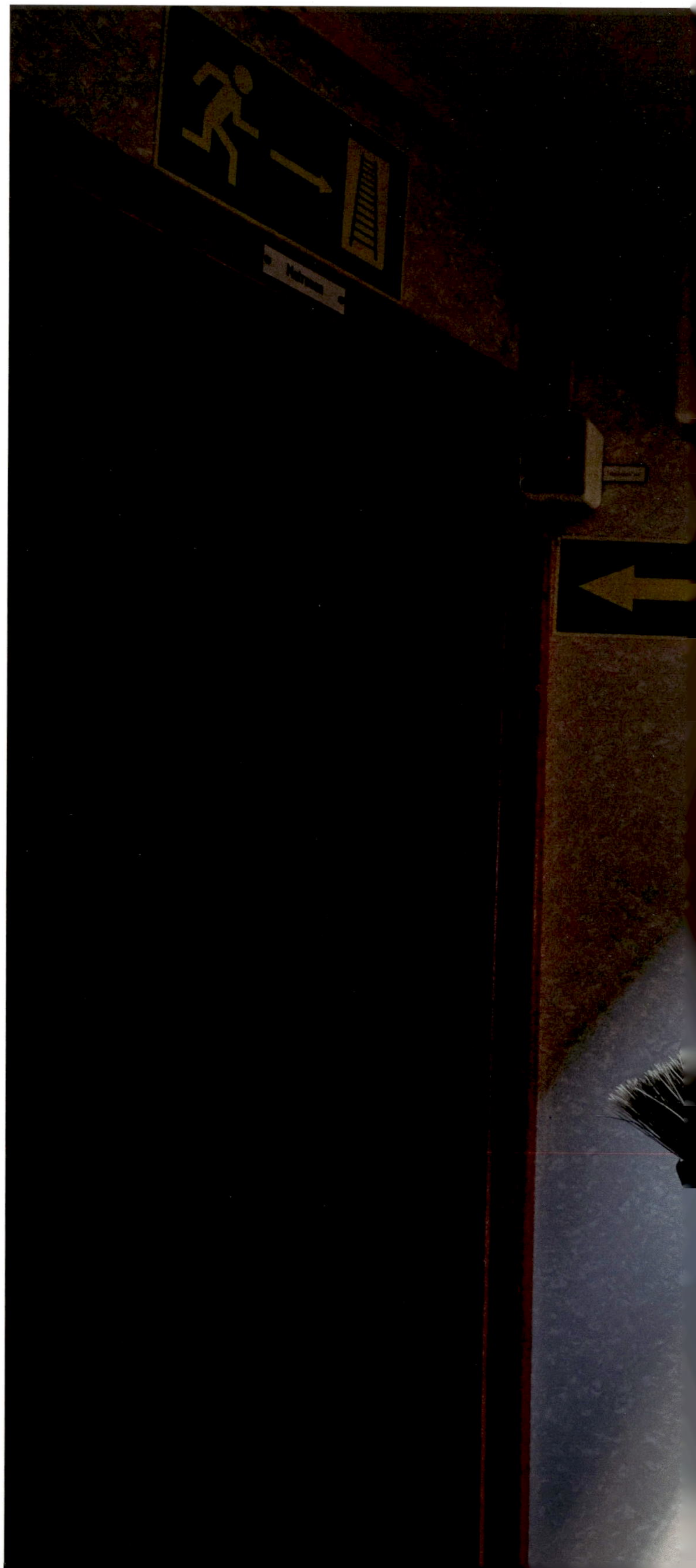

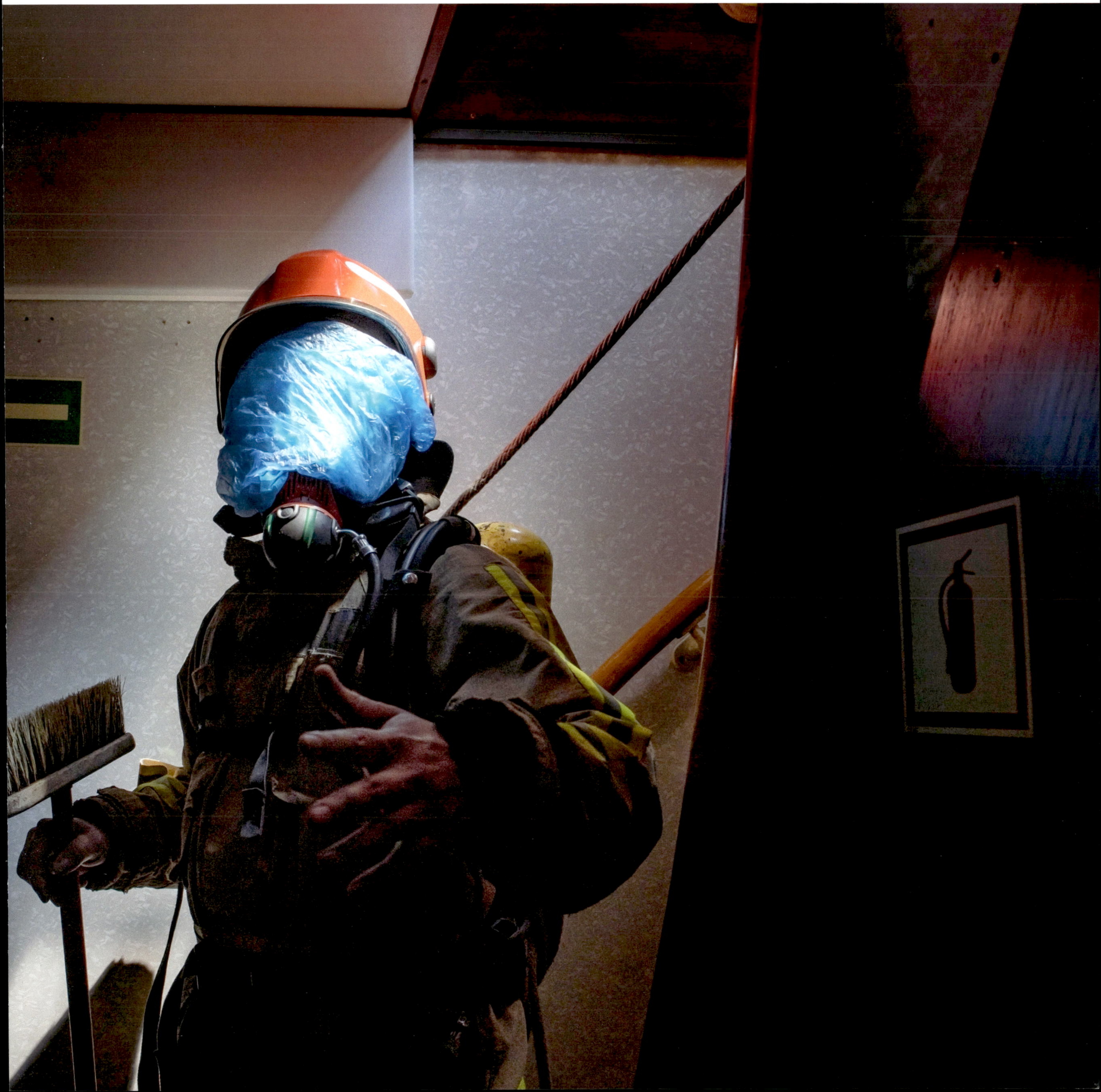

The expedition is entering the realm of the polar bears. Although contact with these predators is to be avoided as far as possible by means of a multilevel security strategy, on the ice floe someone in every group must always carry a loaded firearm. At a shooting range, expedition members practice handling a rifle under stress and in the dark. But above all they learn how to behave so that the weapon never has to be used.

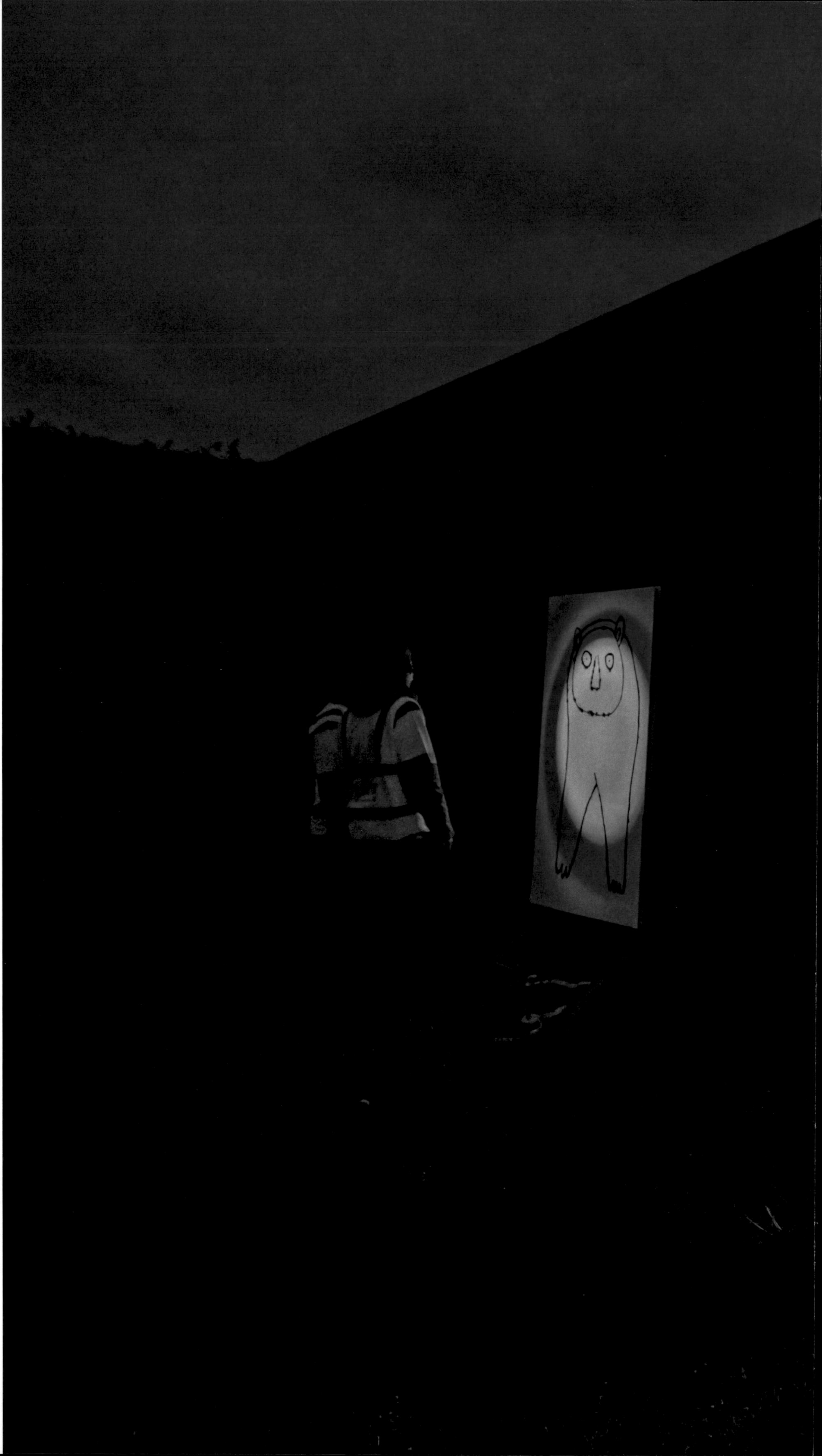

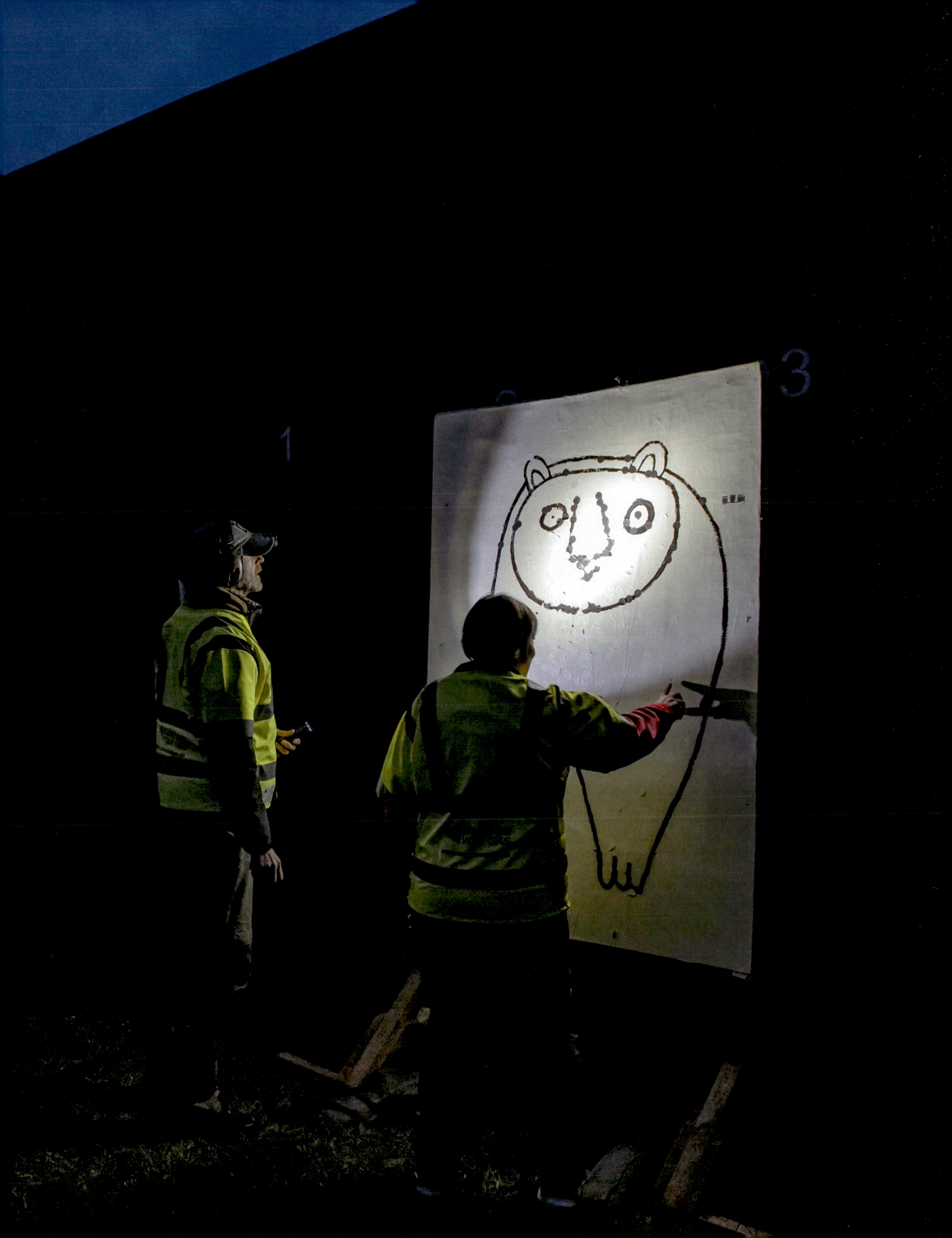
1
3

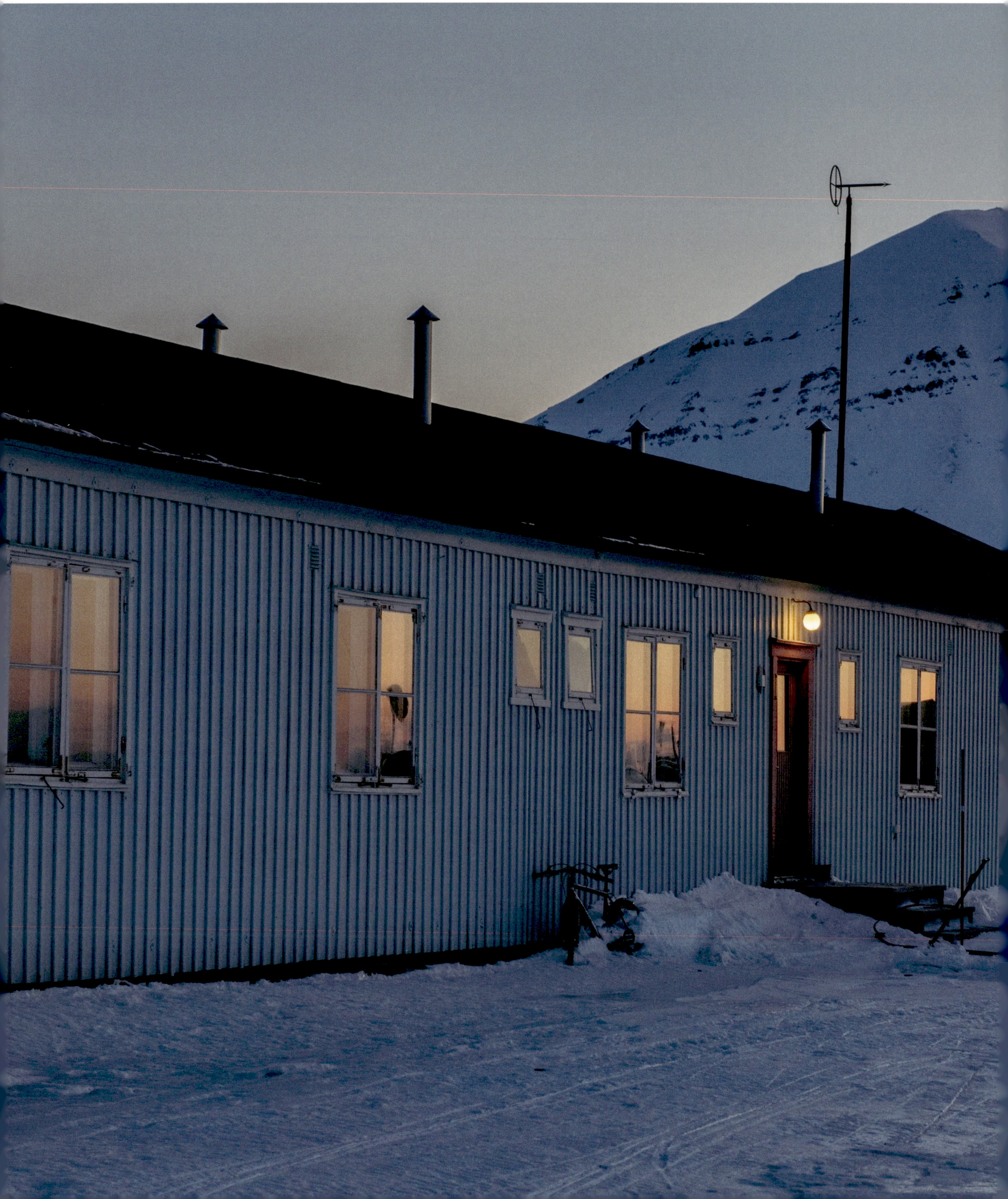

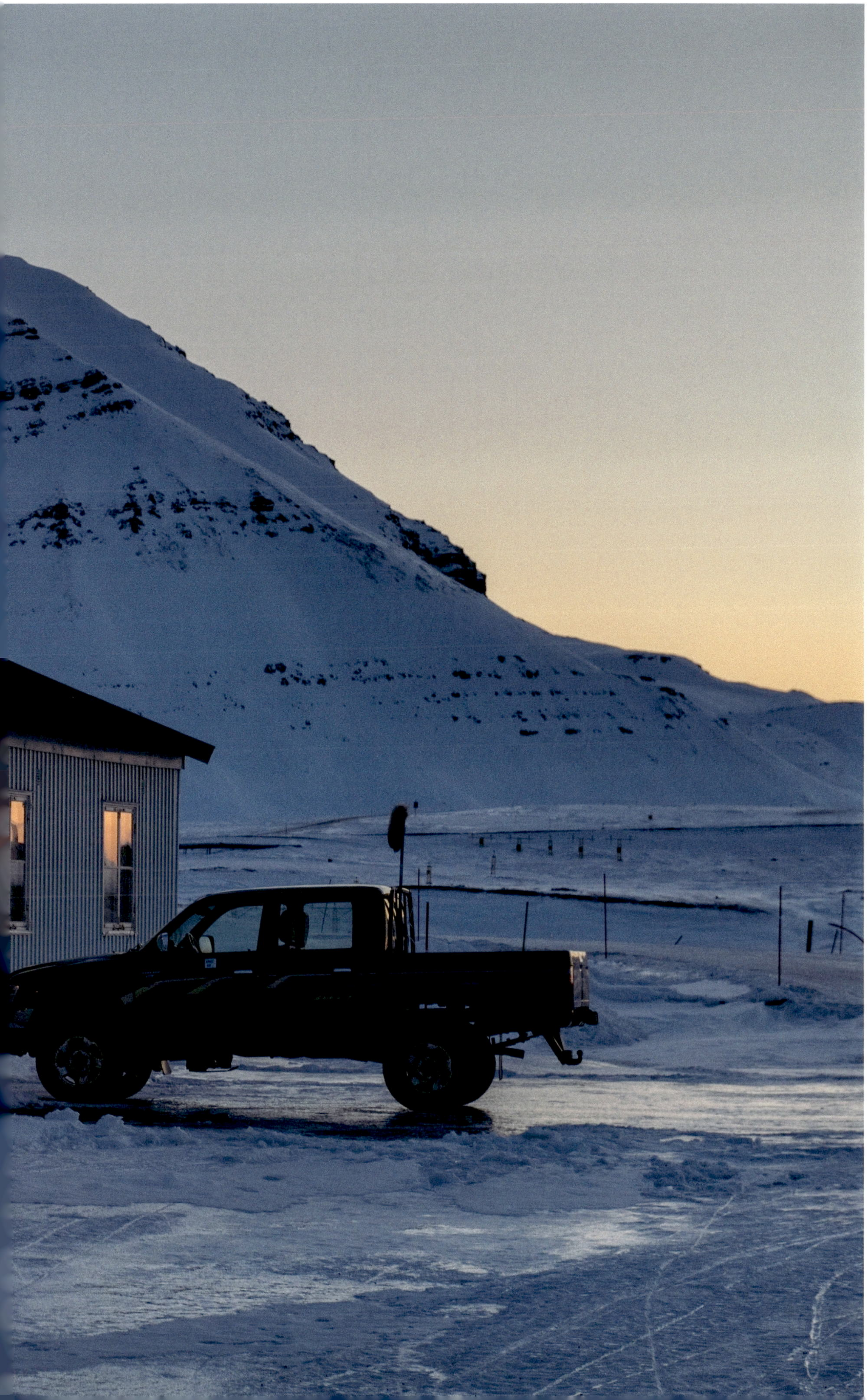

About halfway between Norway and the North Pole, in the middle of the Arctic Ocean, lies the Svalbard archipelago. The research village Ny-Ålesund, one of the northernmost settlements in the world, is situated on the main island. Several nations investigate the Arctic from here. This makes Ny-Ålesund the ideal testing ground for the MOSAiC expedition.

While in the past miners in Ny-Ålesund brought coal up to the surface, today scientists from around the world investigate the transformation of the Arctic as a result of climate change. A few dozen of them stay at the stations in winter as well. No road leads here – the village is accessible only by plane, ship or snowmobile.

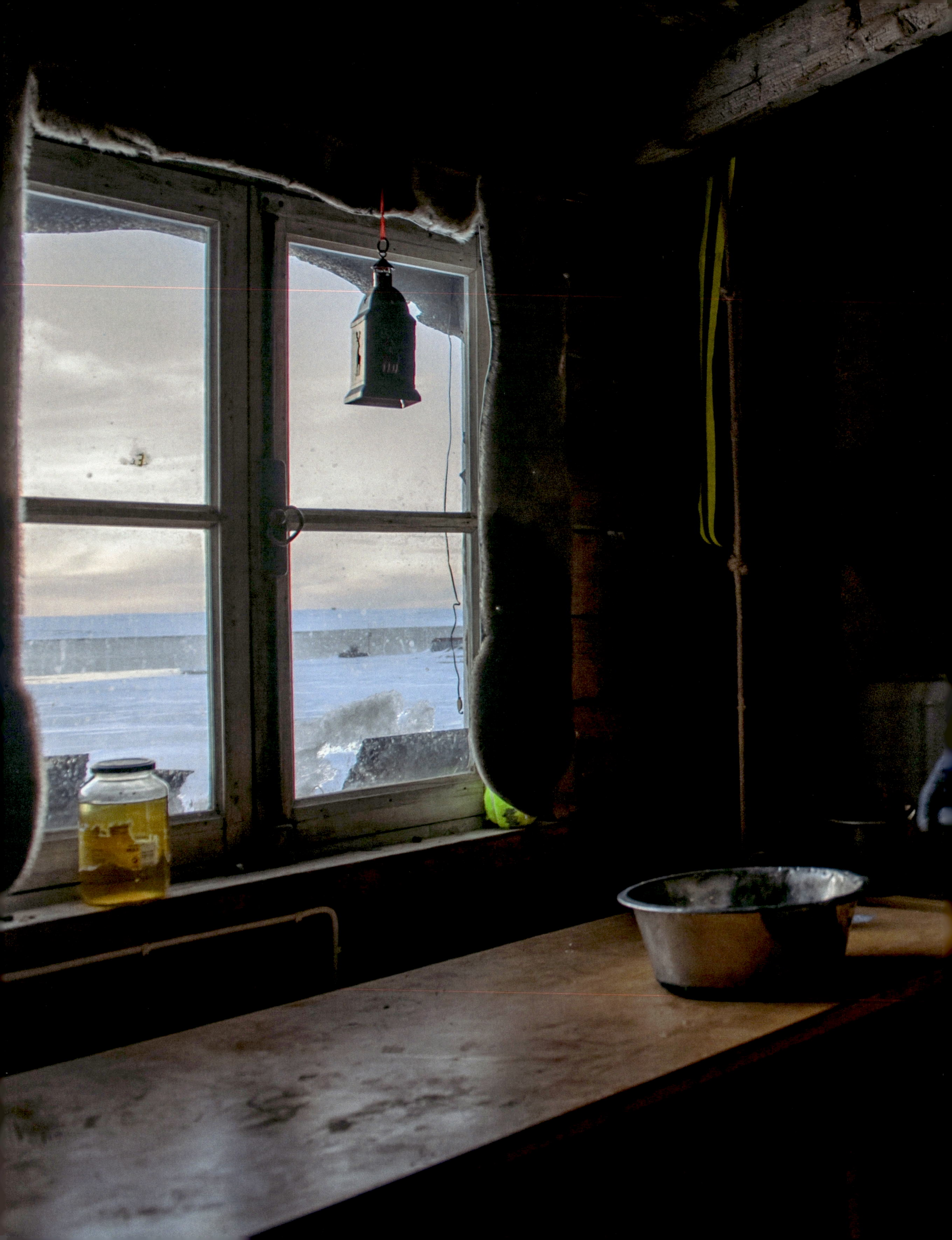

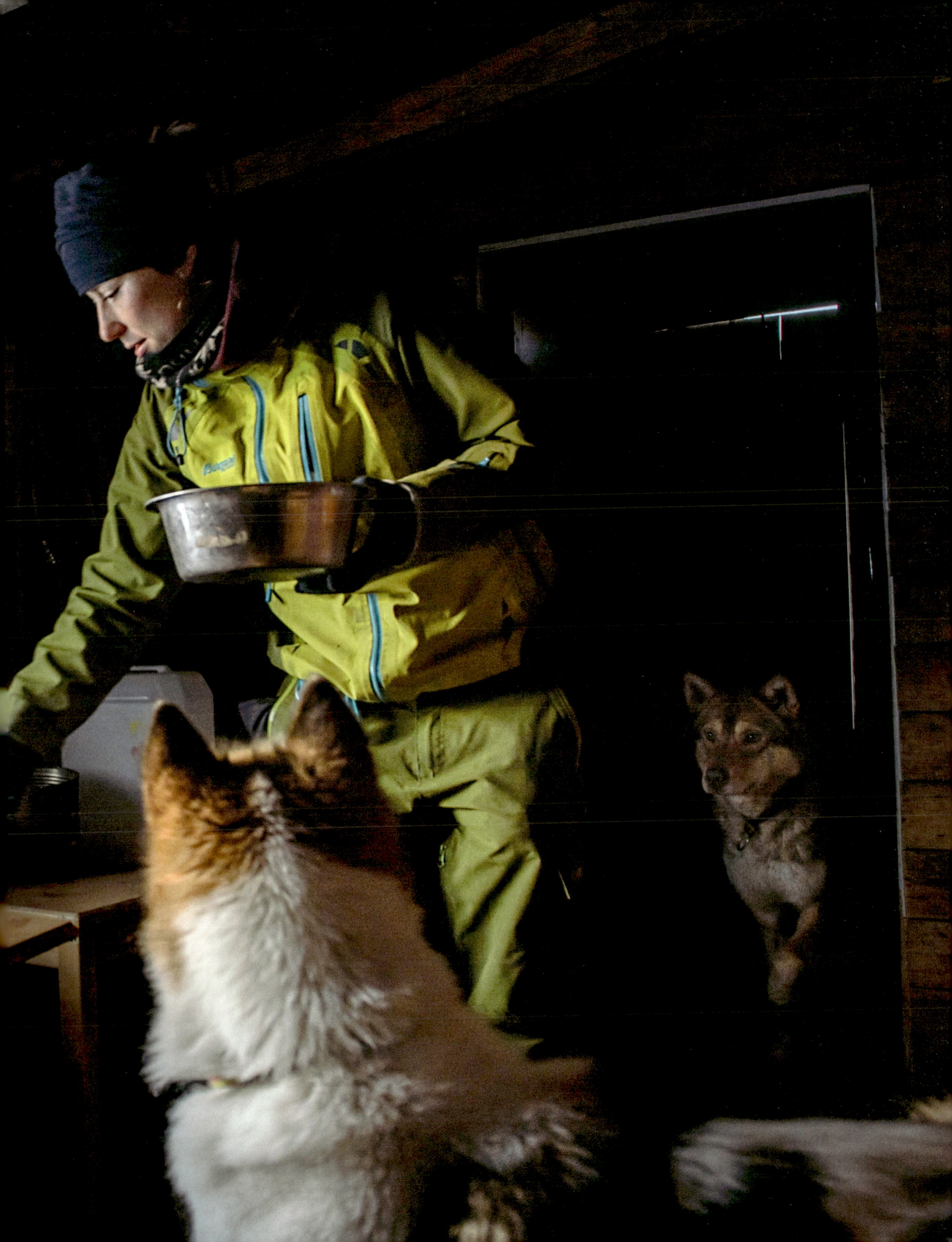

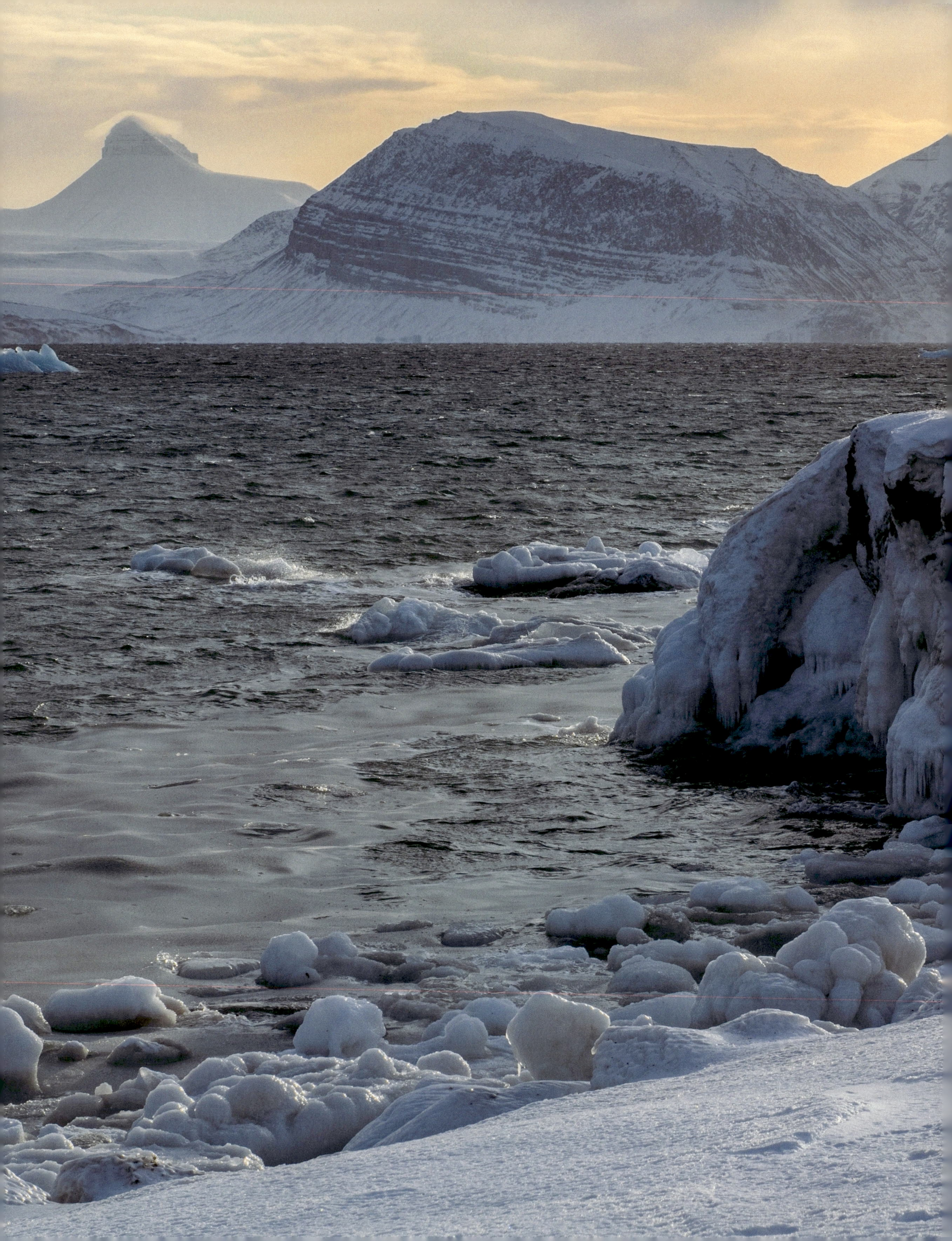

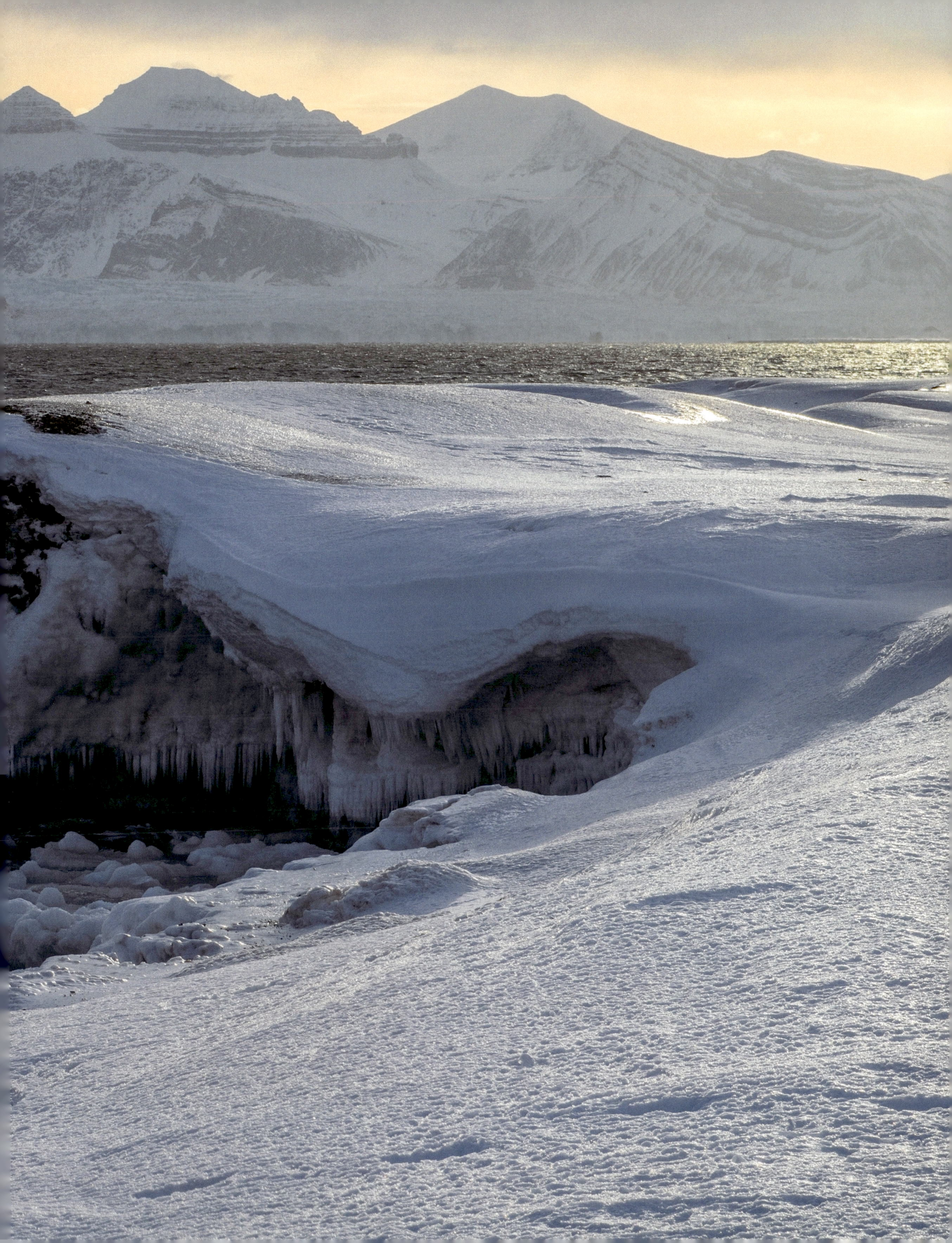

The scientist Anja Sommerfeld prepares the launch of a radiosonde with which meteorological variables such as atmospheric temperature, air pressure and wind speed can be measured. During its flight, the sonde sends data to the station. Since 1993 the scientists in Ny-Ålesund have sent up a weather balloon every day at exactly twelve noon, and have thus accumulated a valuable dataset. For the region near the North Pole, by contrast, hardly any comparable measurements have yet to be made.

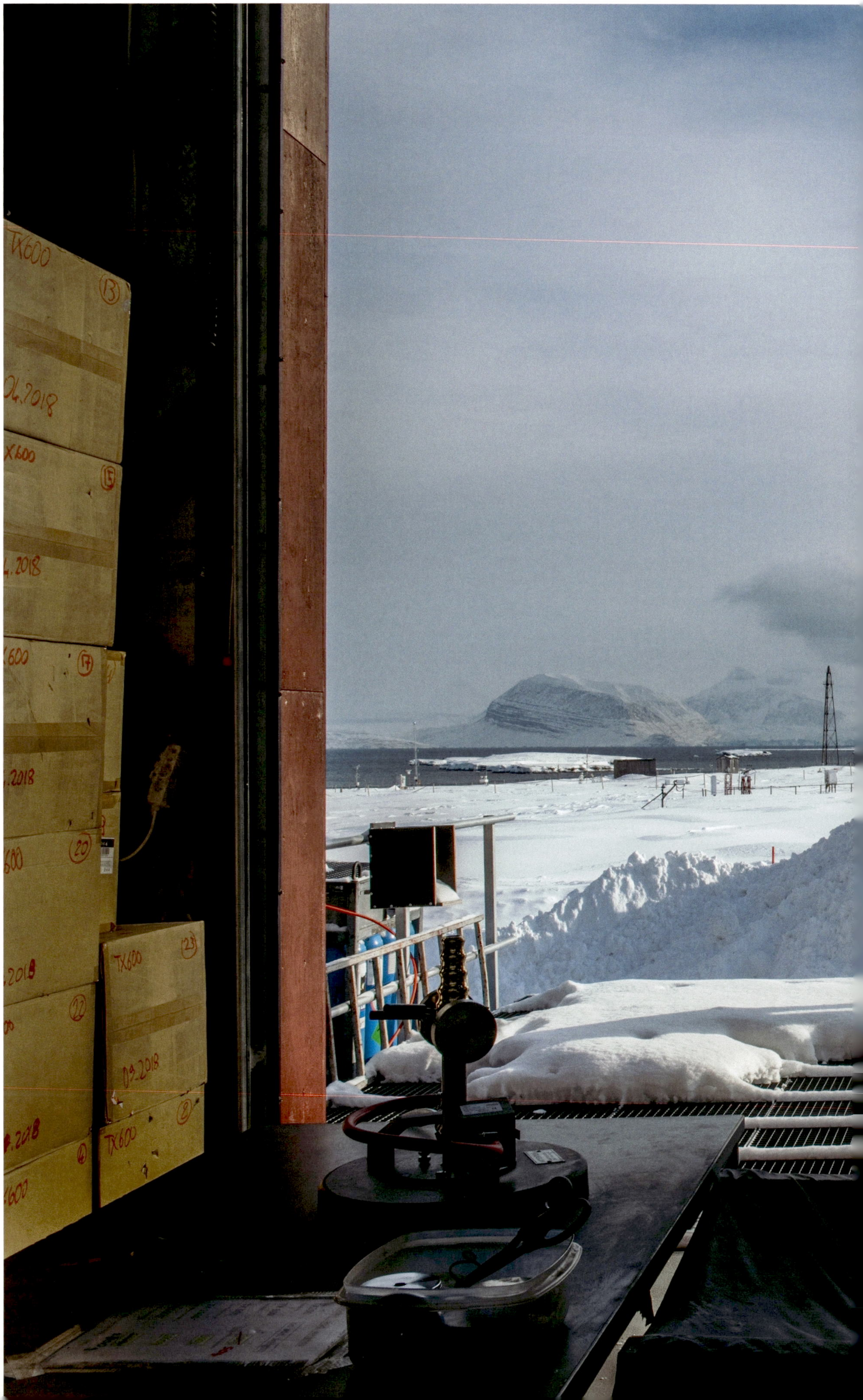

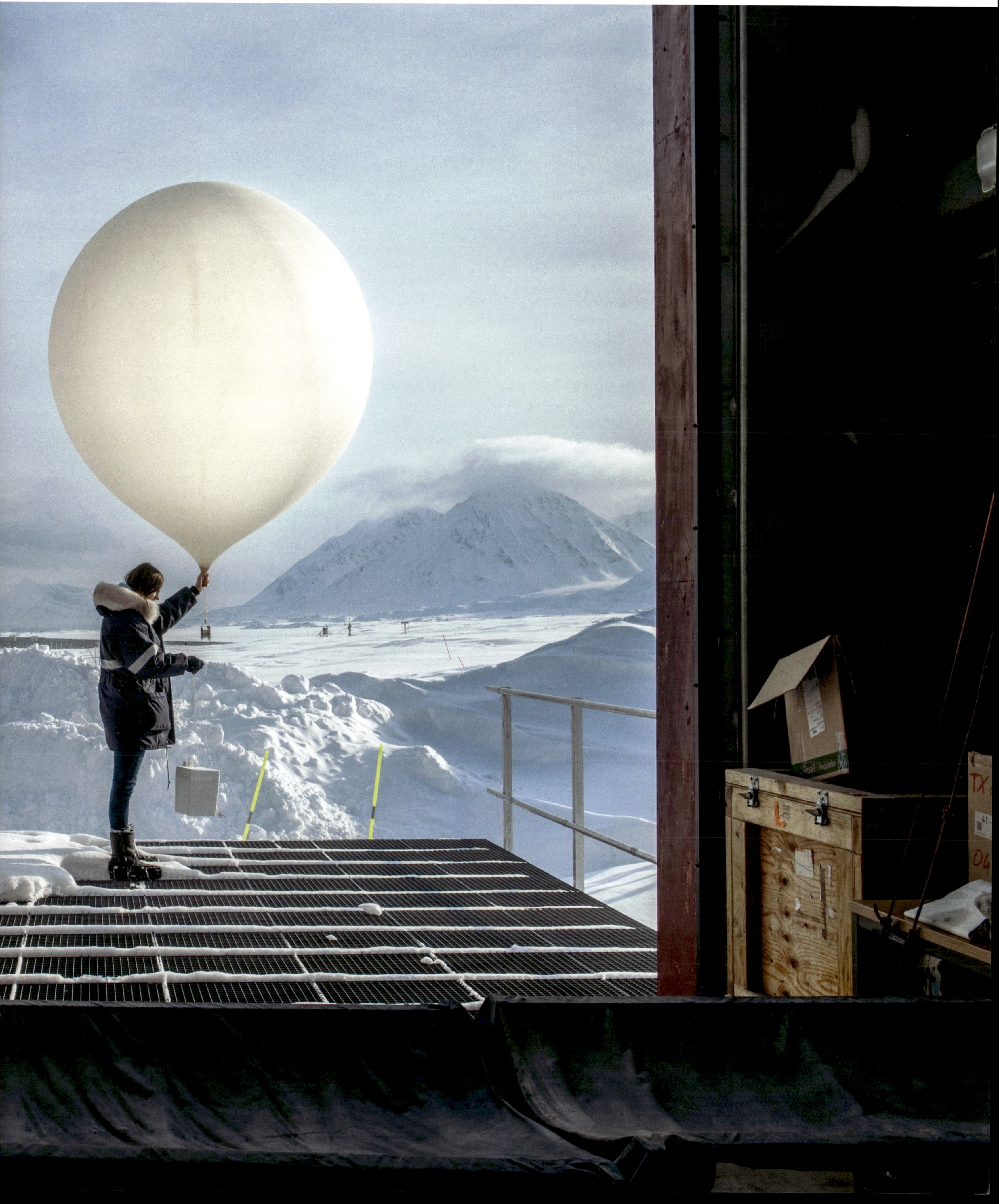

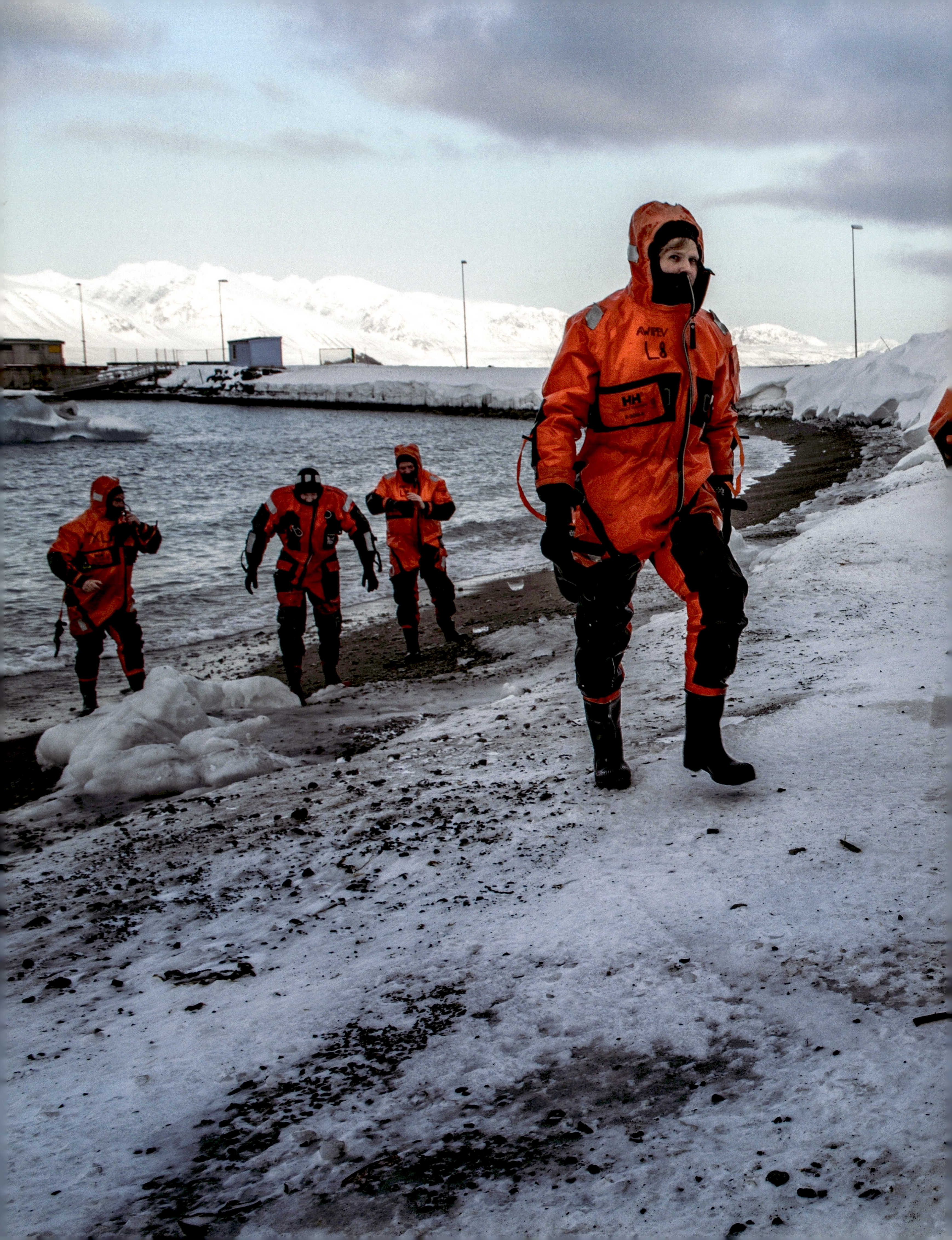
AWIPEV
L8
HH

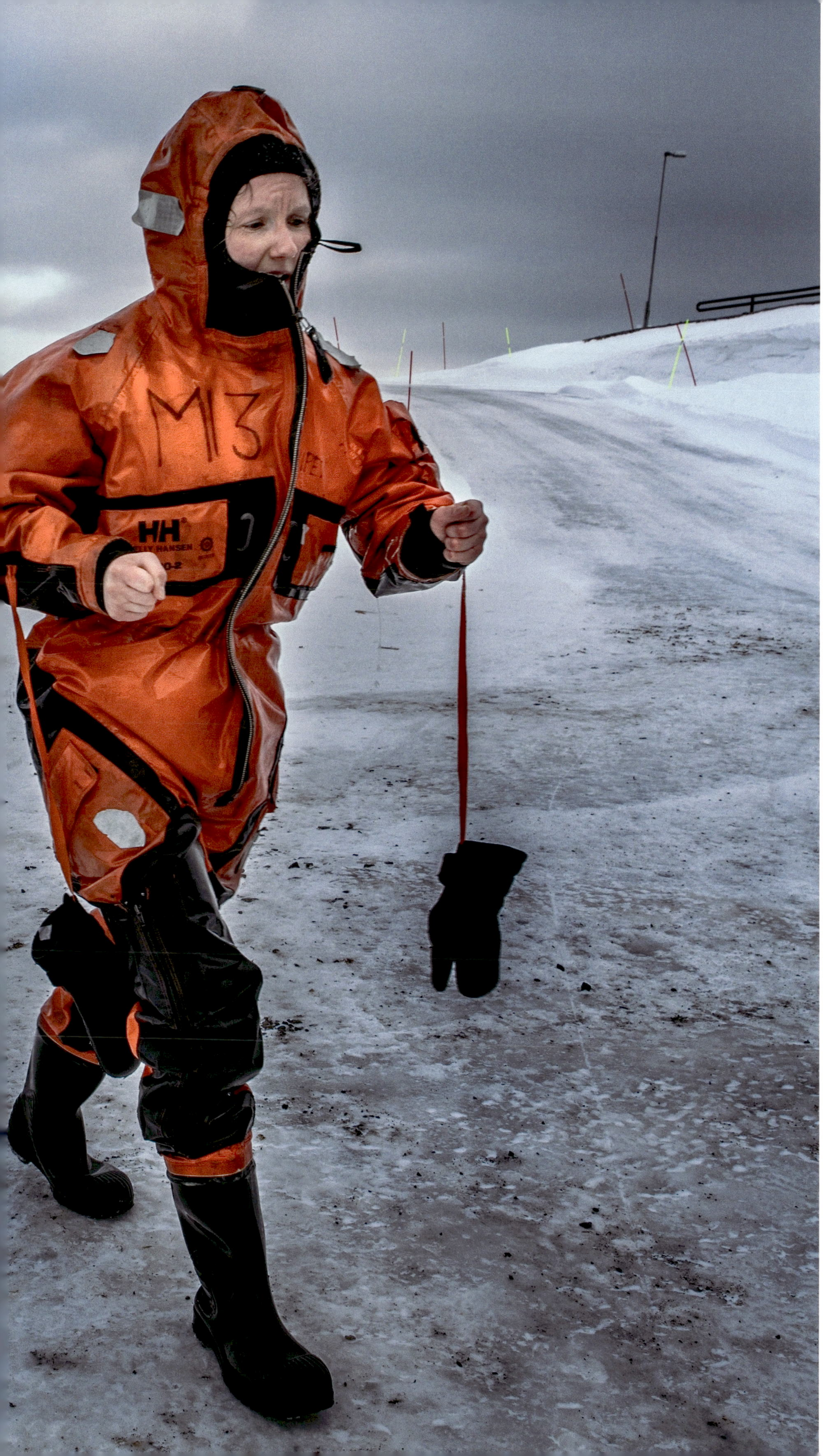

Dressing under extreme conditions. At a water temperature of −1.8 °C (28.8 °F) and an air temperature of −29 °C (−20.2 °F), the future members of the expedition learn how to survive in the water. As some of the research work takes place right at the edge of the ice, the possibility that someone will fall into the water during the expedition cannot be excluded. Flotation suits are therefore standard-issue clothing on the ice.

Following pages: The logistics expert Verena Mohaupt carries out tests with the expedition's most striking research balloon. Given the balloon's shape and colour, the expedition members lovingly call it "Miss Piggy".

"The training in Ny-Ålesund was a good chance for all participants to cooperate with each other – and that's what makes the MOSAiC expedition so special. It is a unique large team effort."

Verena Mohaupt, logistics expert

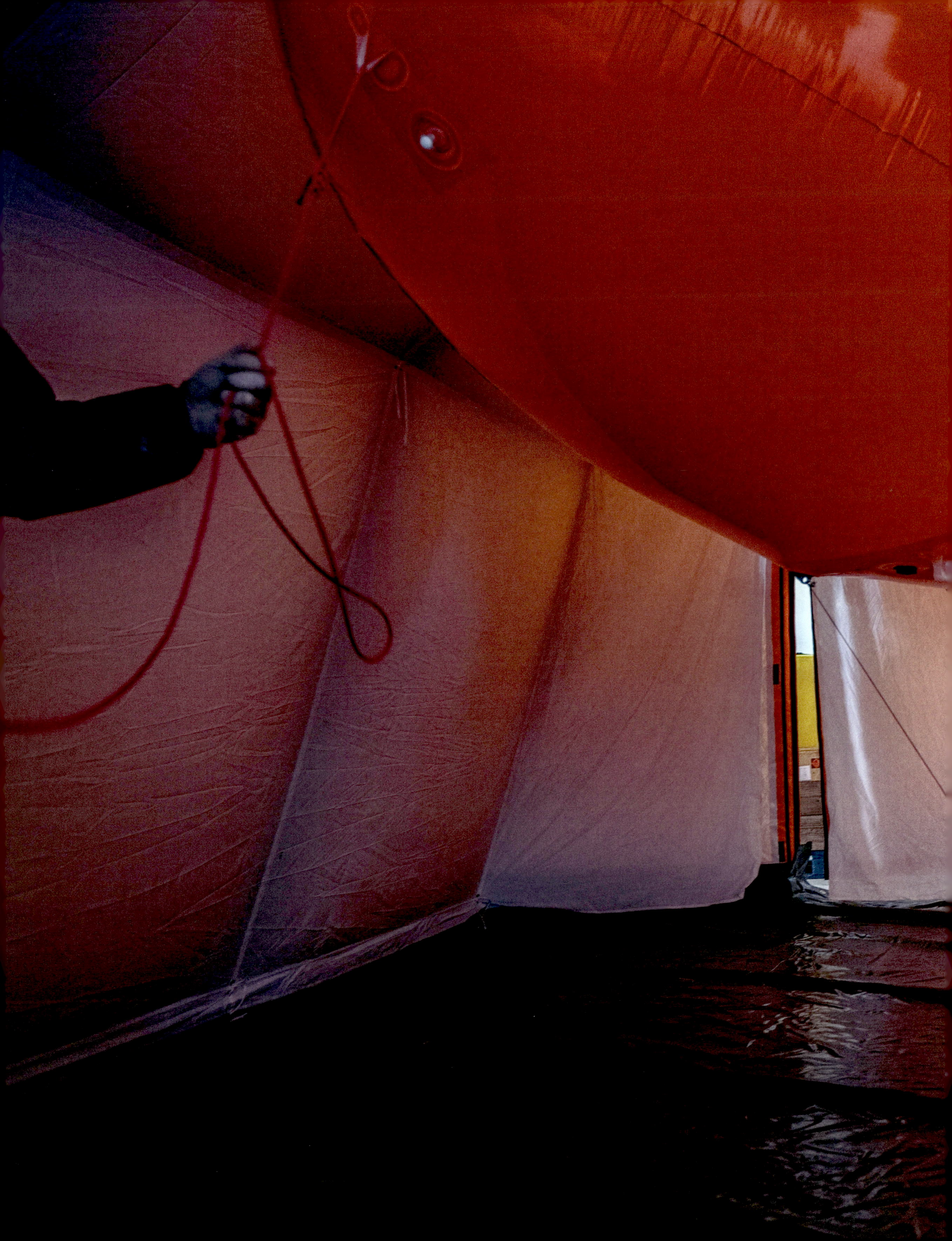

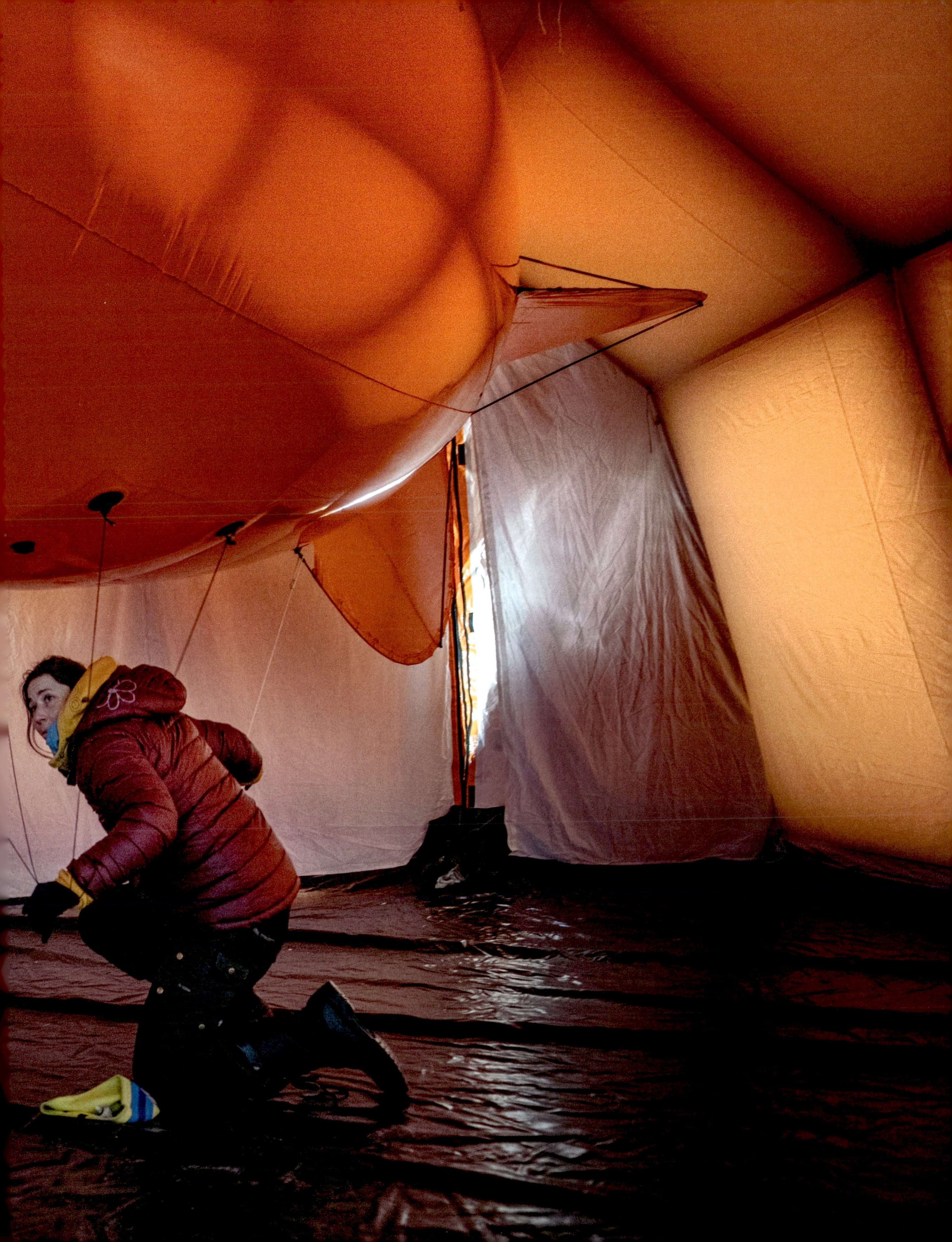

"When I first came to Svalbard in the early 1990s, during March and April the Kongsfjord was still a landscape of ice and snow. In those years we often crossed the ice on skis or in snowmobiles. If I come to this place at the same time of year today, I am standing by open water. Now we travel by boat where we used to have ski tours. There are few places in the world where climate change is so visible."

Markus Rex,
expedition leader

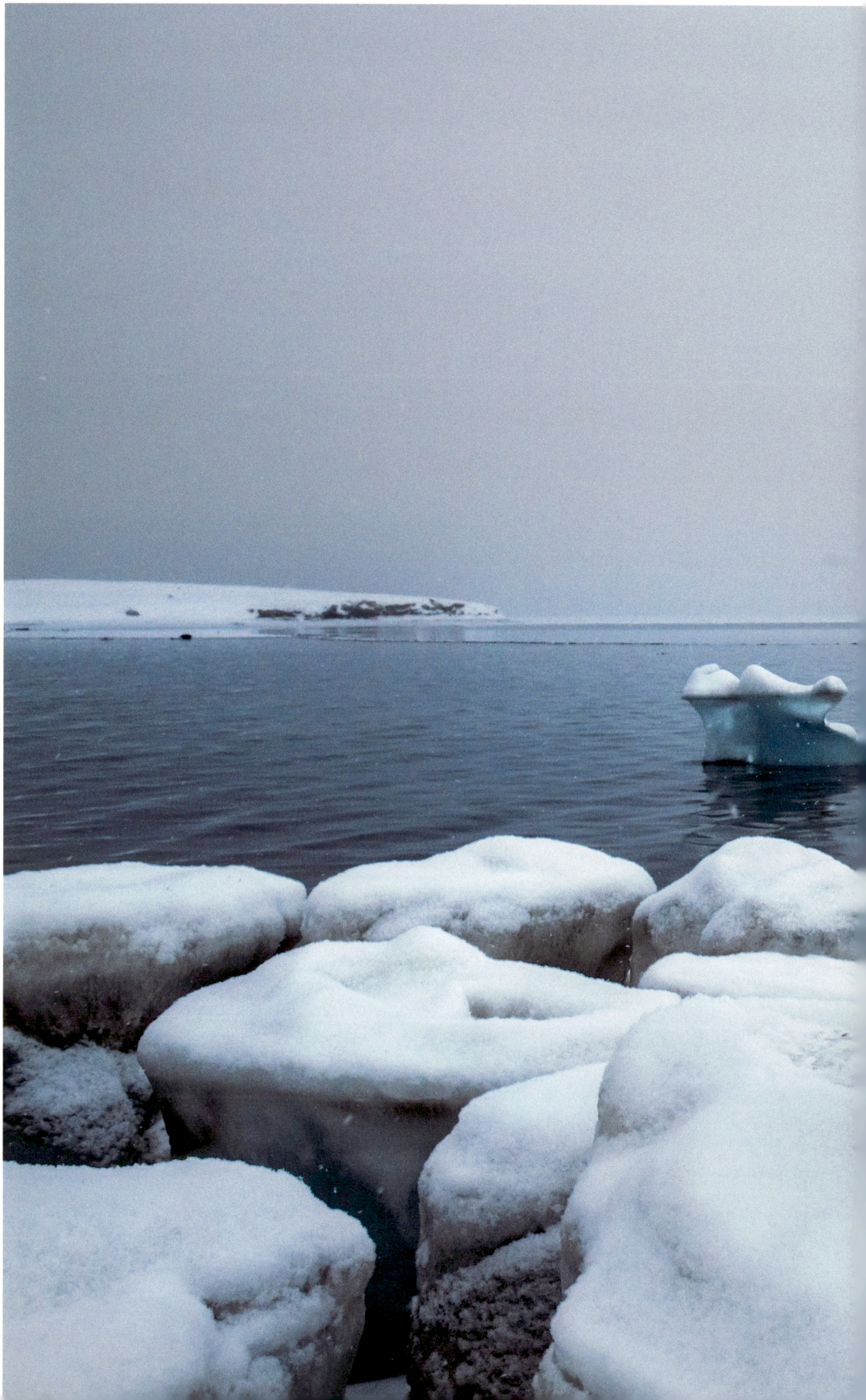

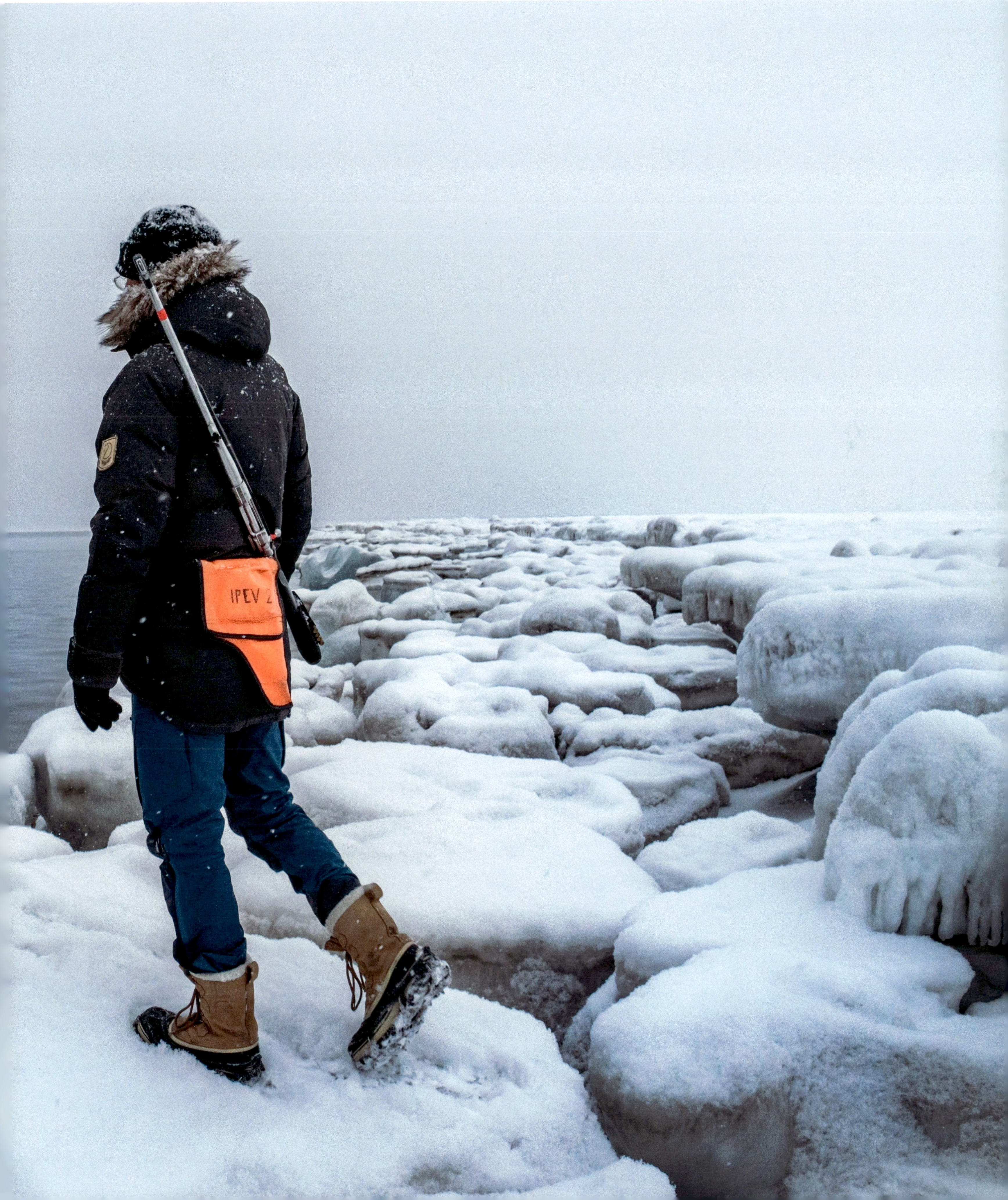
IPEV 2

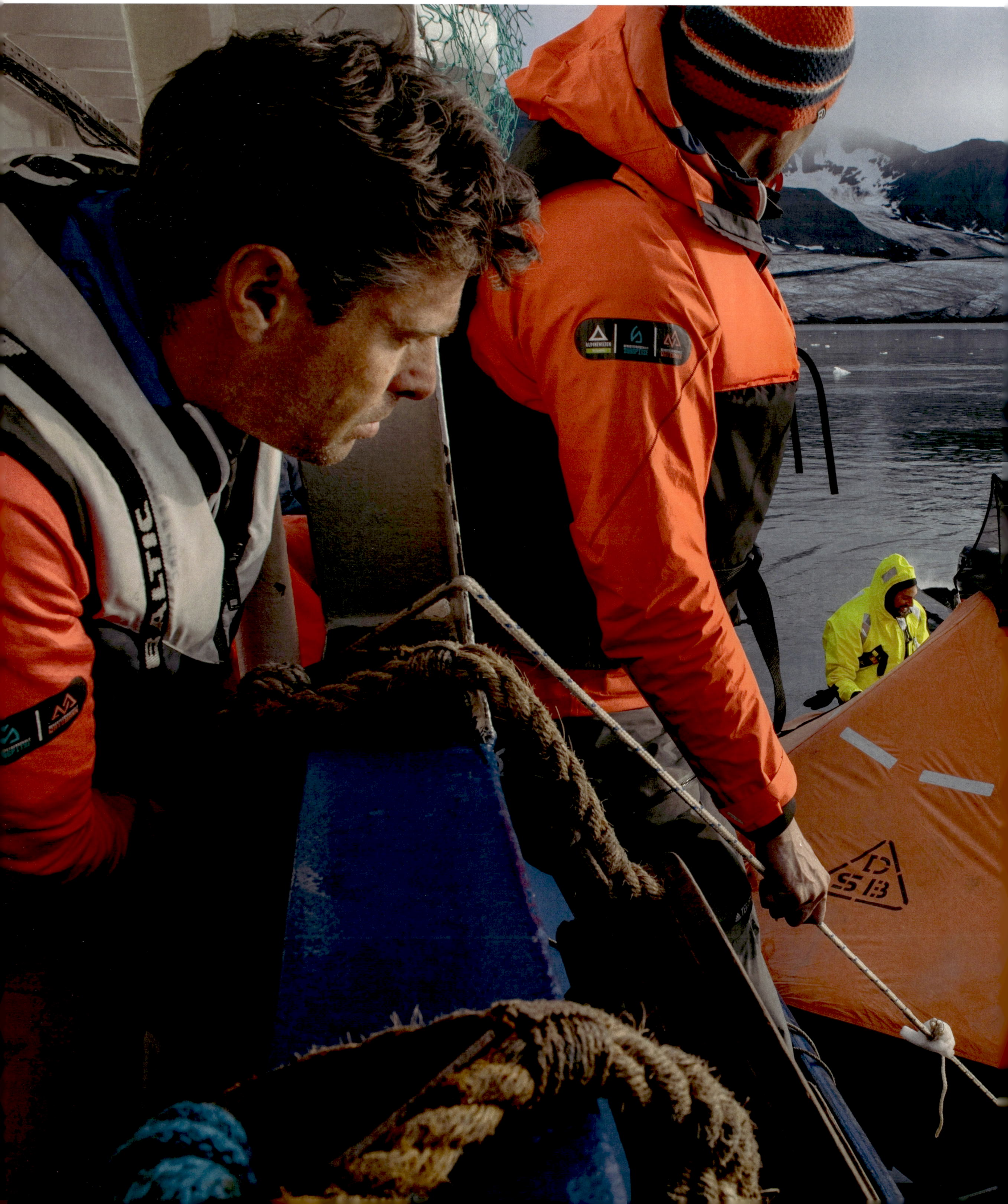
BALTIC
DSB

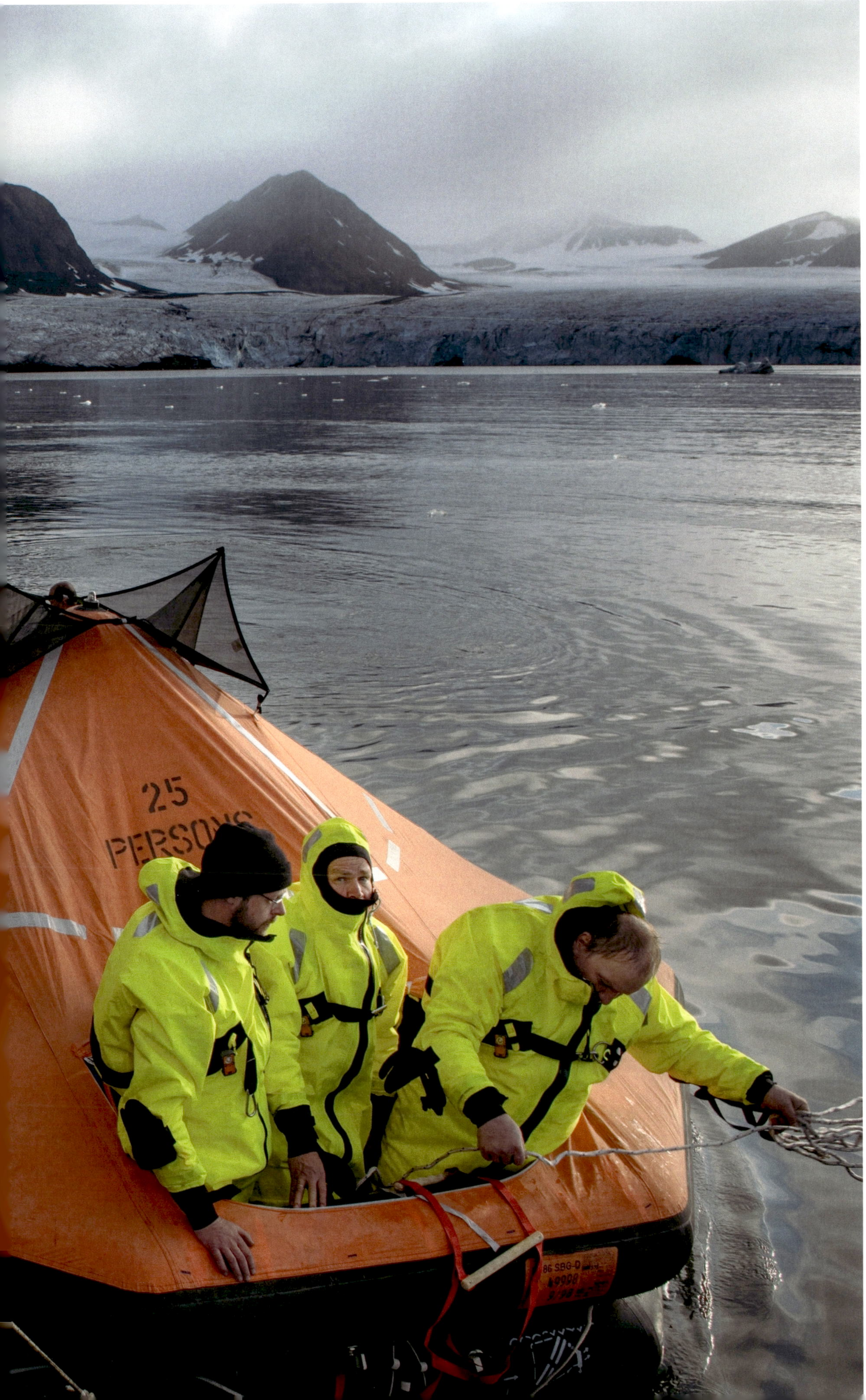

How do you survive if the ship has to be evacuated? The captains and a few officers faced up to this extreme exercise. Their first task was to reach the coast on a life raft. The second was to survive until help arrives. But no one knows how long it would really take.

In the barren fjord landscape, expedition members can get through the exercise only by working as a team. Everyone must be able to rely on everyone else. They carry no drinking water with them and therefore have to search the area for clean ice that they can melt in the camp later. Their food consists of survival biscuits made from flour, oil and sugar. They taste just as you would expect (bottom left).

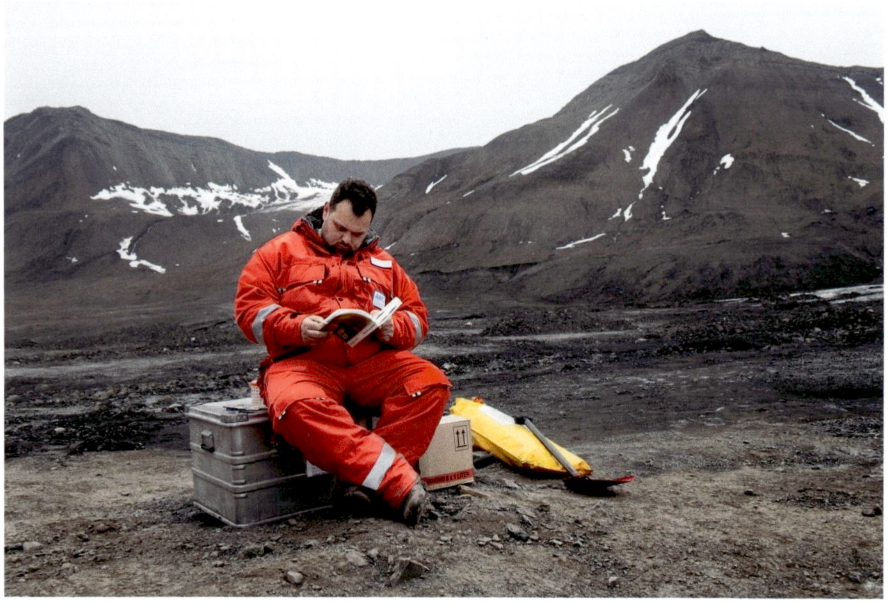

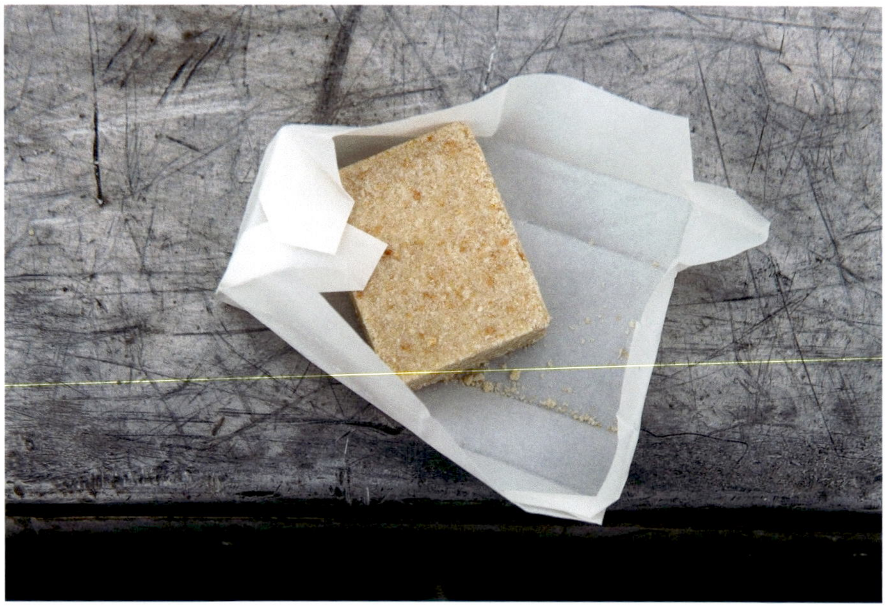

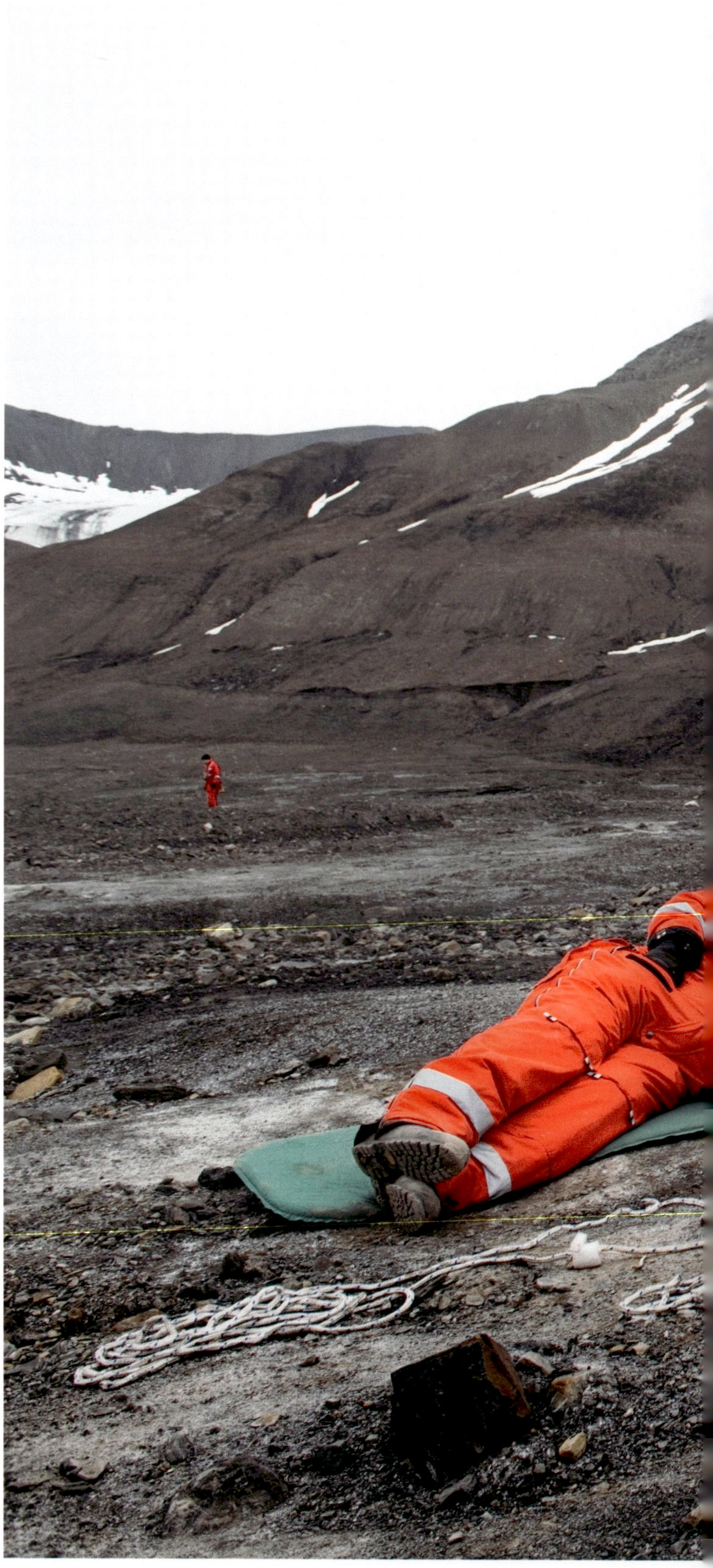

Fourteen stranded people have to share five sleeping bags. They sleep in four-hour shifts. While the first group sleeps, the second keeps a lookout for polar bears and the third carries out what needs to be done in the camp. Sometimes there is even time to read a book.

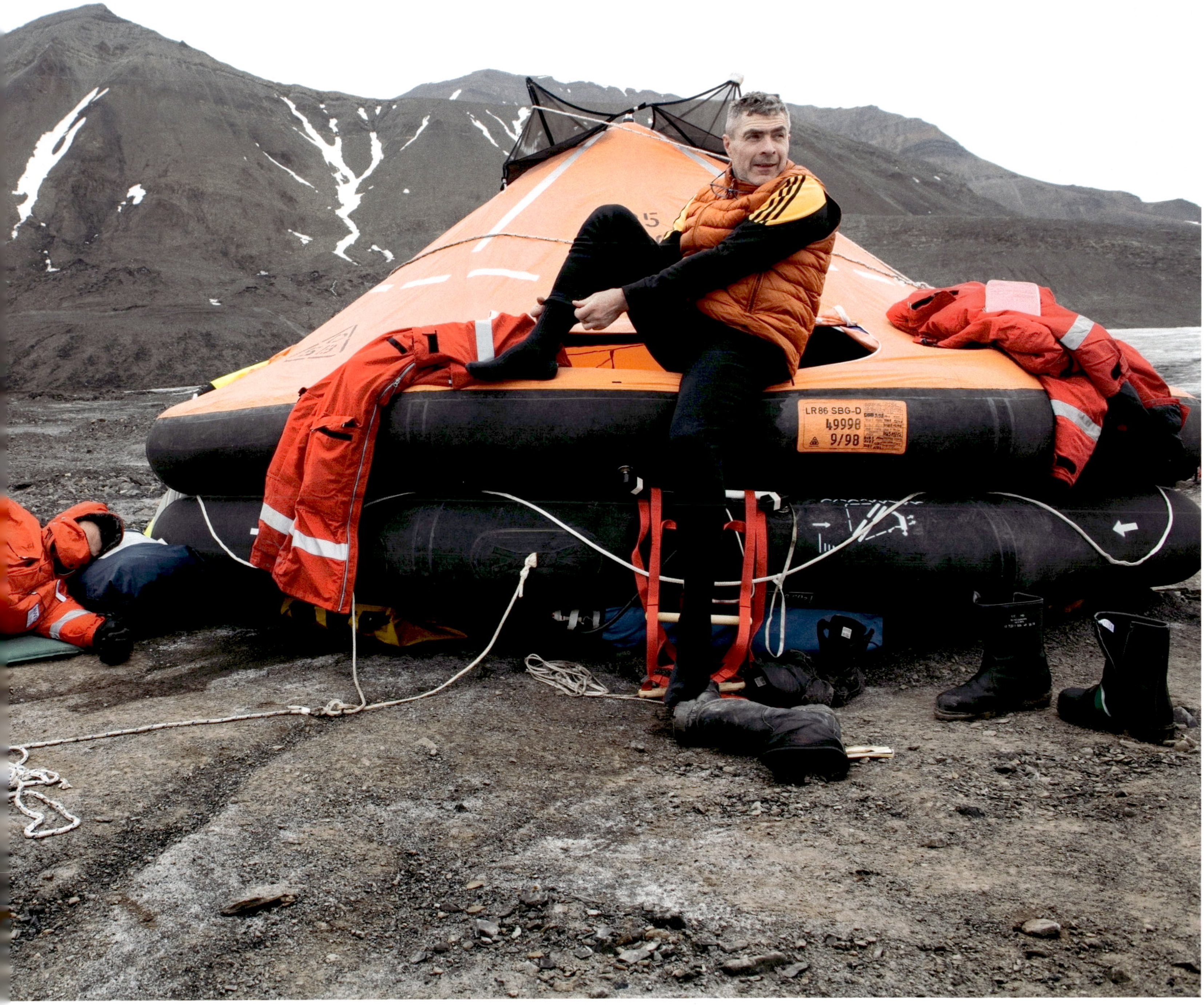

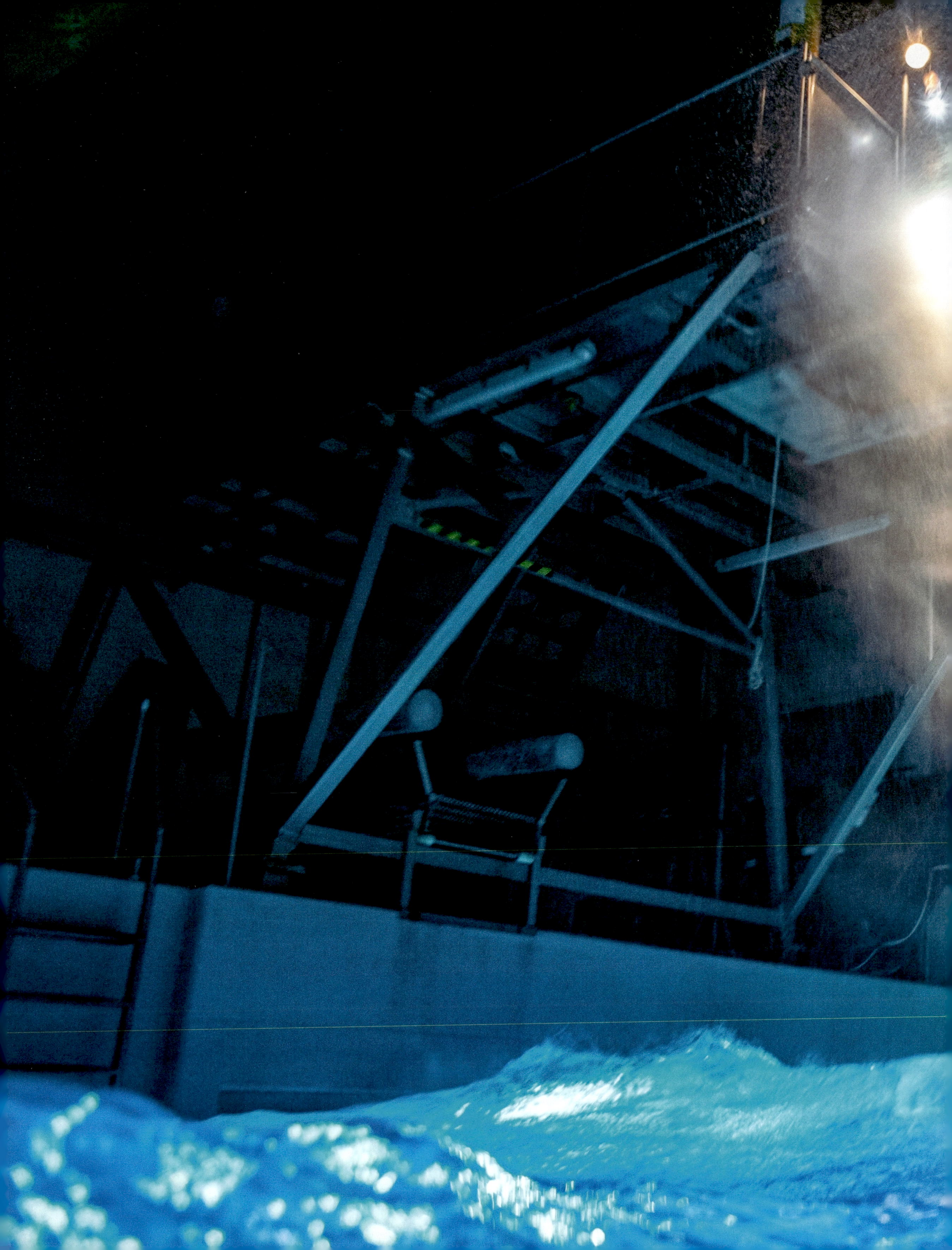

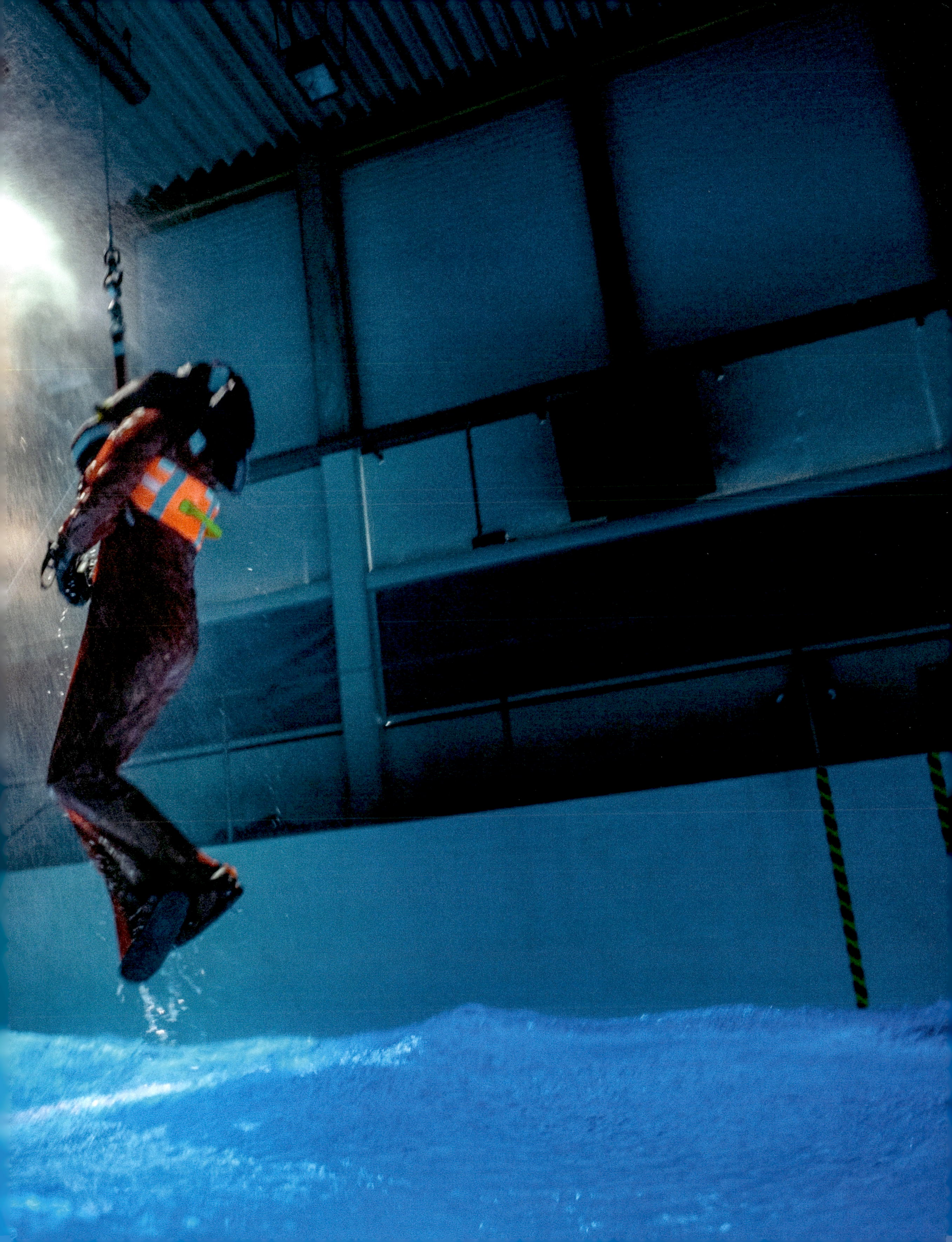

Cast Off

Away into the ice

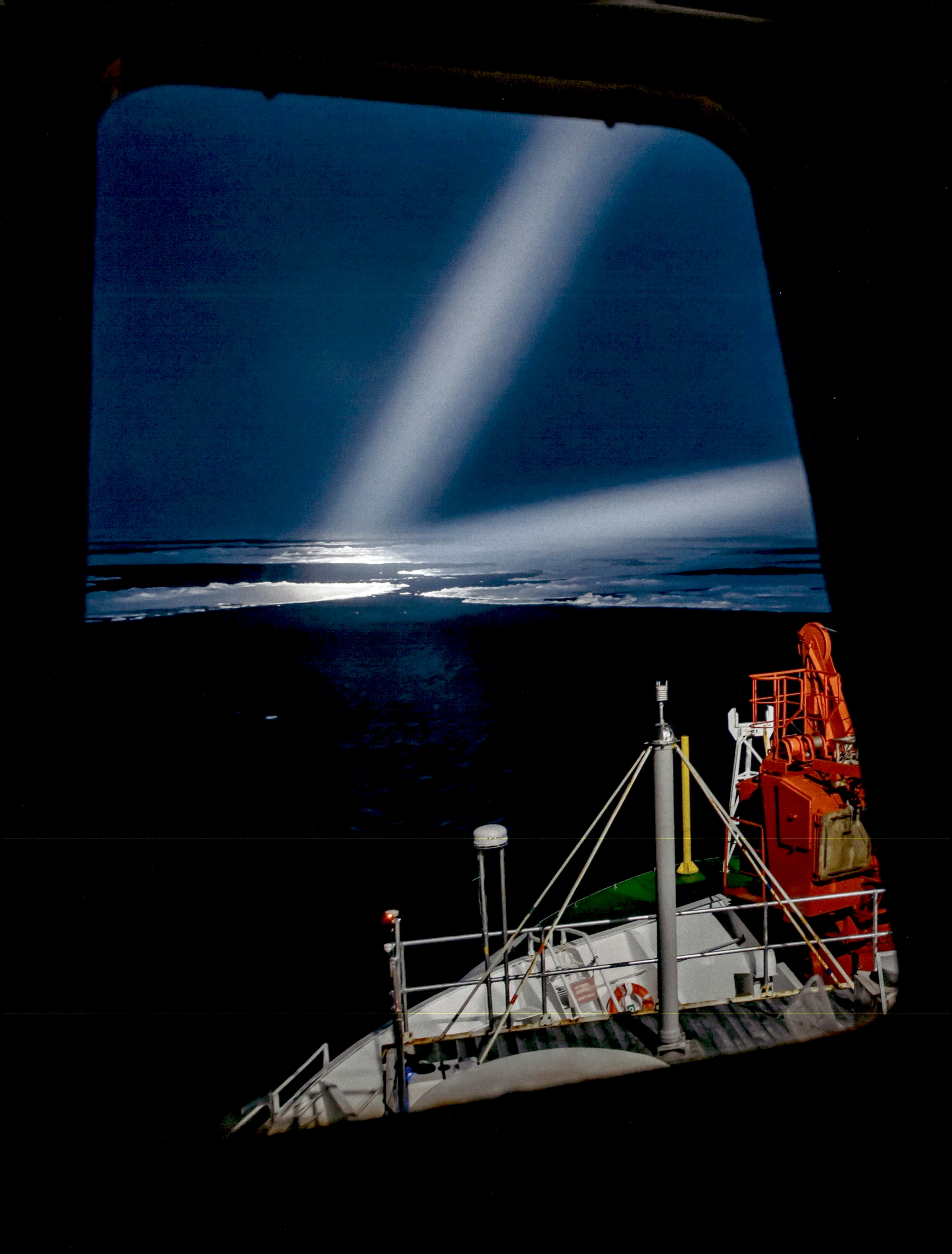

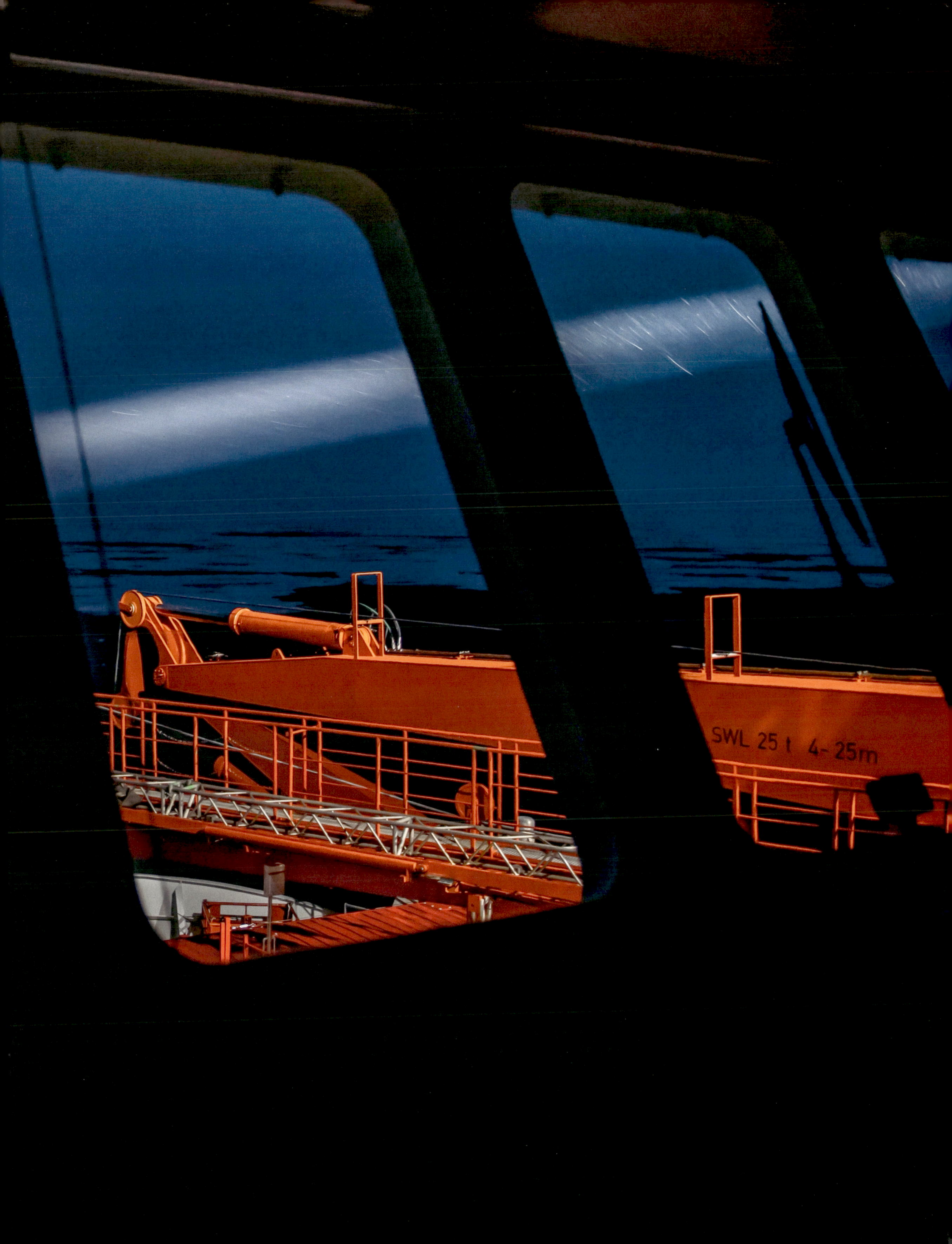
SWL 25 t 4-25m

One of the most exciting moments of the entire expedition begins on the bridge of the *Polarstern* on the evening of 4 October 2019. After only a few days, the scientists have found an ice floe on which they hope to set up their research camp for the year-long drift through the northern polar ocean. As they arrive, the expedition leader Markus Rex and the co-cruise leader Marcel Nicolaus focus on the exact position on the radar. The ship has to penetrate far enough into the stable part of the floe without destroying it.

"The sea ice that we are selecting and that continues to form here will be the basis of the entire MOSAiC expedition. We will examine the ice and its changes in great detail. This floe will largely make scientific history for MOSAiC."

Marcel Nicolaus,
co-leader of the expedition

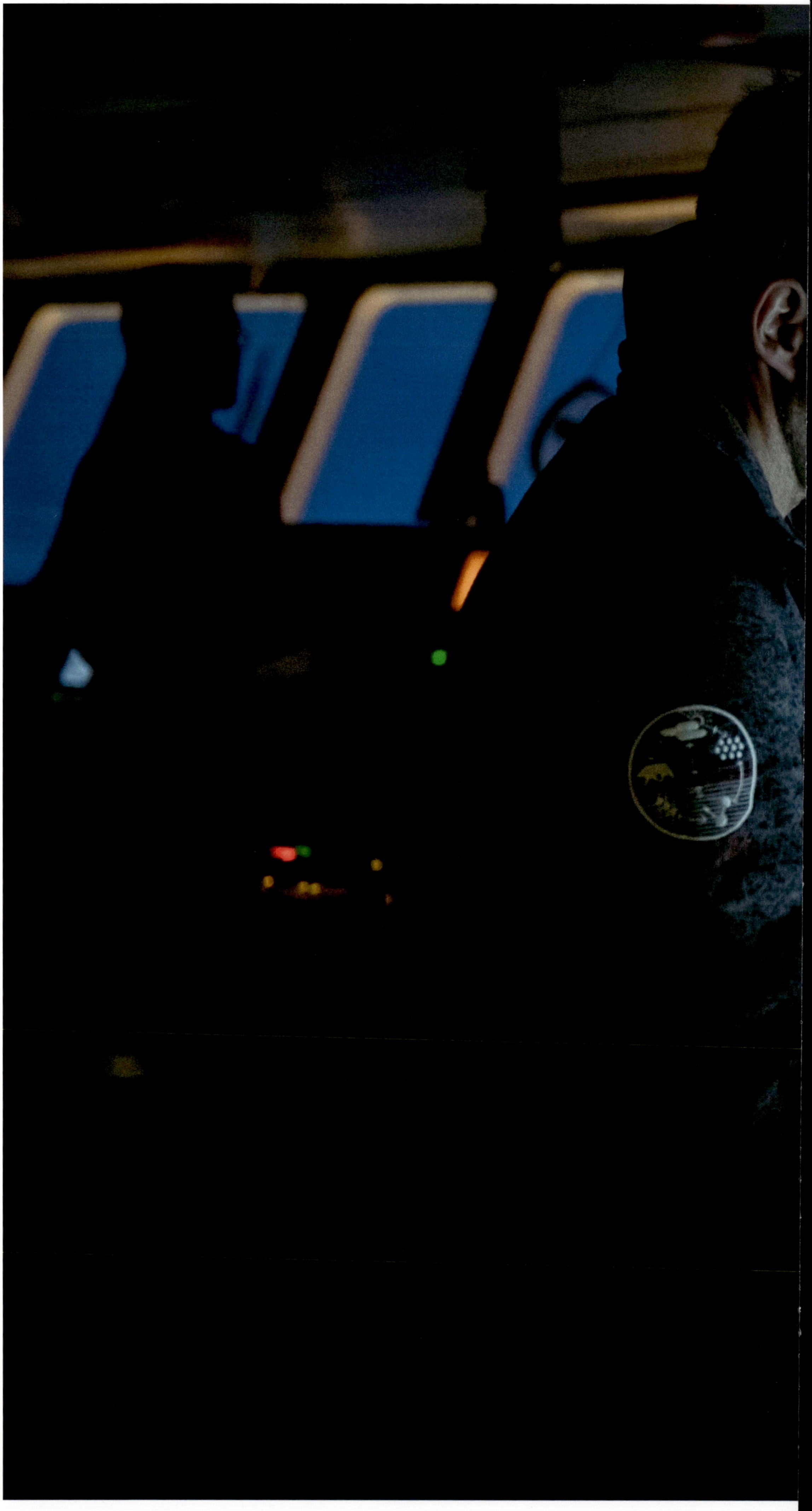

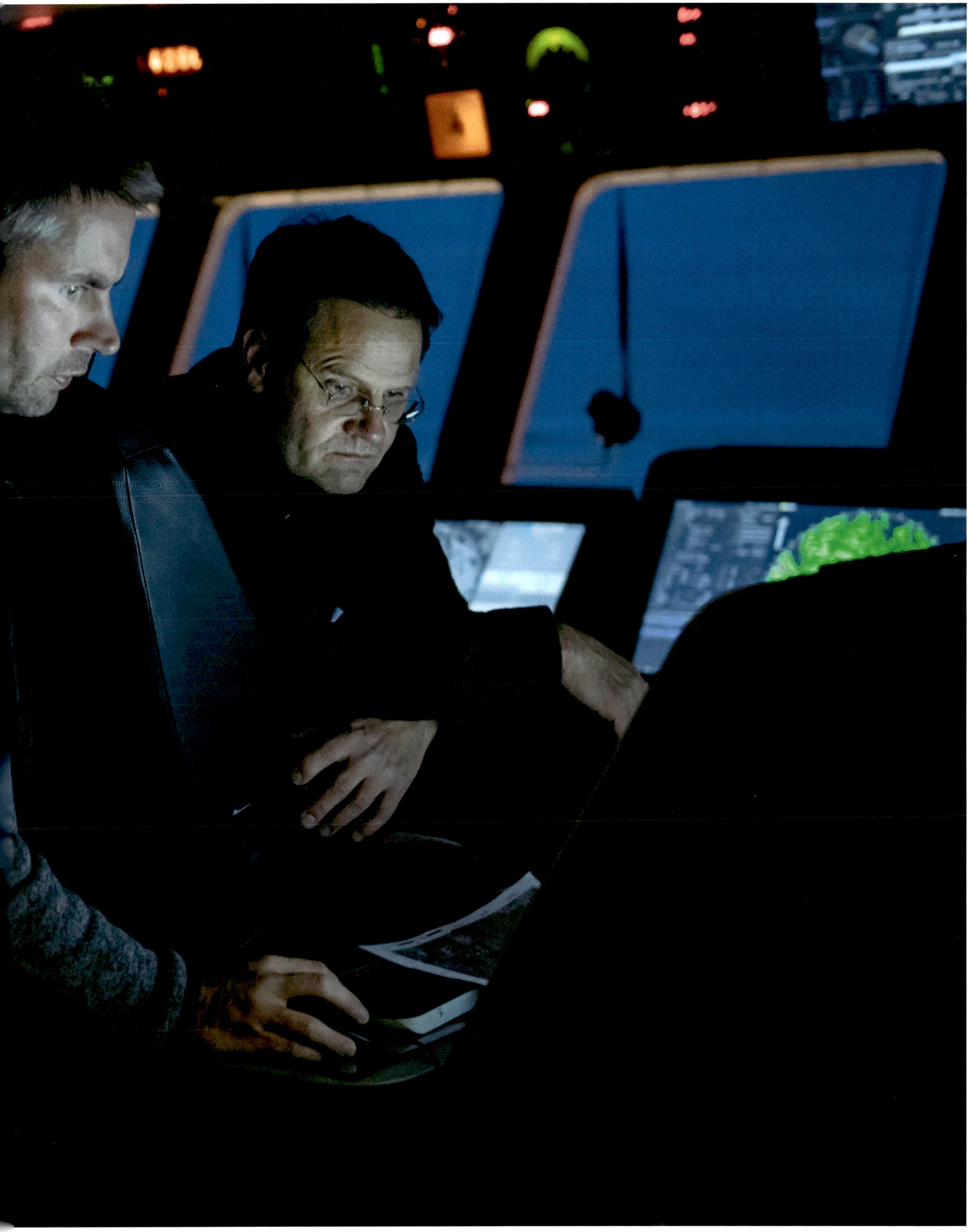

G

Grey clouds shroud the Norwegian port of Tromsø this afternoon of 20 September. It is raining again in the fjord as the first members of the MOSAiC expedition walk the narrow gangway to board the *Polarstern*. Soon, when Arctic precipitation only reaches them in a frozen state, they will almost have forgotten what a raindrop feels like.

20 September is a historic date in maritime exploration, as Ferdinand Magellan began his circumnavigation of the globe exactly 500 years ago to the day. Up on deck, a group forms as a sailor collects passports. This is actually routine, but it still feels like a temporary trade. The members of the expedition are swapping solid ground beneath their feet for life on a ship. Around them, in place of their families, are scientists, old and new colleagues from around the world.

Together they look back to the harbour. Representatives from the worlds of politics, science and the media have assembled on Pier 25 to send the ship off on its journey. This farewell is celebrated properly with a lot of speeches, music and embraces. Then, at a quarter to nine in the evening, the time has finally come: departure. The *Polarstern*'s horn sounds three times. A special mood marks this moment which many in the research team have been looking forward to, some of them for years, and for which they have prepared meticulously. Among the crowd on shore, people are taking out handkerchiefs to wave to the ship and its crew as they disappear into the darkness of the fjord after casting off.

On deck the tranquillity of the ocean slowly cloaks the last shouts from the shore. The expedition has begun. Some of those on board have never been to the Arctic before, while others can look back on decades of polar research. Yet all agree on one thing: this will be the expedition of their lives.

Exactly two weeks later, an excited silence falls on the bridge of the *Polarstern*. Next to the captain, Markus Rex and the officers, curious expedition members have gathered and are crowding at the windows. Their faces are hardly identifiable in the darkness. Only a few lamps on the bridge give off a dim red light. This shortwave light enables their eyes to focus on the essentials: outside, three spotlights illuminate the expanse of the Arctic sea ice. The coming minutes will be decisive for the future course of the expedition.

At this moment it feels like an eternity since the first night on board, when the ship left the last fjords behind and on day one of the expedition its members awoke to find themselves surrounded only by the sea. They laid eyes on their last piece of land at the North Cape of Norway, on their way to the Barents Sea, above which the Northern Lights were dancing. On its north-easterly course, the ship passed through six time zones. The voyage took the *Polarstern* through the Kara Sea, past the Russian archipelagos of Novaya Zemlya and Severnaya Zemlya, on course for the Arctic sea ice, the outer fringes of which they reached after almost a week. At this point, a race against time began, as the sun was retreating further and further. Before the start of the polar night they had to find an ice floe for their mission. With satellite images, helicopters and their first scouting trips on foot, they searched indefatigably for a patch of ice that was suitable for their expedition. Here the *Polarstern* was finally reunited with the Russian support vessel *Akademik Fedorov*, which left Tromsø one day later and now assisted in the search.

On the bridge of the *Polarstern* the decisive moment has now arrived. The cracking noise as the ship slowly breaks through the ice is growing louder and louder, until finally the vessel reaches a floe measuring approximately 2.5 by 3.5 kilometres with an unusually stable zone at its centre. After a short but thorough search,

Following successful mooring manoeuvres, at half past ten in the evening the expedition reaches its destination at 85° N and 134° E. From here on the two engines are idle, but the watch on the bridge continues for the duration of the drift. First Officer Felix Lauber starts his shift at midnight. In the logbook he and his colleagues note down the distance travelled, from now on in drift miles rather than nautical miles.

a conclusion is reached: this will be the starting point of the ice drift. The icebreaker moves carefully into the floe. Nothing can go wrong now, as the ice is fragile in many places, in spite of its solid core. It almost seems as if the expedition is digging the ground from beneath its own feet. One false manoeuvre could irrevocably destroy the expedition members' new home. At the same time the icebreaker has to get as close as possible to the stable area, as only here is there a realistic chance of setting up the research camp. At half past ten in the evening the ship reaches its destination, and the engines can idle. The captain enters the coordinates 85°04.582′ N and 134°25.769′ E into the logbook. This position is transitory, however, and will already have changed by the following day. The location of the expedition's base will now be determined by the movements of the Arctic sea ice alone.

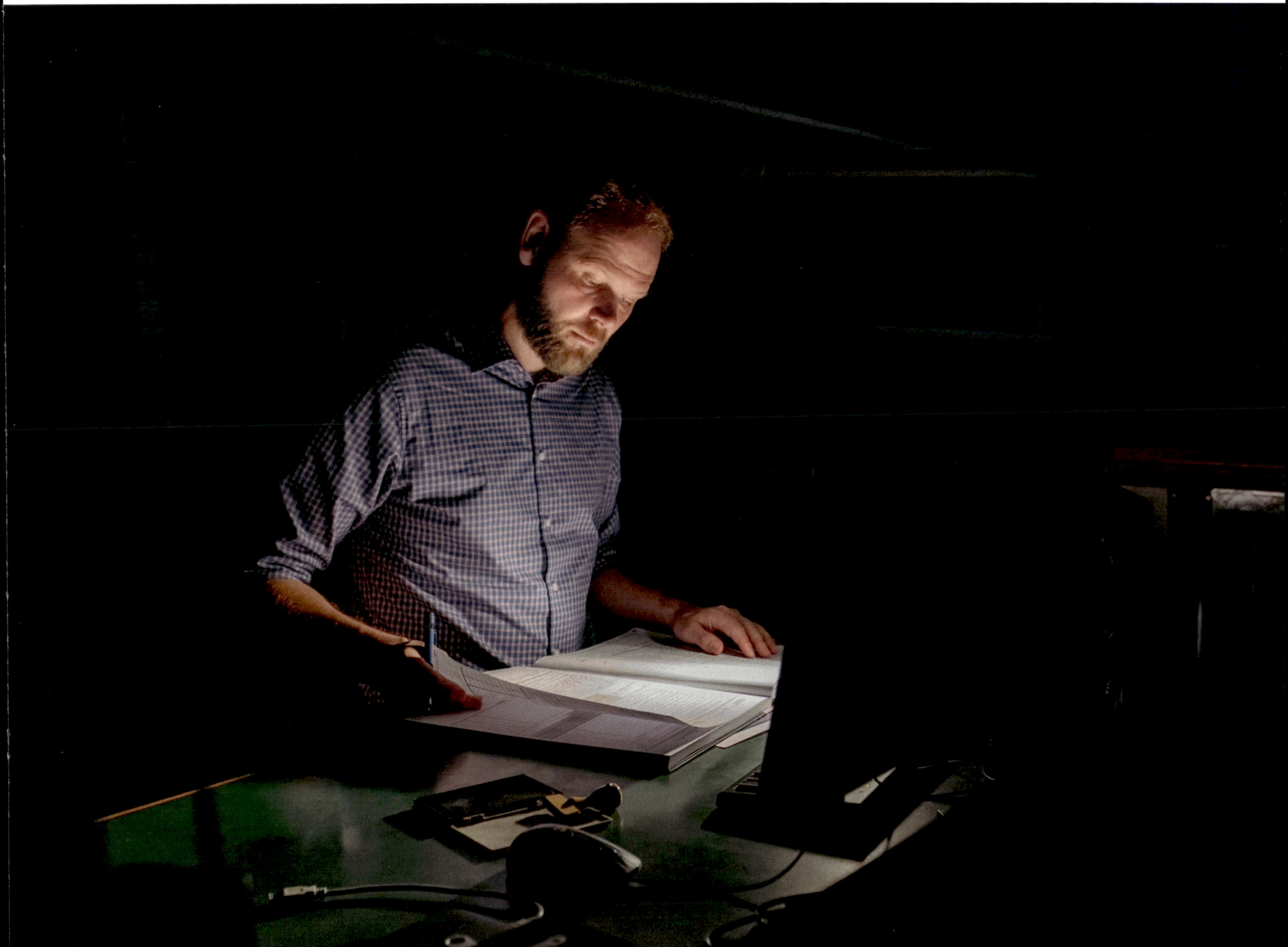

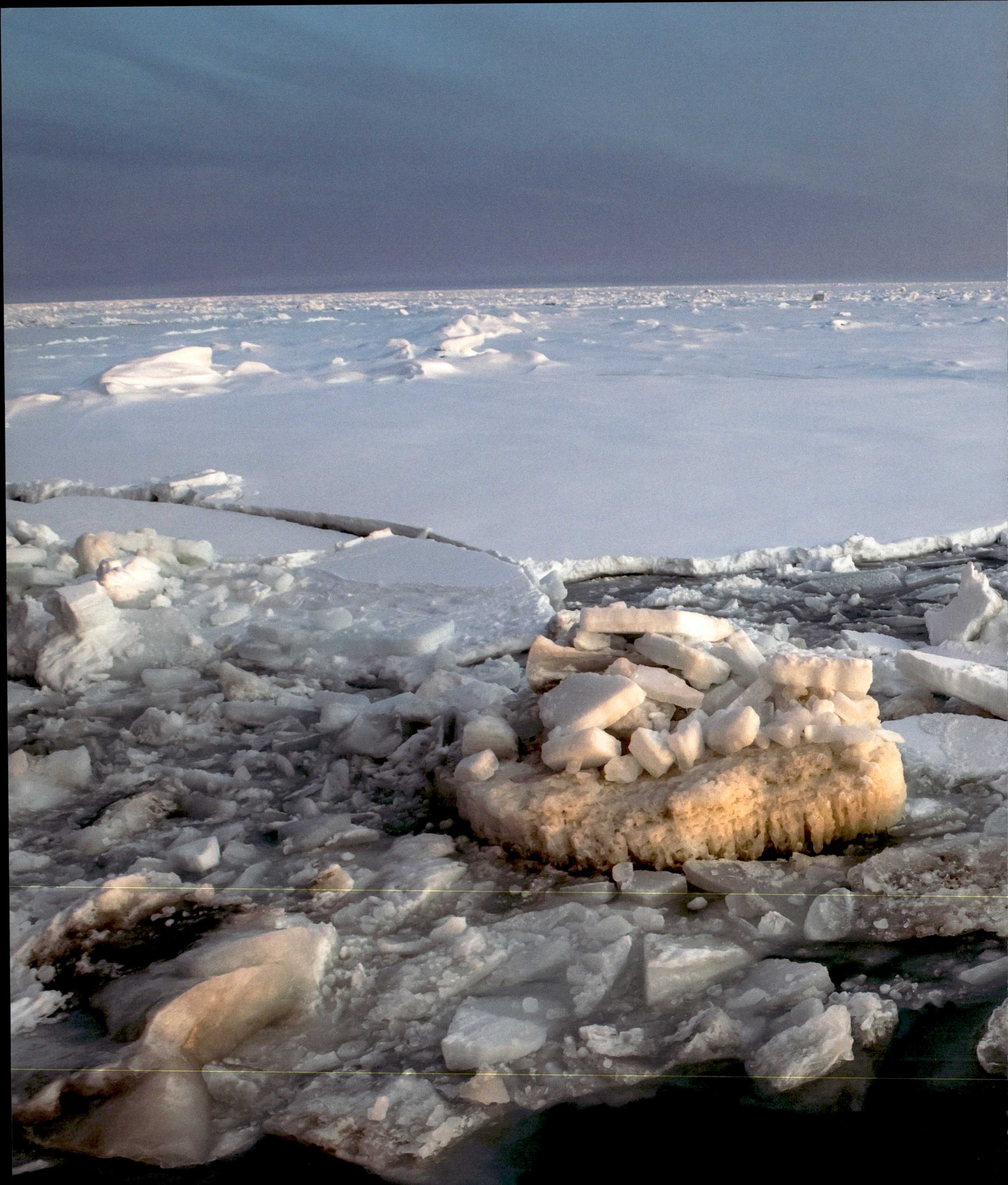

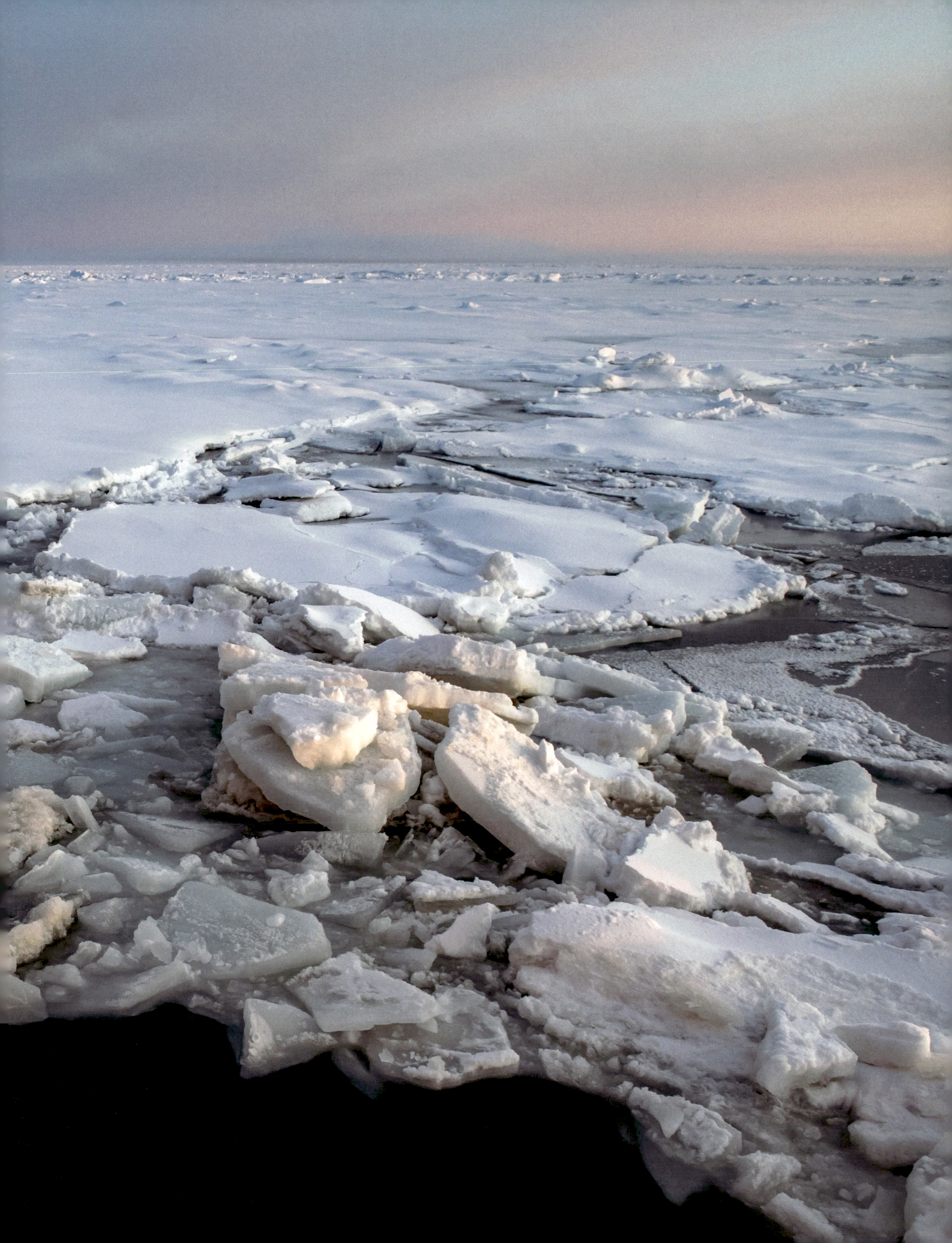

Only a few days earlier, the two research ships *Polarstern* and *Akademik Fedorov* were reunited in the central Arctic. In a well-coordinated manoeuvre, the captains brought their vessels to a standstill right next to each other. This means that people and cargo can change ships with the use of a crane. The scientists from the two vessels discuss the results to date of the search for an ice floe.

Following pages: During this time two inquisitive polar bears approach the ships. The mother bear and her cub seem to be in good condition.

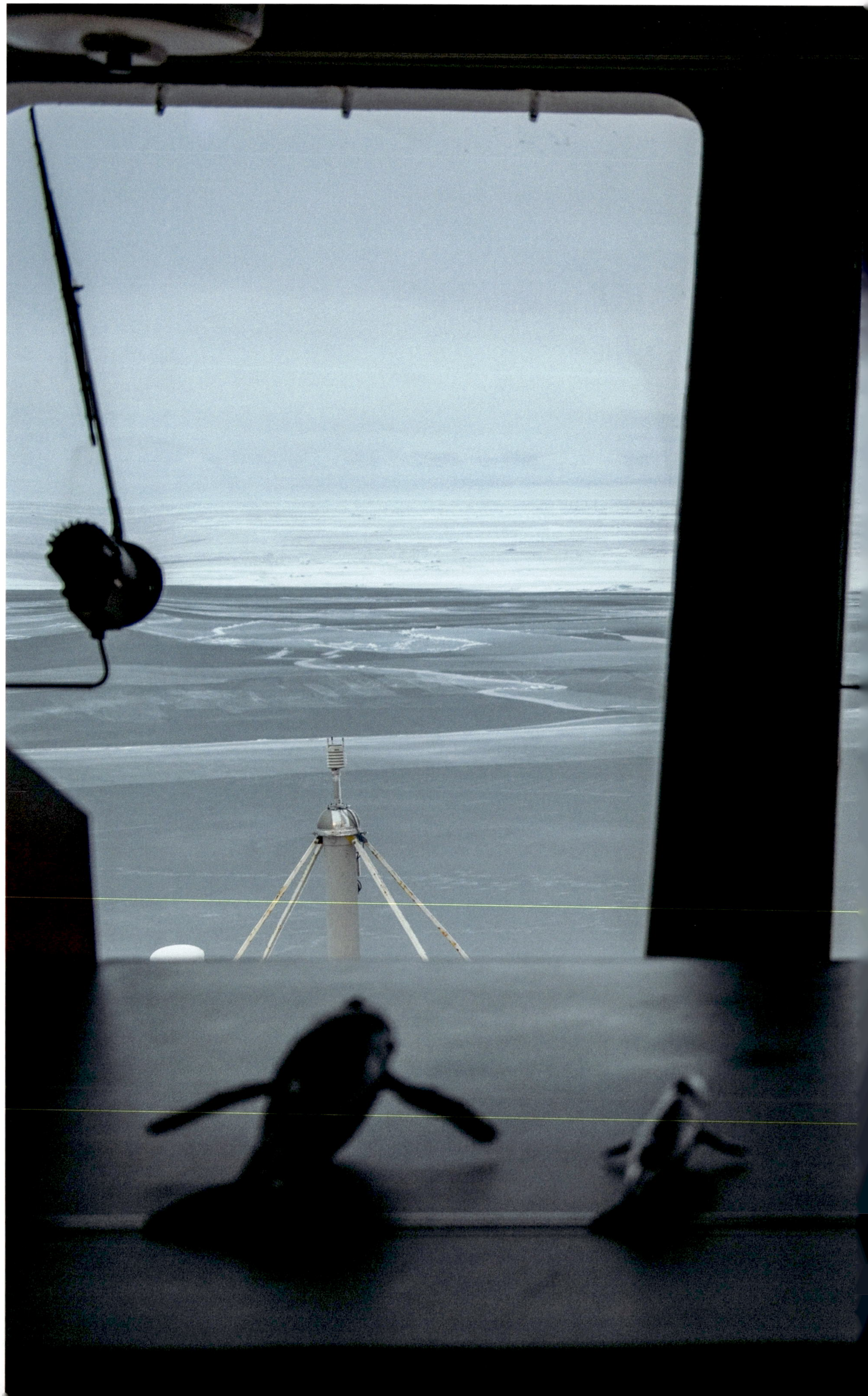

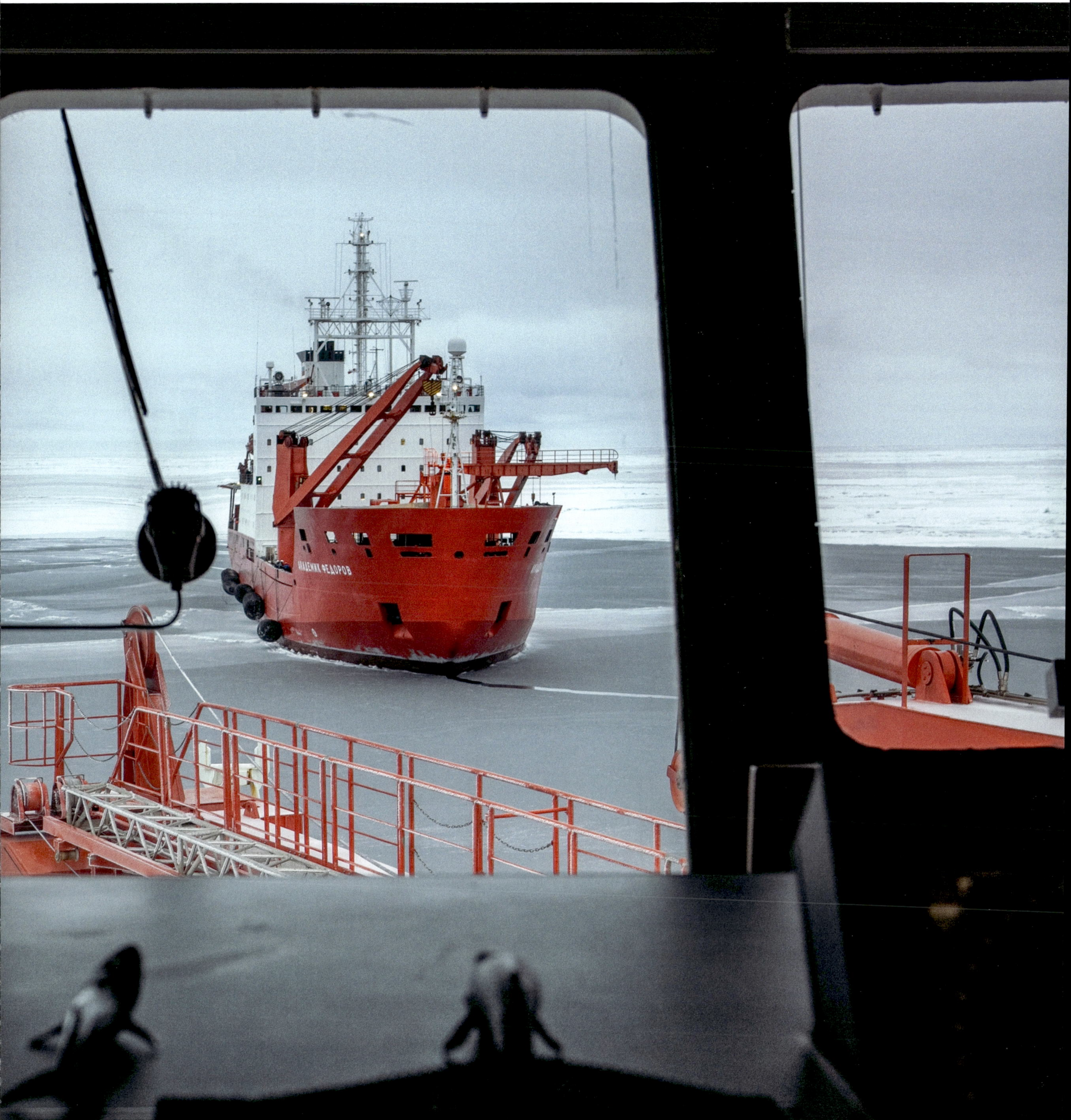
ФЕДОРОВ

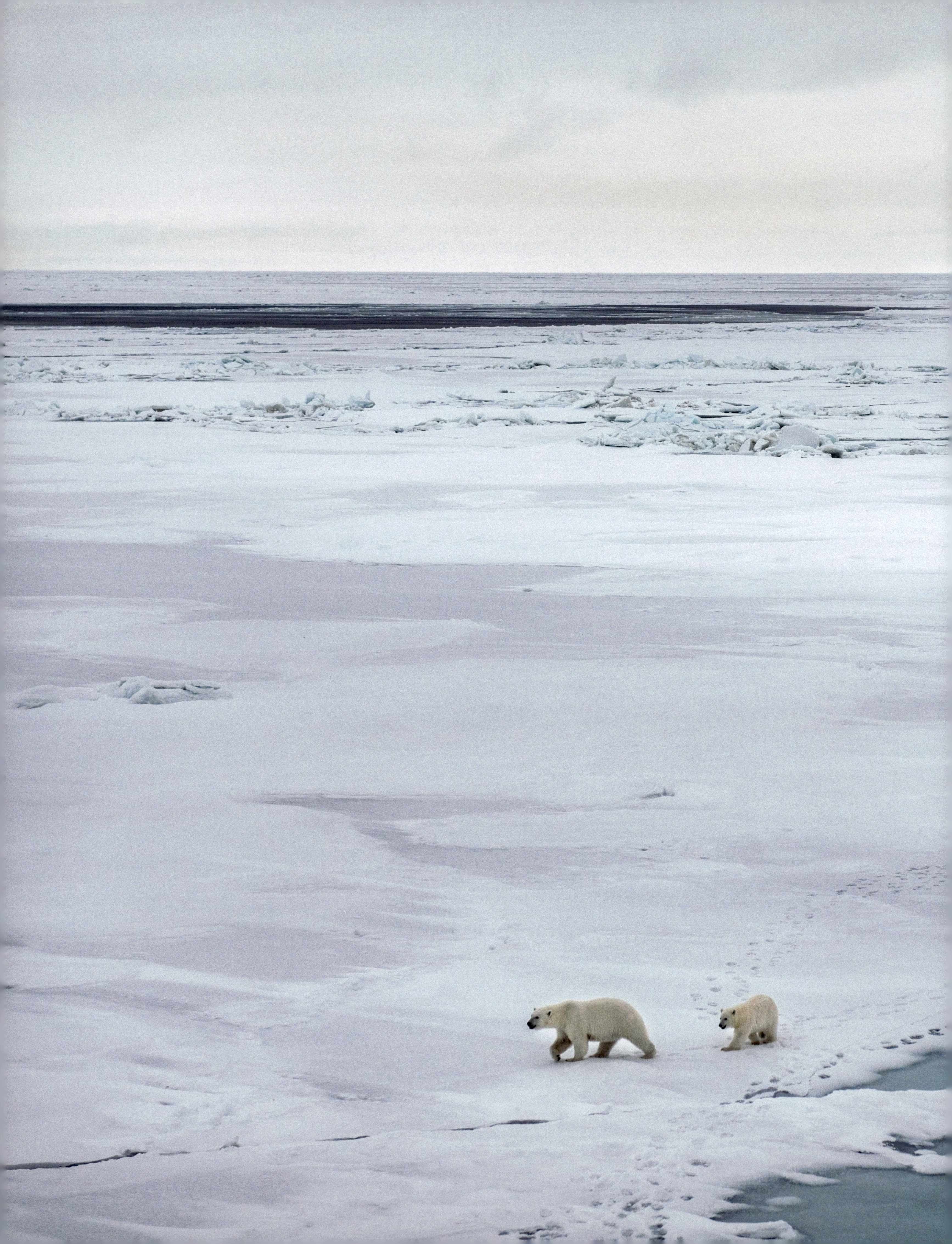

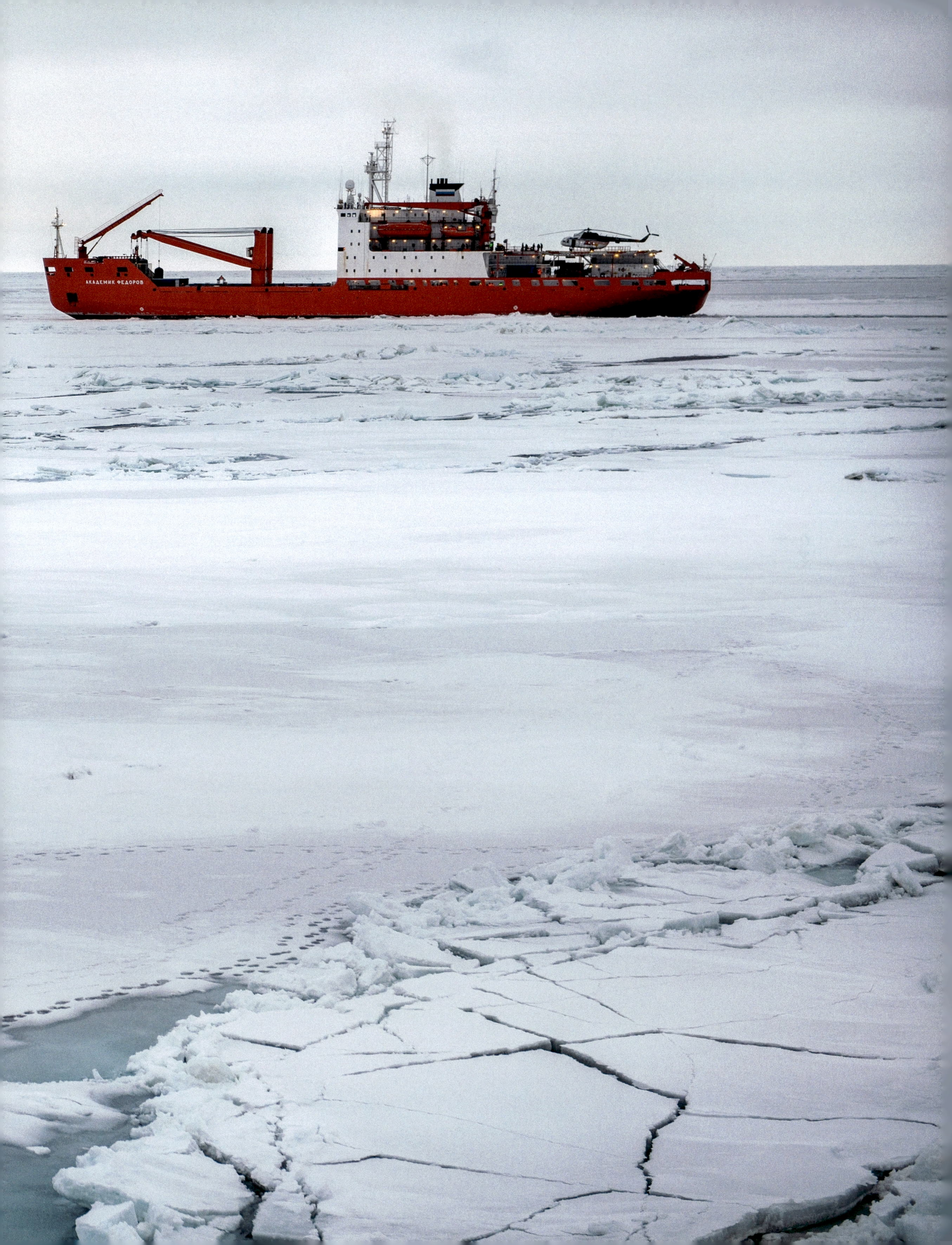
АКАДЕМИК ФЕДОРОВ

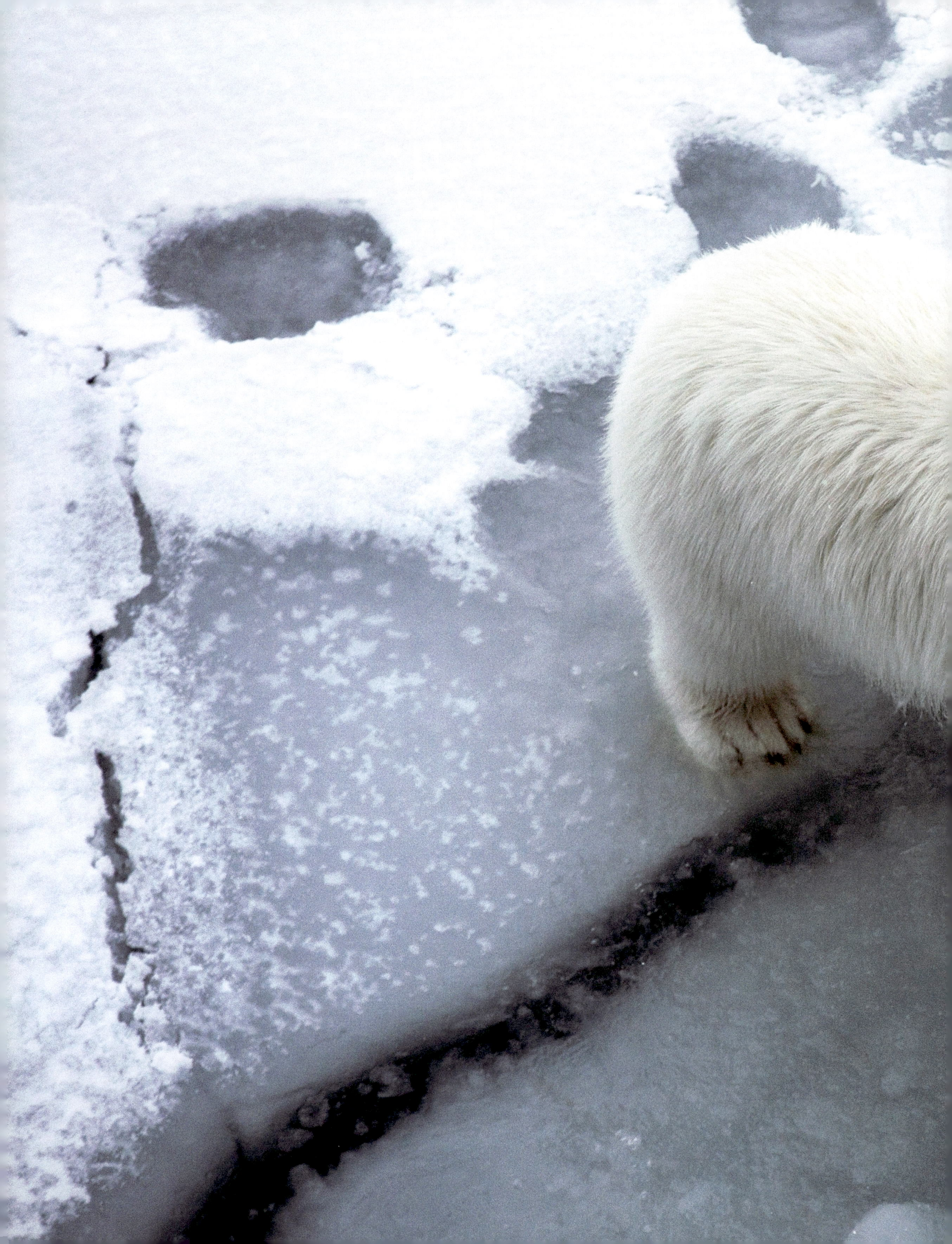

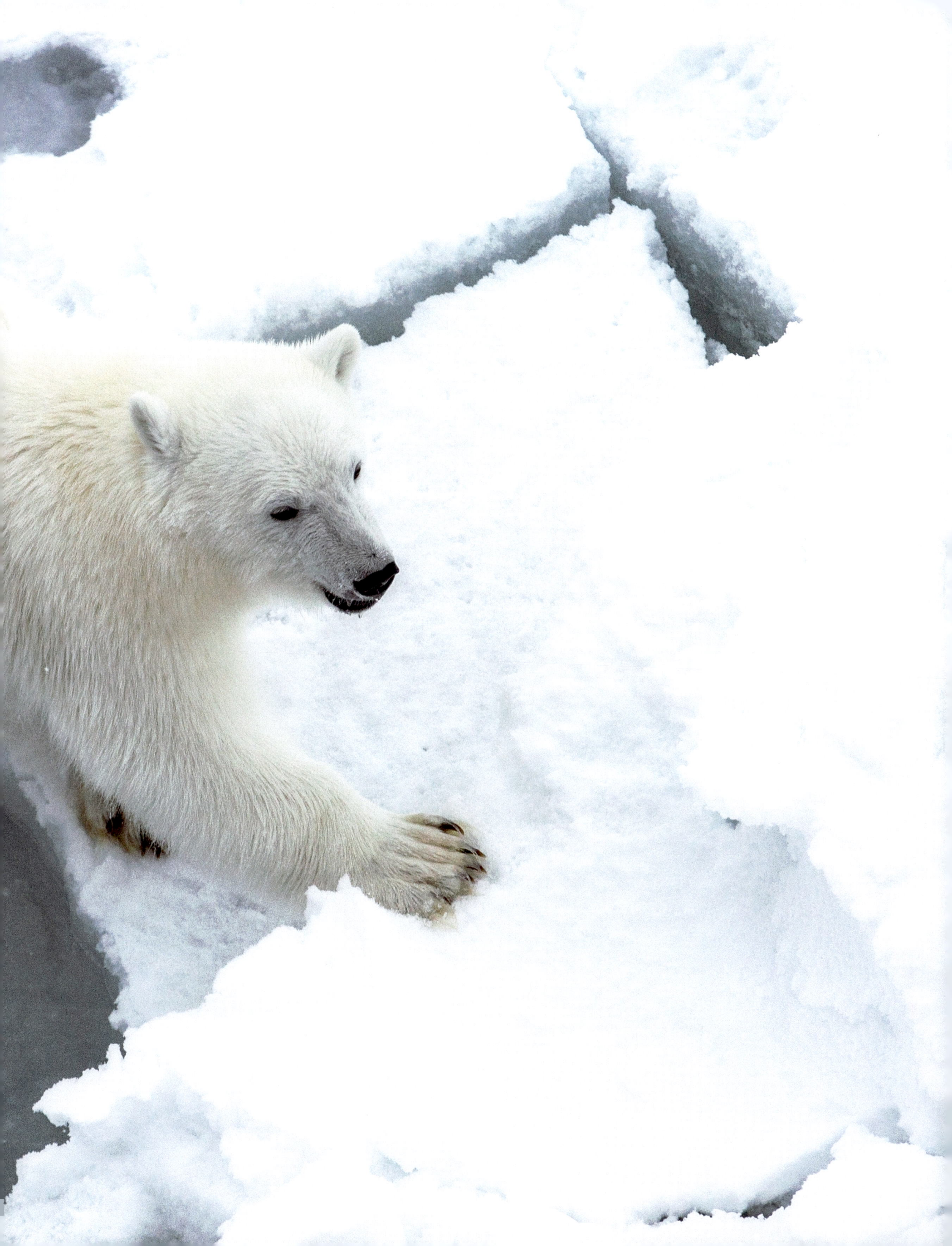

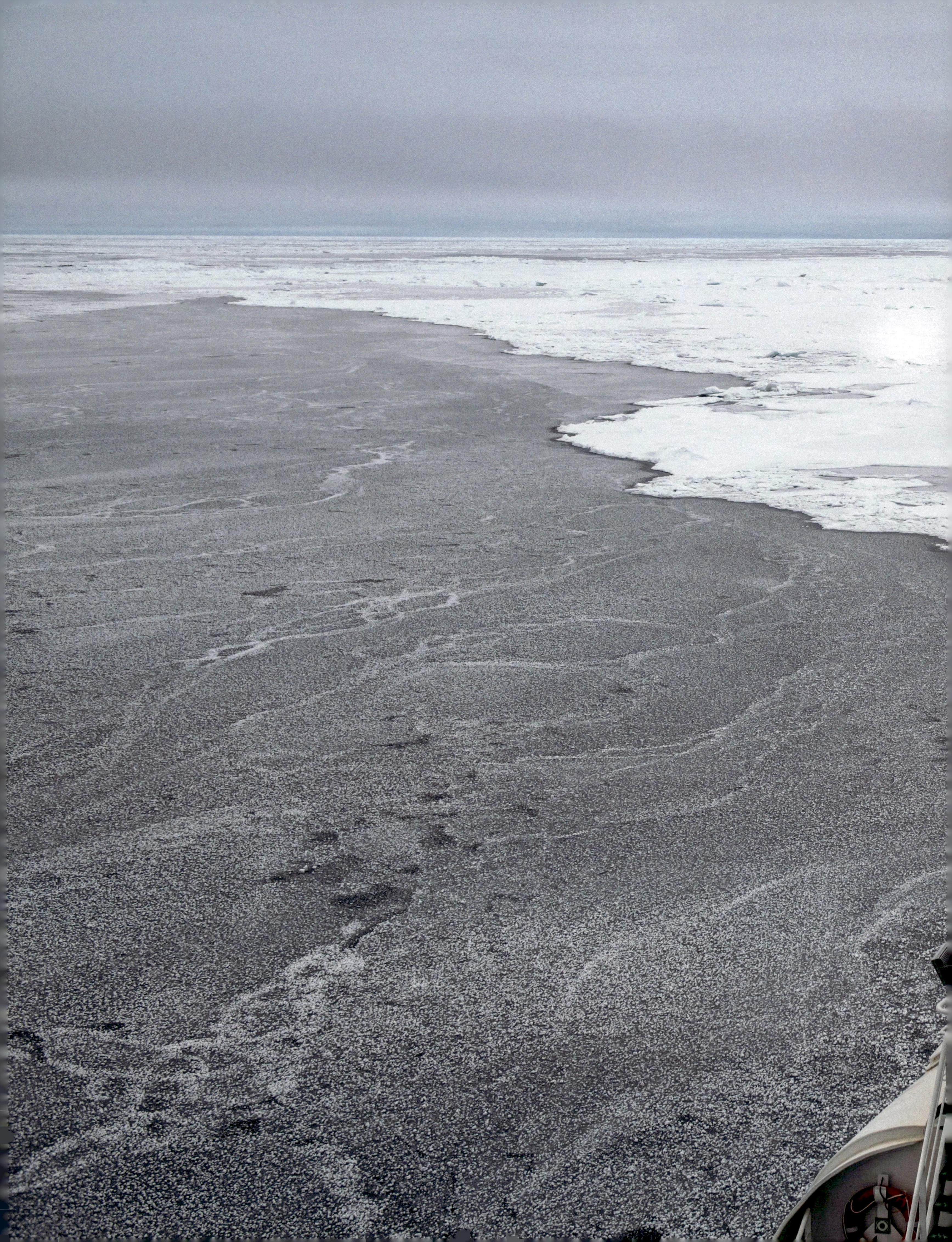

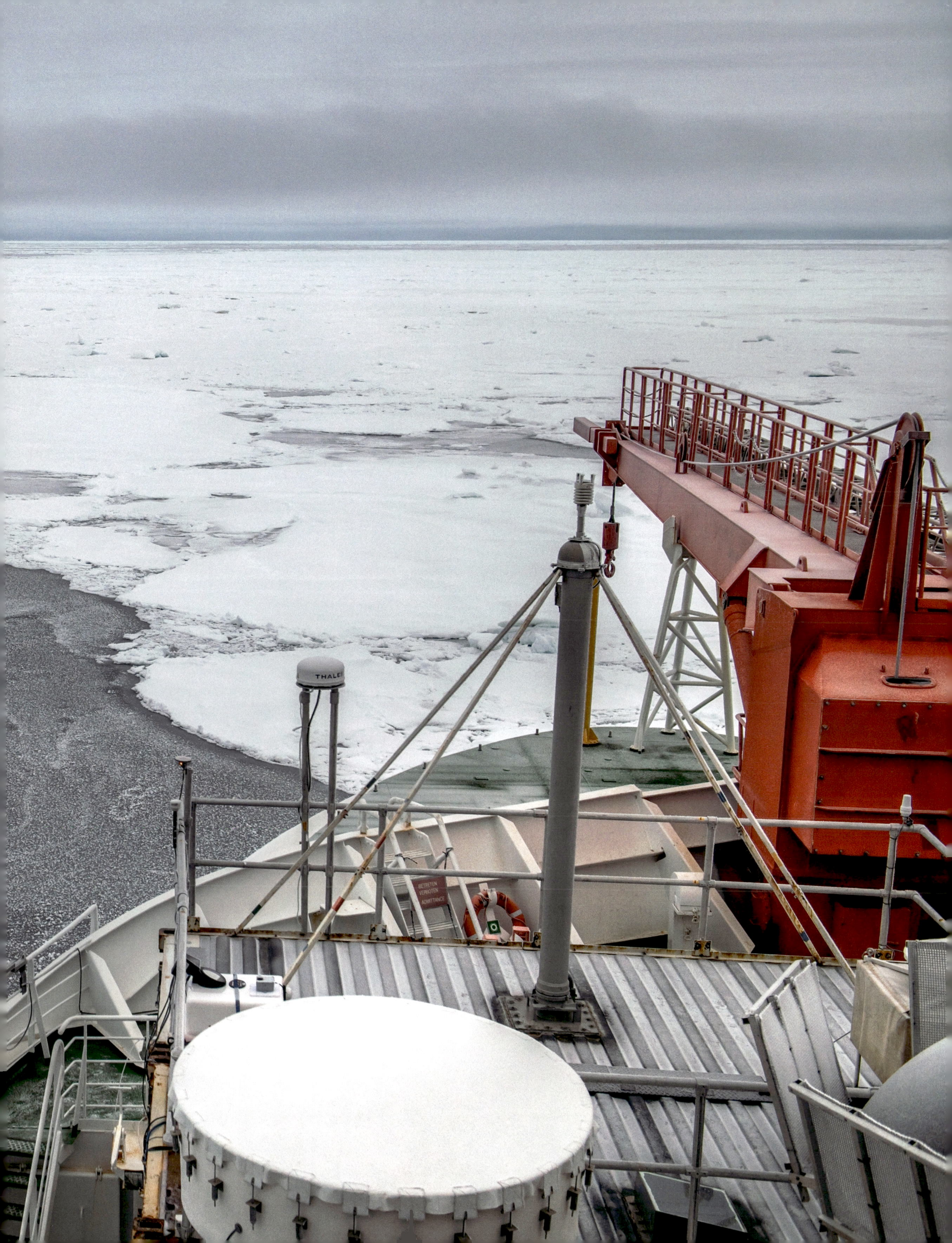
THALES

Bjela König has joined the expedition as a safety expert. For more than a year she has considered all scenarios that could pose a threat in the Arctic Ocean and has developed counter-strategies. Many dangers arise from the weather conditions. Fog can come in suddenly, making orientation on the ice difficult. Frostbite can occur on unprotected areas of the skin within minutes. The expedition members must also be prepared for polar bears everywhere and at all times. Now it will become clear how well the safety systems fare in practice.

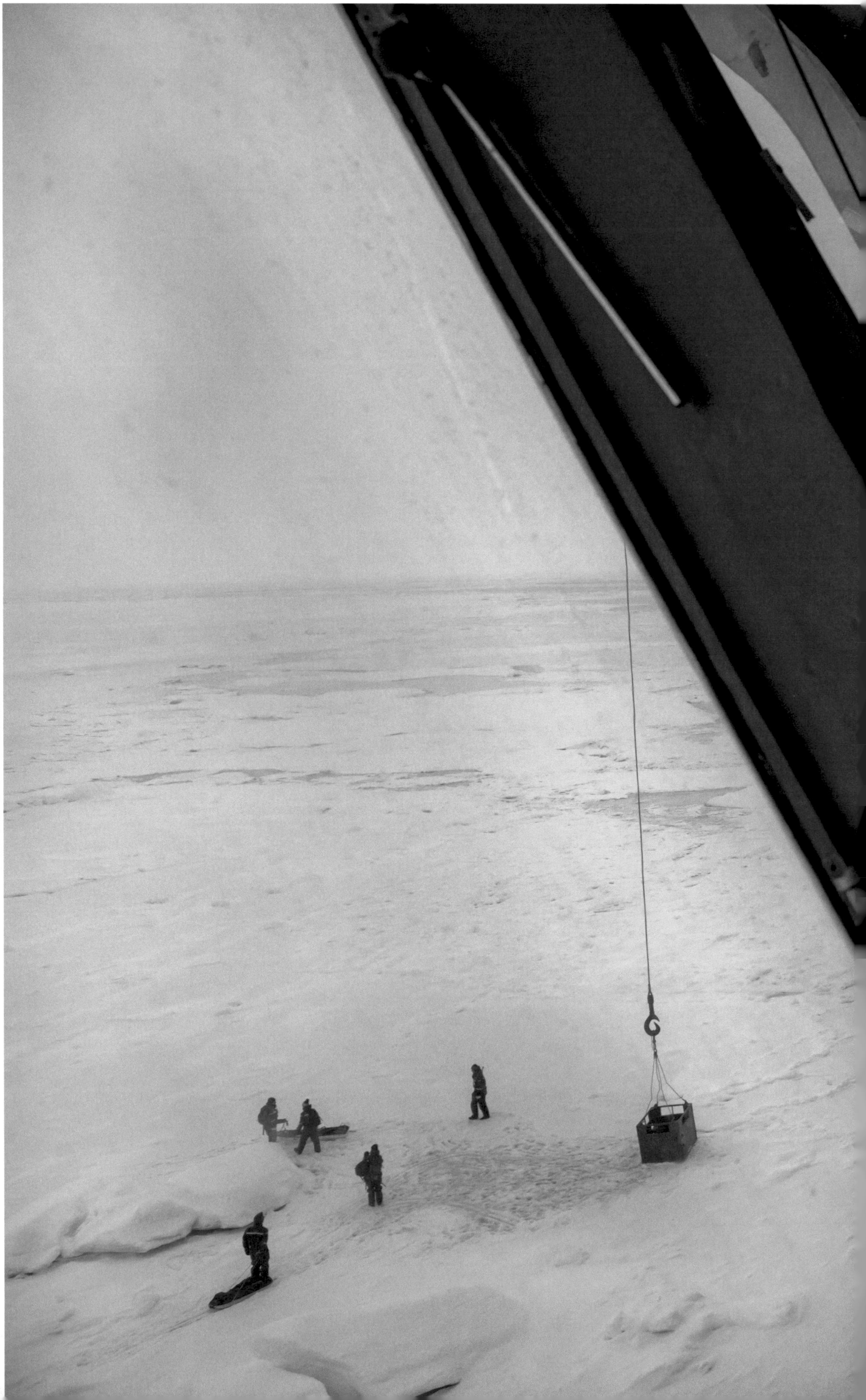

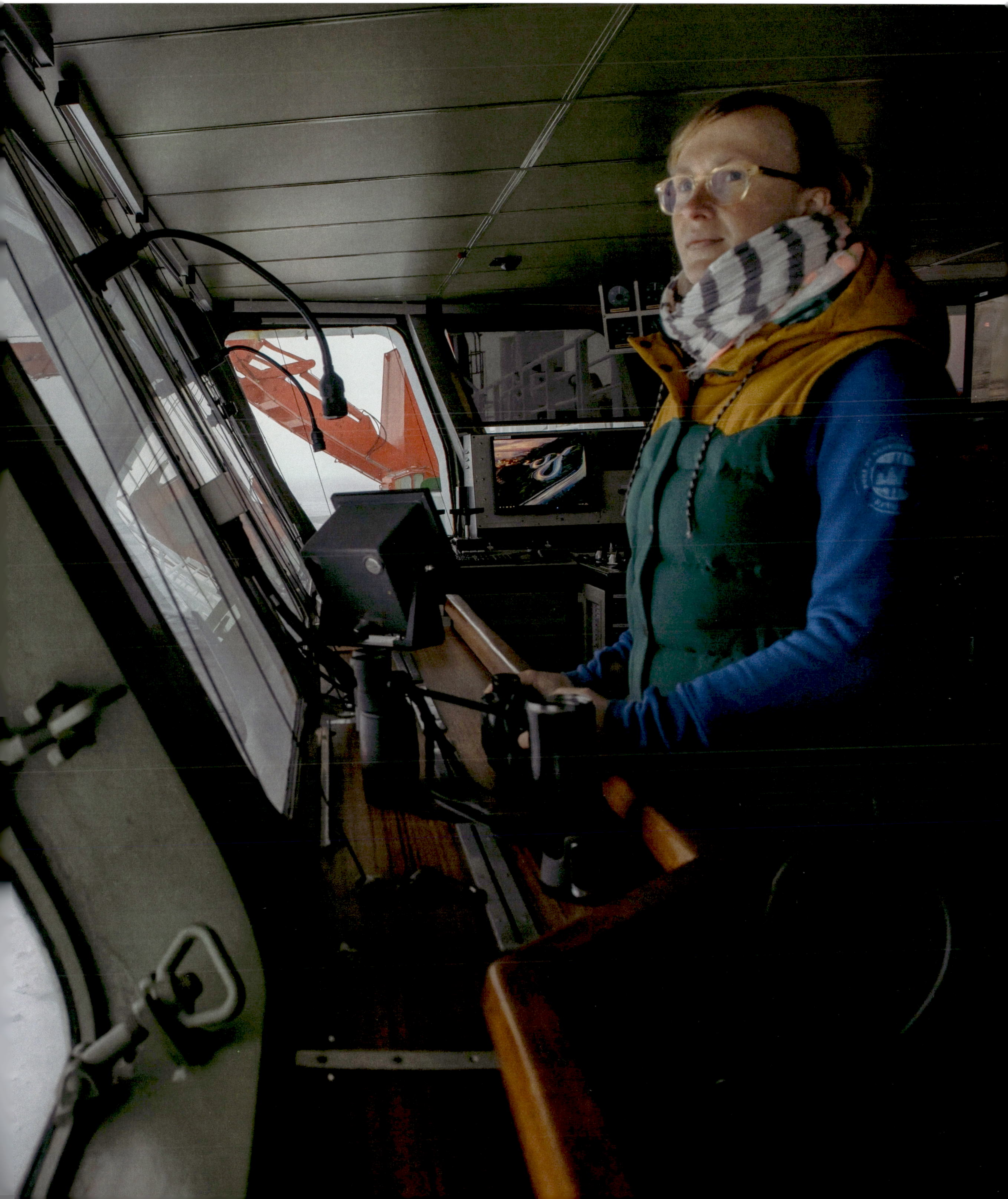

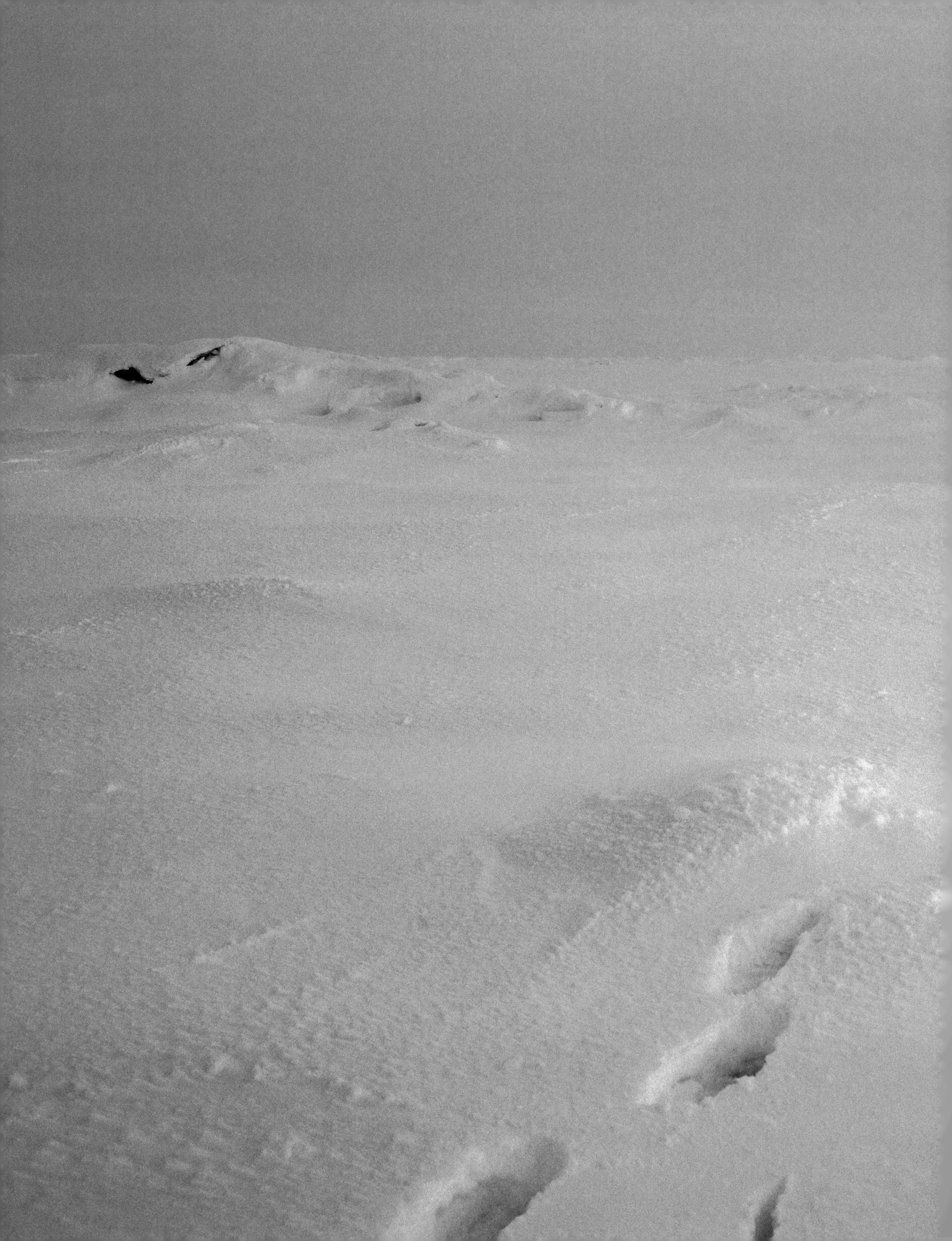

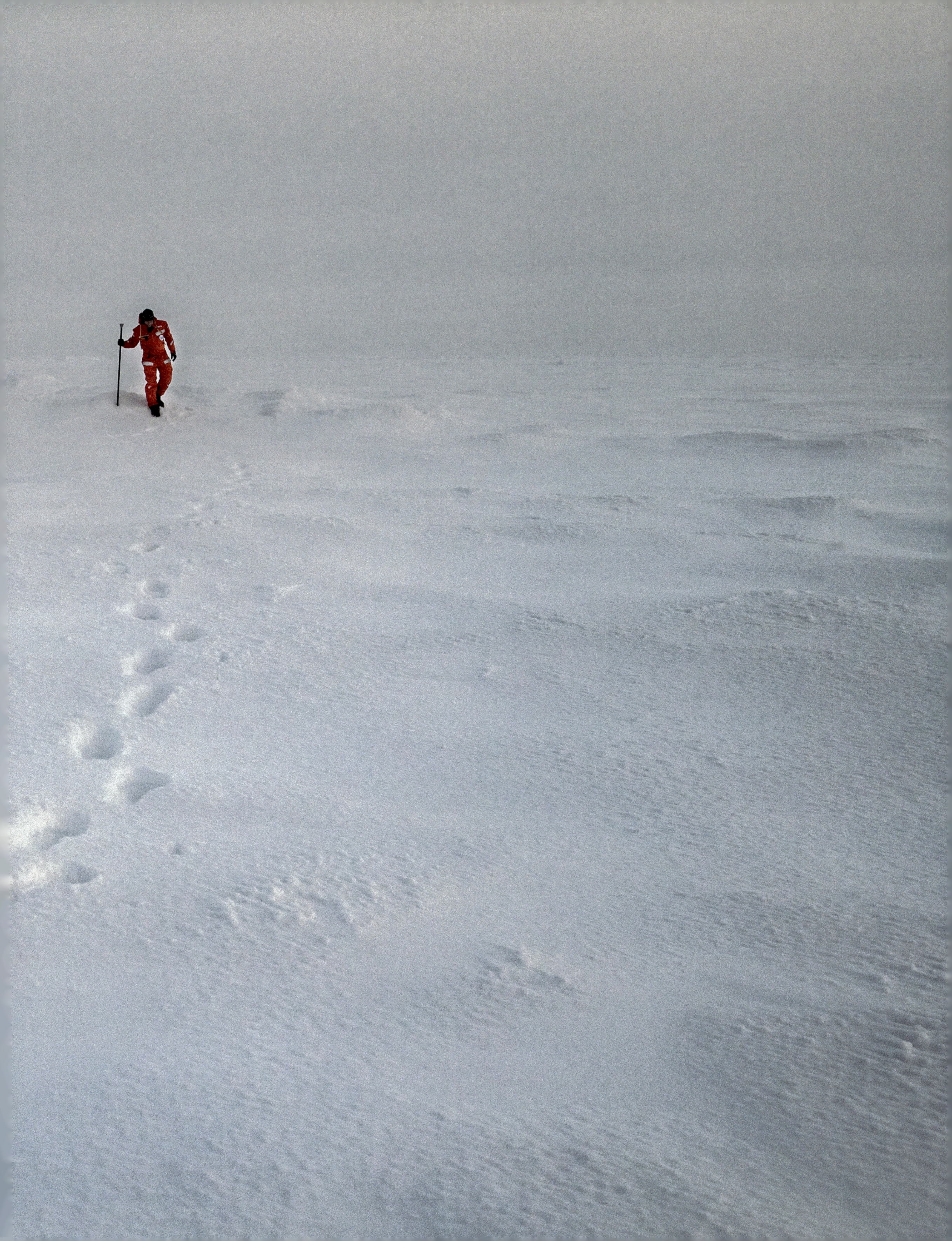

"It is not the perfect floe, but it is the best one in this area of the Arctic, and it provides us with better conditions than we could have expected following a warm Arctic summer." Markus Rex, expedition leader

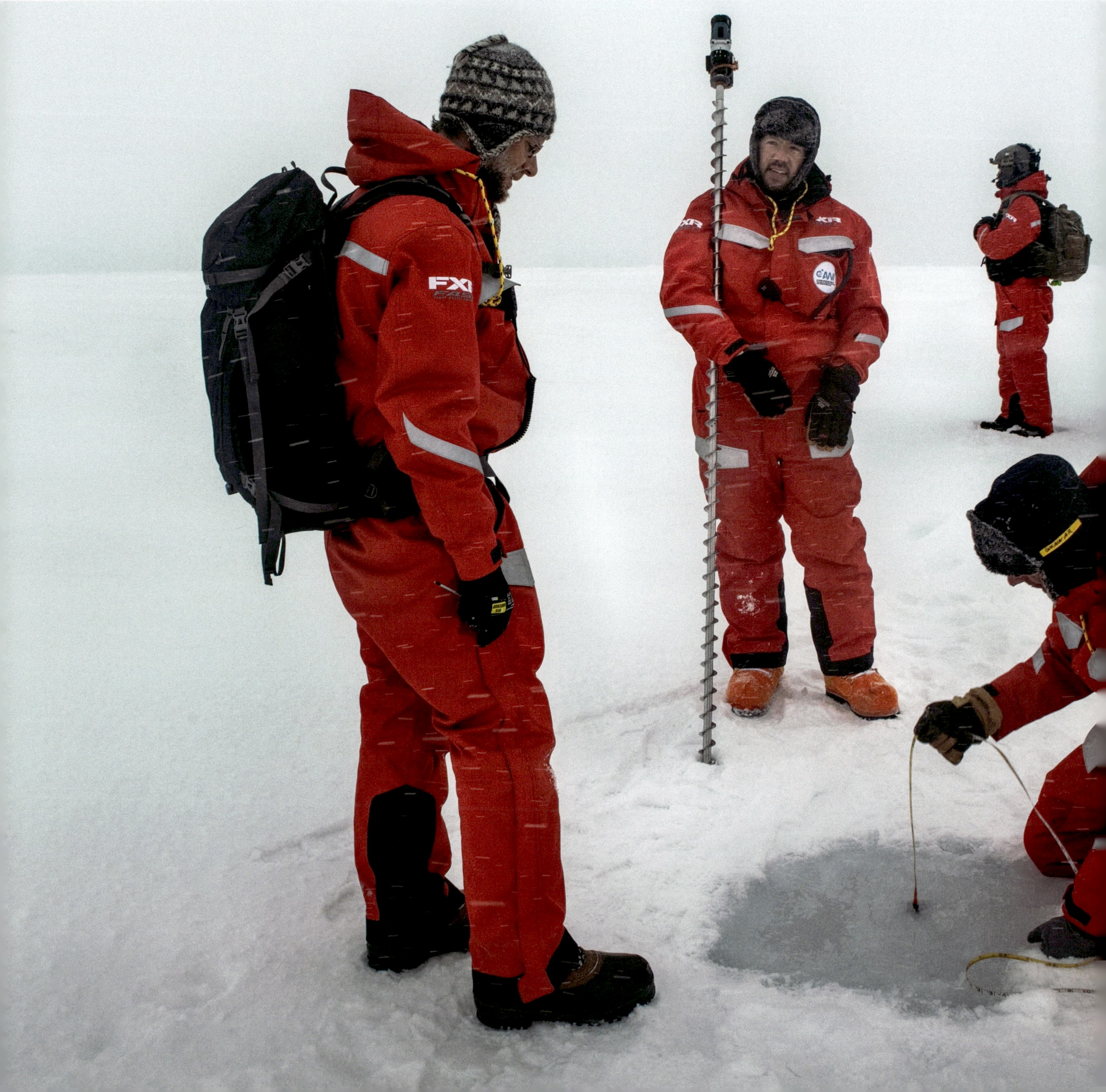

Since 28 September the first teams have been landing from the *Polarstern* by crane on the ice floe which, given analysis of satellite data, has long been seen as the favourite. They cannot know the exact condition of the ice at this location until, accompanied by polar bear guards, they have mapped the floe for several days and nights with an electromagnetic sensor and repeatedly taken ice samples (top right). Their investigations are coordinated from the bridge and monitored with infrared cameras.

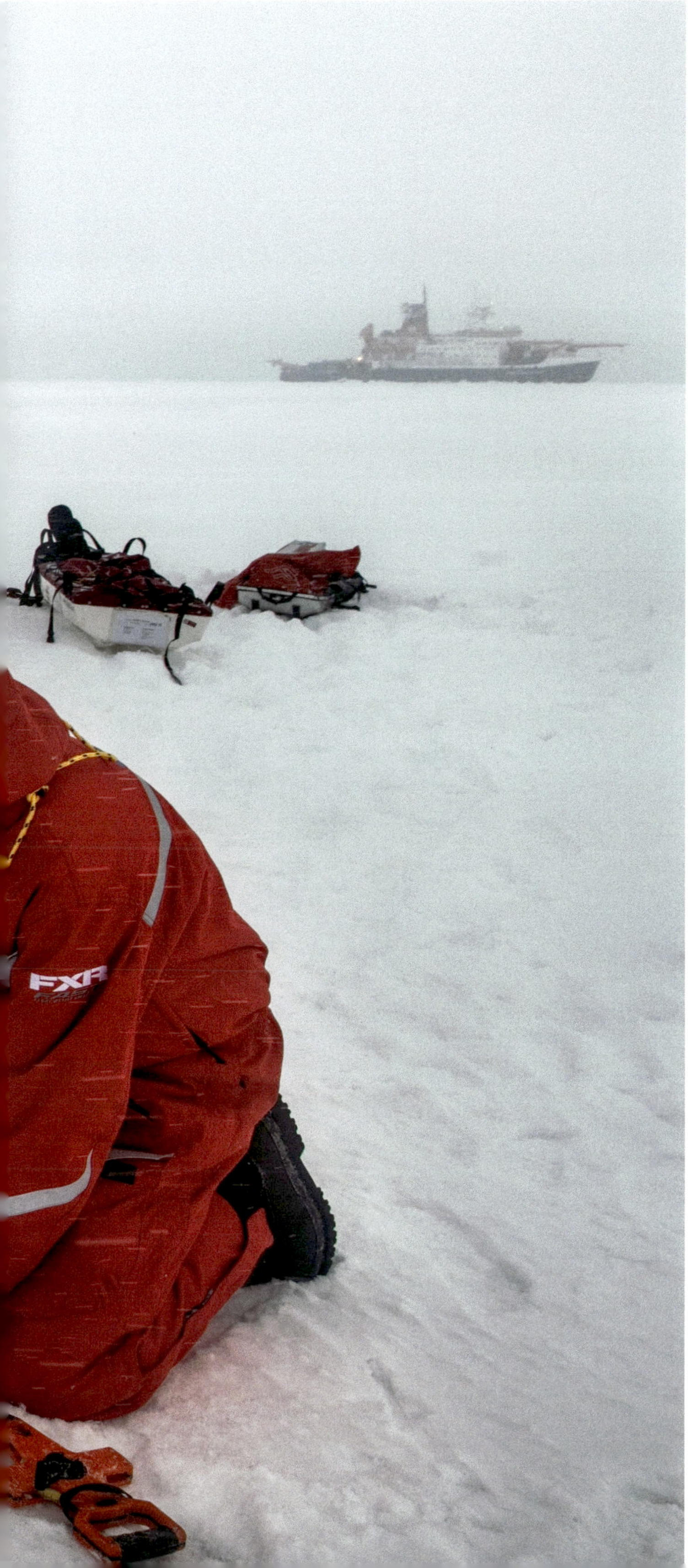

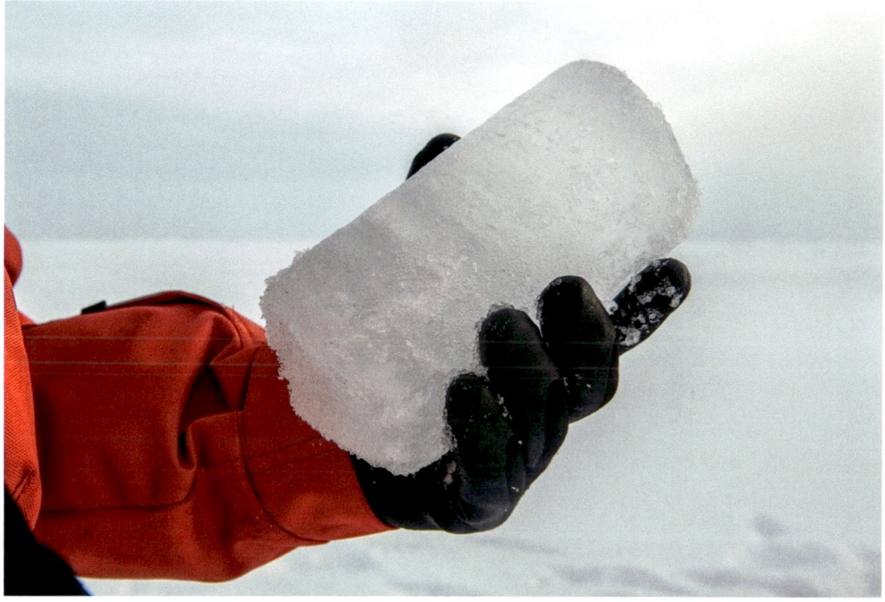

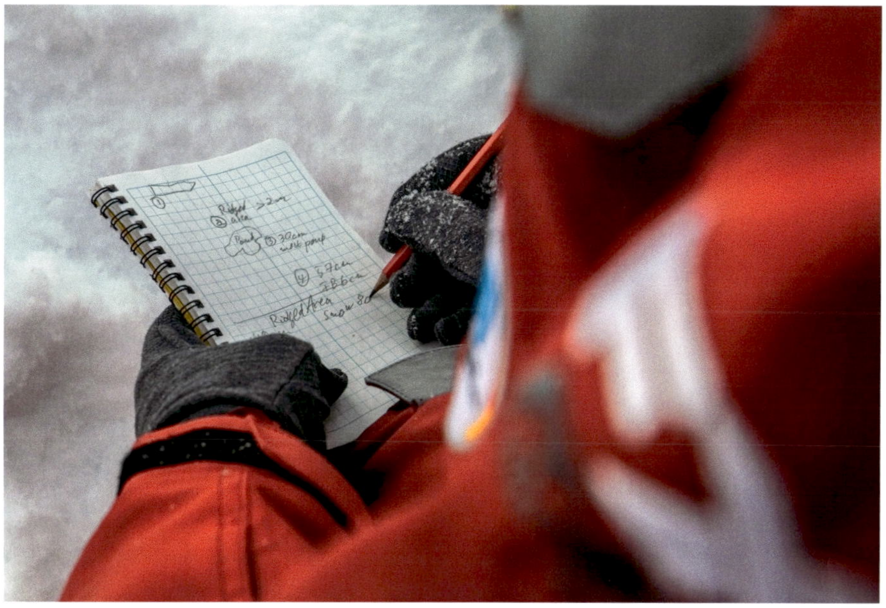

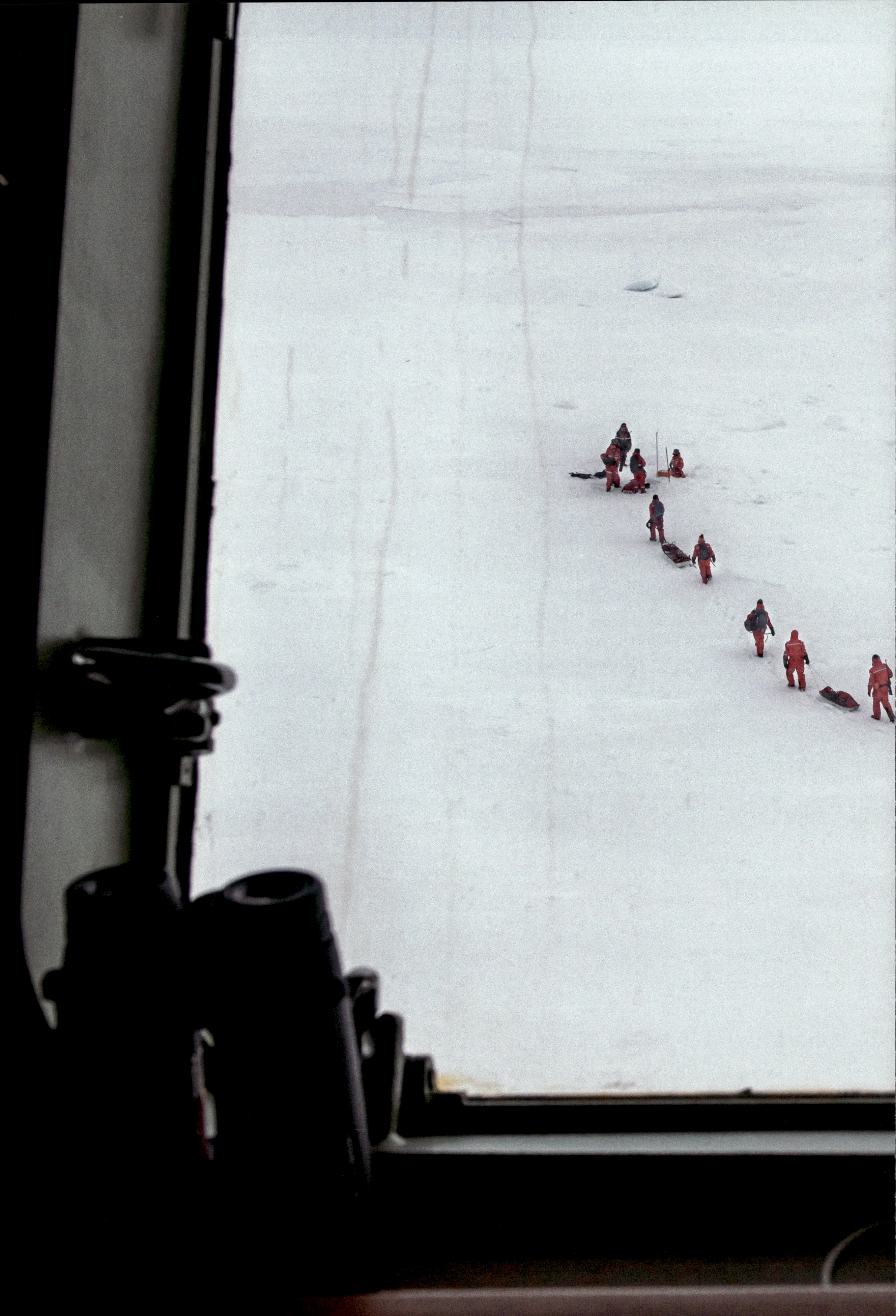

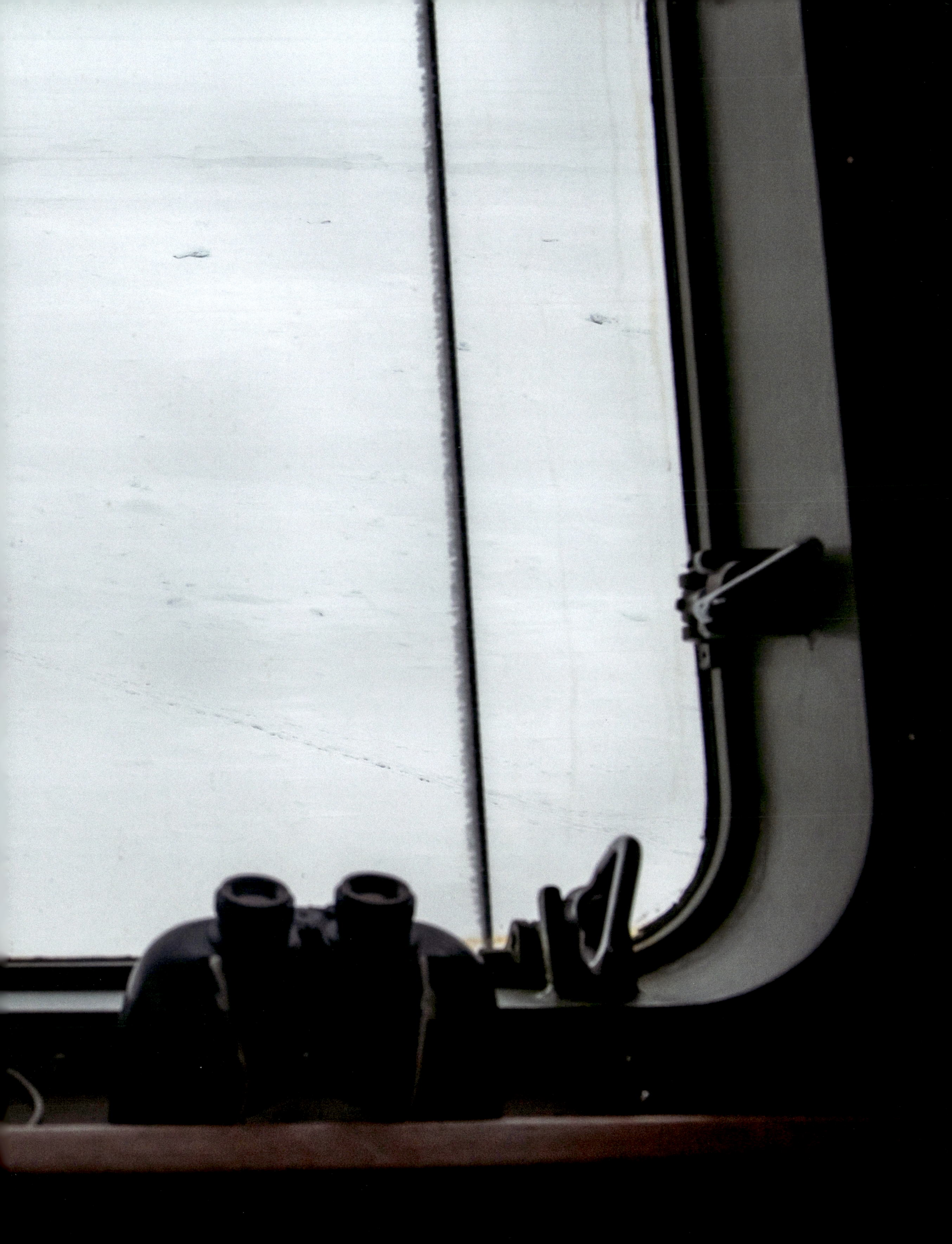

The Invention of the Ice Drift

Fridtjof Nansen and the Norwegian polar expedition

For Fridtjof Nansen, April 1895 began with a realization that probably saved his life. Far from home, after days of going over it in his head, the Norwegian decided to abandon his ambitious goal of being the first person to reach the North Pole. Locked into pack ice with twelve companions on the expedition ship *Fram*, he had drifted through the northern polar seas for almost two years but had missed the North Pole by hundreds of kilometres. In order to reach his goal in spite of this, he had left the ship a few weeks before, accompanied by Hjalmar Johansen, to cover the last degrees of latitude on sledges and snowshoes. However, endless chains of tall pressure ridges of compacted ice repeatedly forced them to carry their sledges. They hardly made any progress, and every kilometre was an ordeal. There was so little snow on the ice blocks that their snowshoes were useless. Everything was white on white: visibility was so poor they could barely spot bumps and cracks in the ice. Their desperation finally reached a limit. Their lives were worth more to them than the North Pole – and they turned back.

Despite this setback, the drift expedition of the *Fram* was a monumental achievement in the history of polar exploration. In his comprehensive account of the journey, *Farthest North*, Nansen explained how this daring expedition came into being and how it came to an end.

According to this account, he came upon the idea in autumn 1884, when he happened upon an astonishing article by the Norwegian scientist Henrik Mohn: on the west coast of Greenland, it stated, a few items had been found that undoubtedly belonged to the American naval ship *Jeannette*, which had sunk off the Siberian coast three years prior during an Arctic expedition. But how did this detritus reach Greenland? This find, according to Mohn, could only be explained by the theory of transpolar drift: the wind and sea currents, he assumed, pushed the ice slowly from Siberia to Greenland, passing the North Pole. Since at that time pack ice was an insurmountable obstacle for ships, the North Pole was practically inaccessible. And since the Arctic coasts were too far away, there was no suitable base camp nearby for an expedition on skis.

Nansen saw his chance: what worked for flotsam must also be possible for an expedition with a specially built ship. He conceived a vessel as small and strong as possible, just large enough for a crew of a dozen sailors and provisions for five years. The *Fram*, in contrast to previous expedition ships, was given a rounded hull that would provide almost no surface for the forces of the ice to assail. Instead of being crushed by masses of ice, the pressure would push the *Fram* upwards. People at the time undoubtedly found this idea odd, ignoring as it did numerous rules of shipbuilding. And most experts thought Nansen's plan foolhardy, but his enthusiasm won out. The idea of the Norwegian flag being the first to fly at the North Pole was simply too enticing. Thanks in part to public attention, considerable sums were raised to finance the expedition and countless applications arrived from experienced seamen. It was entirely uncertain whether the crew would ever return alive.

In 1893, on Midsummer's Day, 24 June, the hour had come: it was time to say goodbye. In Kristiania, as the Norwegian capital Oslo was then called, people flocked to the quayside, joyfully waving their hats and handkerchiefs. Nansen later noted that the view through the telescope back to his homeland was the most difficult moment of the whole journey. In the months and years to follow, he and his crew were to face indescribable hardships. After voyaging along the coasts of Norway and Russia, on 20 September they finally reached the edge of the ice at a latitude of 78°N. In his report, Nansen described what happened four days later:

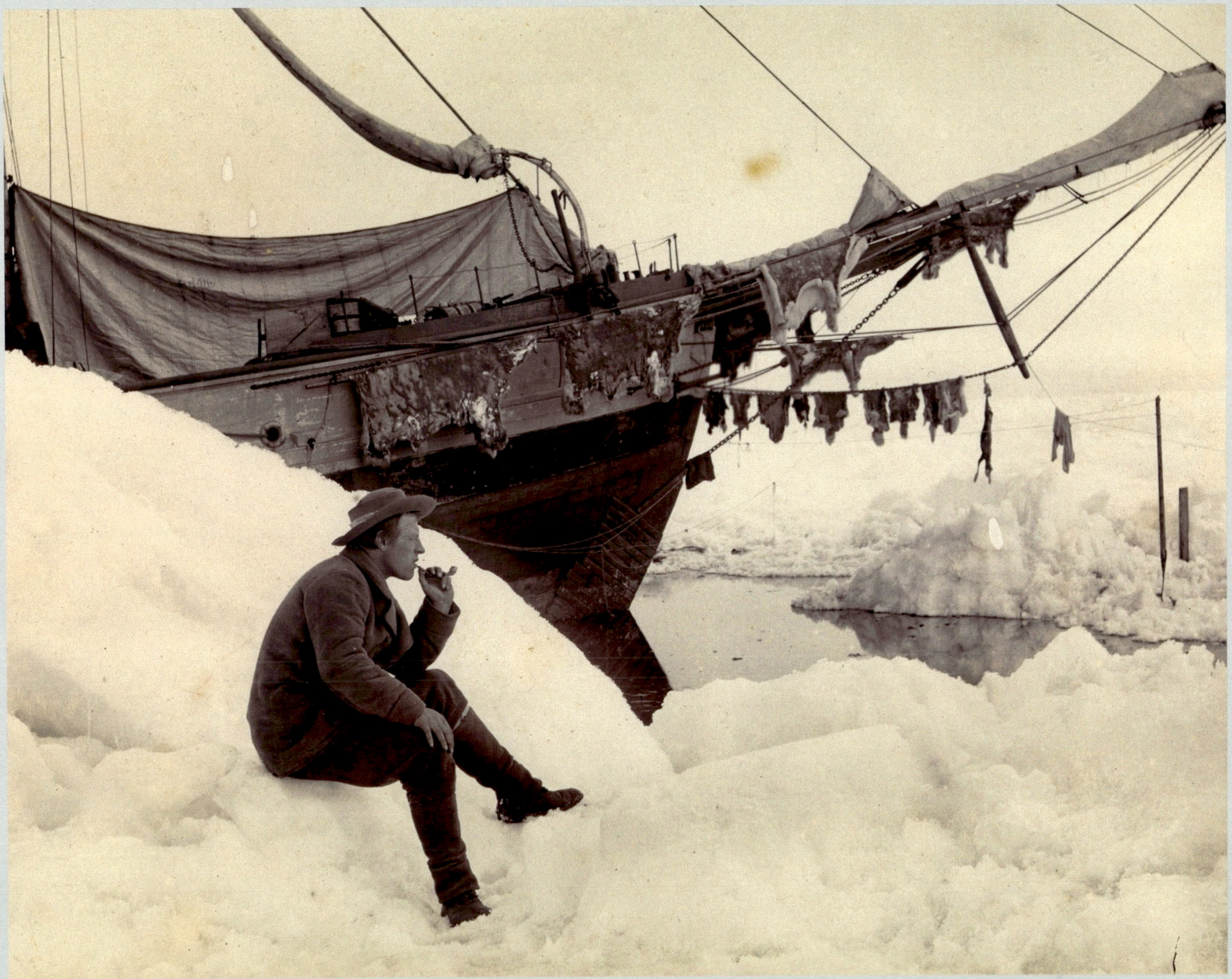

Fog in the morning, which cleared off as the day went on, when we discovered that we were closely surrounded on all sides by tolerably thick ice. Between the floes lies slush-ice, which will soon be quite firm. There is an open pool to be seen to the north, but not a large one. From the crow's-nest, with the telescope, we can still descry the sea across the ice to the south. It looks as if we were being shut in. Well, must e'en bid the ice welcome.

The polar night was approaching and in the following days Nansen and his team converted the *Fram* into winter quarters. This winter was to last almost three years. The force of the ice made the ship shudder with a loud noise. Nansen was delighted, as the *Fram* behaved exactly as he had conceived it: it was pushed upwards by the ice. Nansen meticulously noted his observations; for example, he realized that the pressure of the ice was particularly strong at full moon and new moon. Over the course of the winter, however, the initial euphoria turned to frustration, as the drift hardly took them northwards: every time they drifted a little to the north, a setback followed shortly afterwards. On 18 February 1894 he wrote:

We are now in about 80°N. lat., in September we were in 79°; that is, let us say, one degree for five months. If we go on at this rate, we shall be at the Pole in forty-five, or say fifty, months, and in ninety or one hundred months at 80°N. lat. on the other side of it, with probably some prospect of getting out of the ice and home in a month or two more. At best, if things go on as they are doing now, we shall be home in eight years.

The less north they drifted, the more serious his crew's faces became. Every day it was clearer to Nansen that he would only be able to reach the North Pole, if at all, with sledges. Finally, on 16 November 1894, after a walk in snowshoes by the moonlight, he announced his decision to set out in spring. For Nansen and

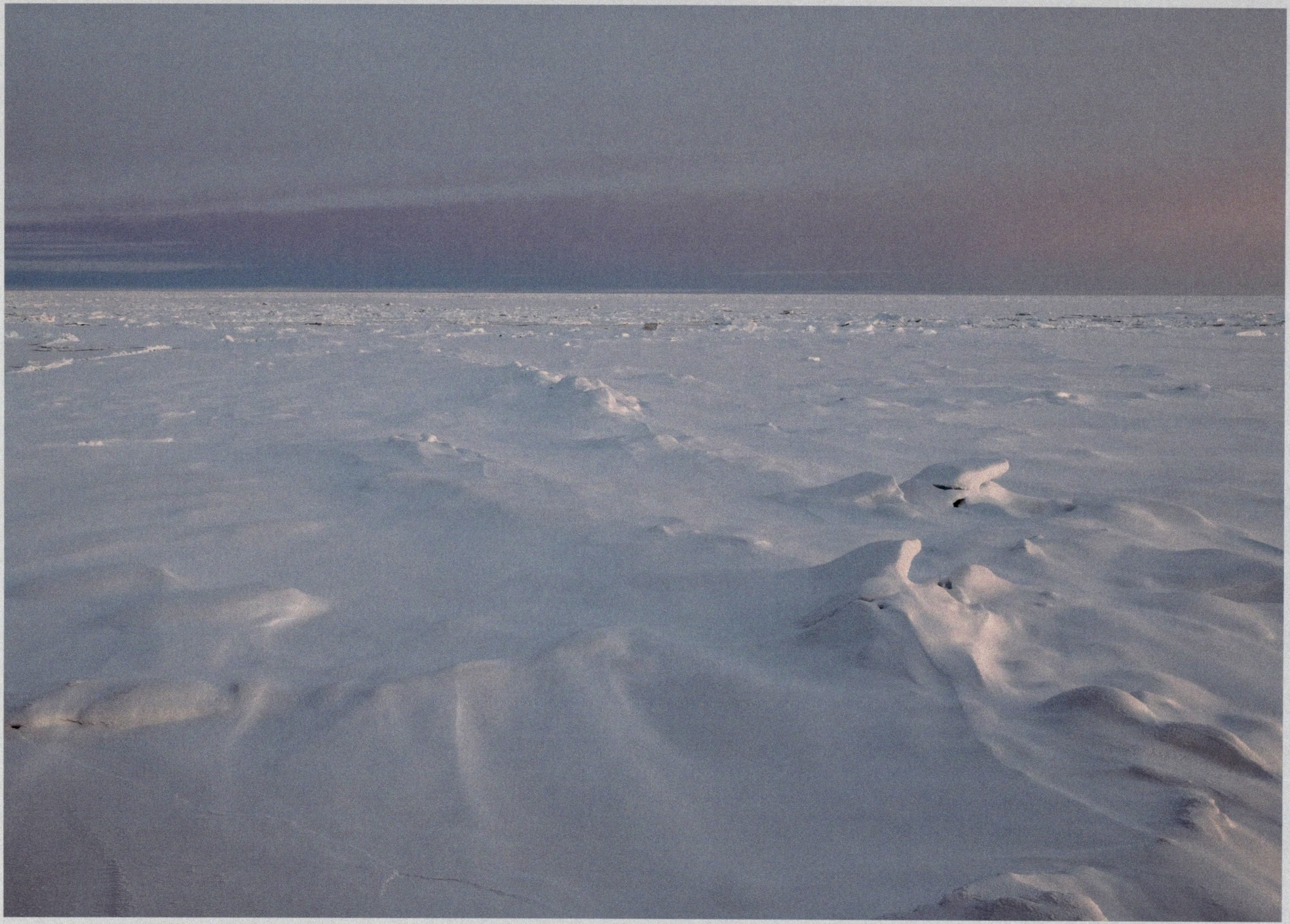

his team, preparing for this part of the expedition meant a longed-for turning point. They calculated what they needed to cover the 780 kilometres to the North Pole: two men, twenty-eight dogs, 1,050 kilograms of provisions. When his eyes roamed across the map, a shiver ran down Nansen's spine.

Nansen and Johansen had to quickly abandon the first two attempts and return to the safety of the ship after their sledges were damaged. But on 14 March 1895, accompanied by the thunder of gunfire, it was at last time to bid farewell for an extended period. Their expedition began at about 84°N and at first led them across smooth formations in good weather, with only the occasional ice ridge rising rose from the surface. Yet this was to change only a few days later. On 6 April 1895 Nansen noted:

The ice grew worse and worse. Yesterday it brought me to the verge of despair, and when we stopped this morning I had almost decided to turn back. I will go on one day longer, however, to see if the ice is really as bad further northward as it appears to be from the ridge, 30 feet [9.1 m] in height, where we are encamped. We hardly made 4 miles [6.4 km] yesterday. Lanes, ridges and endless rough ice, it looks like an endless moraine of ice-blocks; and this continual lifting of the sledges over every irregularity is enough to tire out giants. Curious this rubble-ice; for the most part it is not so very massive, and seems as if it had been forced up somewhat recently, for it is incompletely covered with thin, loose snow, through which one falls suddenly up to one's middle. And thus it extends miles after mile northward.

Two days later, Nansen and Johansen finally gave up. Their new destination was now the south. Since they did not know where the *Fram* had drifted to, they aimed for the Franz Josef Land archipelago. At first they progressed far better than in previous weeks, but as time passed the large number of small and large channels running through the ice complicated their journey. Food for their sledge dogs also began to run short and they were forced to put a

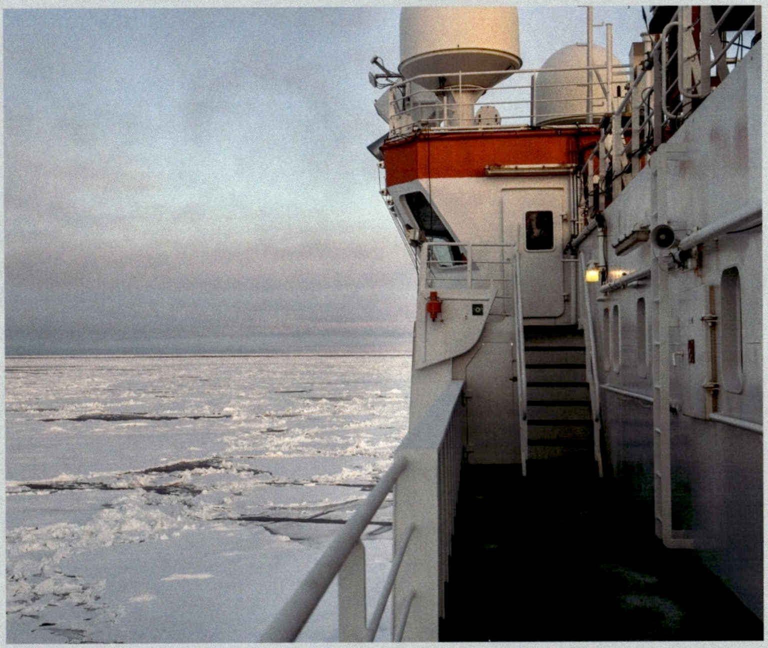

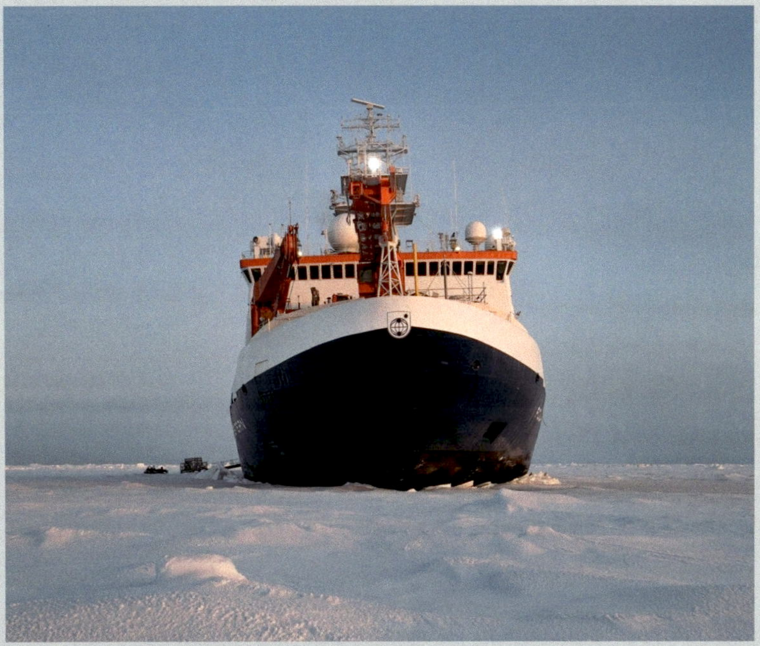

number of them down. The further south they came, the more they encountered animals such as seals, whales and seabirds, and their hopes rose of soon reaching land. It was two years after they had set sail with the *Fram*. On 24 July 1895 Nansen wrote:

At last the marvel has come to pass. Land, land! and after we had almost given up our belief in it! After nearly two years, we again see something rising above that never-ending white line on the horizon yonder – a white line which for millennium after millennium has stretched over this sea, and which for millenniums to come shall stretch in the same way. We are leaving it, and leaving no trace behind us, for the track of our little caravan across the endless plains has long ago disappeared. A new life is beginning for us; for the ice it is ever the same. It has long haunted our dreams, this land, and now it comes like a vision, like fairy-land! Drift-white, it arches above the horizon like distant clouds, which one is afraid will disappear every minute.

Although they now had terra firma beneath their feet, Nansen and Johansen were still far from their destination. They established primitive winter quarters on the archipelago, little more than a cave in the ice that provided some protection against the weather. They found enough animals to hunt, but the seemingly endless time was now their greatest enemy. At times they slept as much as twenty hours a day in order to endure as little as possible of the winter. Time passed, week by week, month by month, until finally in May 1896 they set off again. In June there took place what must be one of the strangest encounters in history. At first Nansen thought he heard a voice, then he saw a dog between the hills, and a little later a dark figure. He said the figure was speaking English to the dog, and even thought he had seen this person before. Indeed, it was the British polar explorer Frederick George Jackson, who had been surveying Franz Josef Land for several years. They thus came face to face with one another: on one side Jackson in a plaid suit, scented with perfume; on the other side Nansen, dressed in rags and with a face that, under the patina of past winters, hardly resembled that of a person any more. Jackson was thrilled to see Nansen, who was long given up for lost.

With Jackson's help the two men reached the northern Norwegian port of Vardø on 13 August 1896. That same day Nansen headed to the local post office and put a large bundle of telegrams on the desk which he wanted to be sent as soon as possible. On that day, before Nansen and Johansen were reunited with their companions from the *Fram*, which by then had broken free of the ice, the world learned that two members of the Norwegian polar expedition had returned safely and, although they had not achieved their goal, in crossing the 86th parallel had reached further north than any human in history.

The Drift

Setting up the world's northern-most research camp

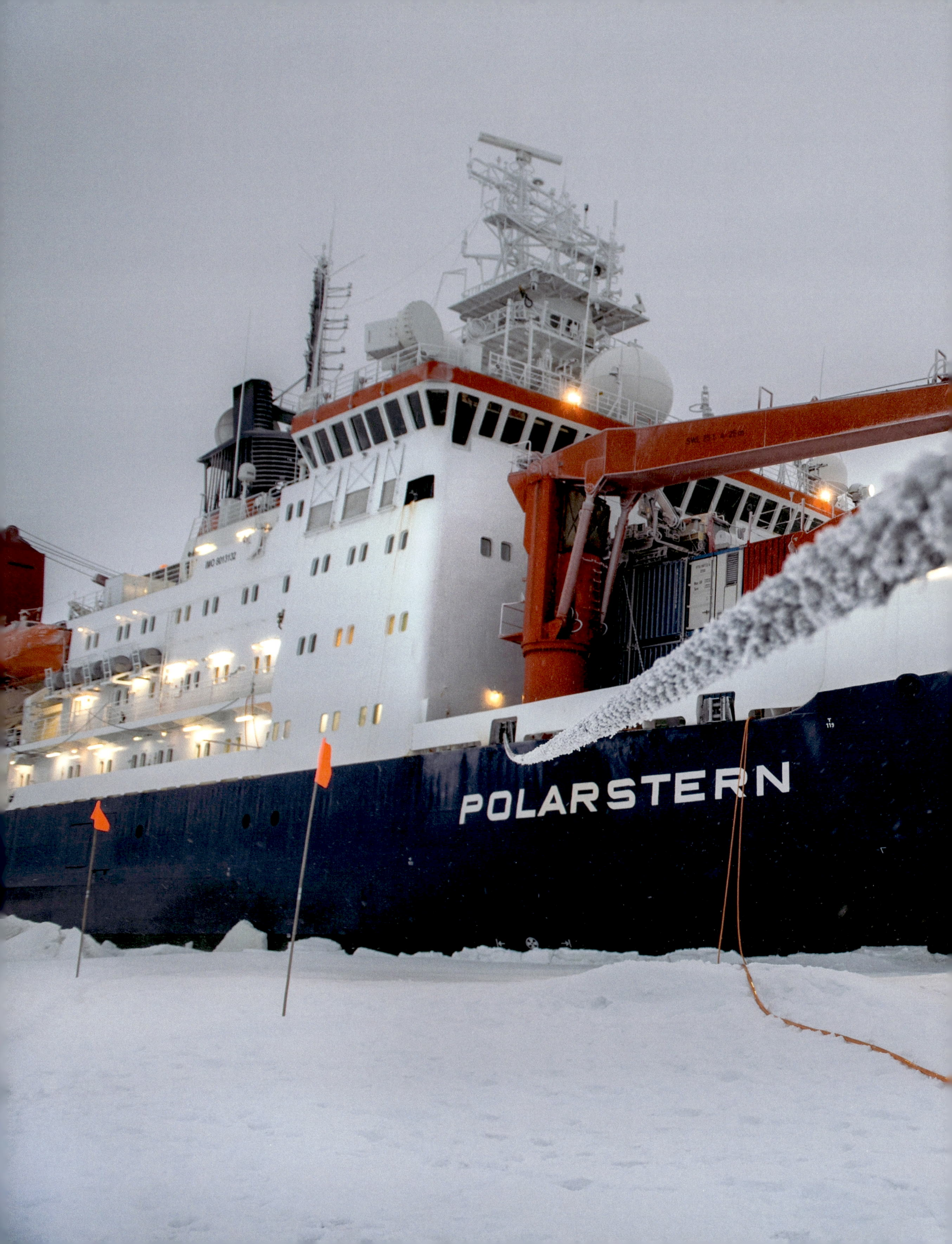
POLARSTERN

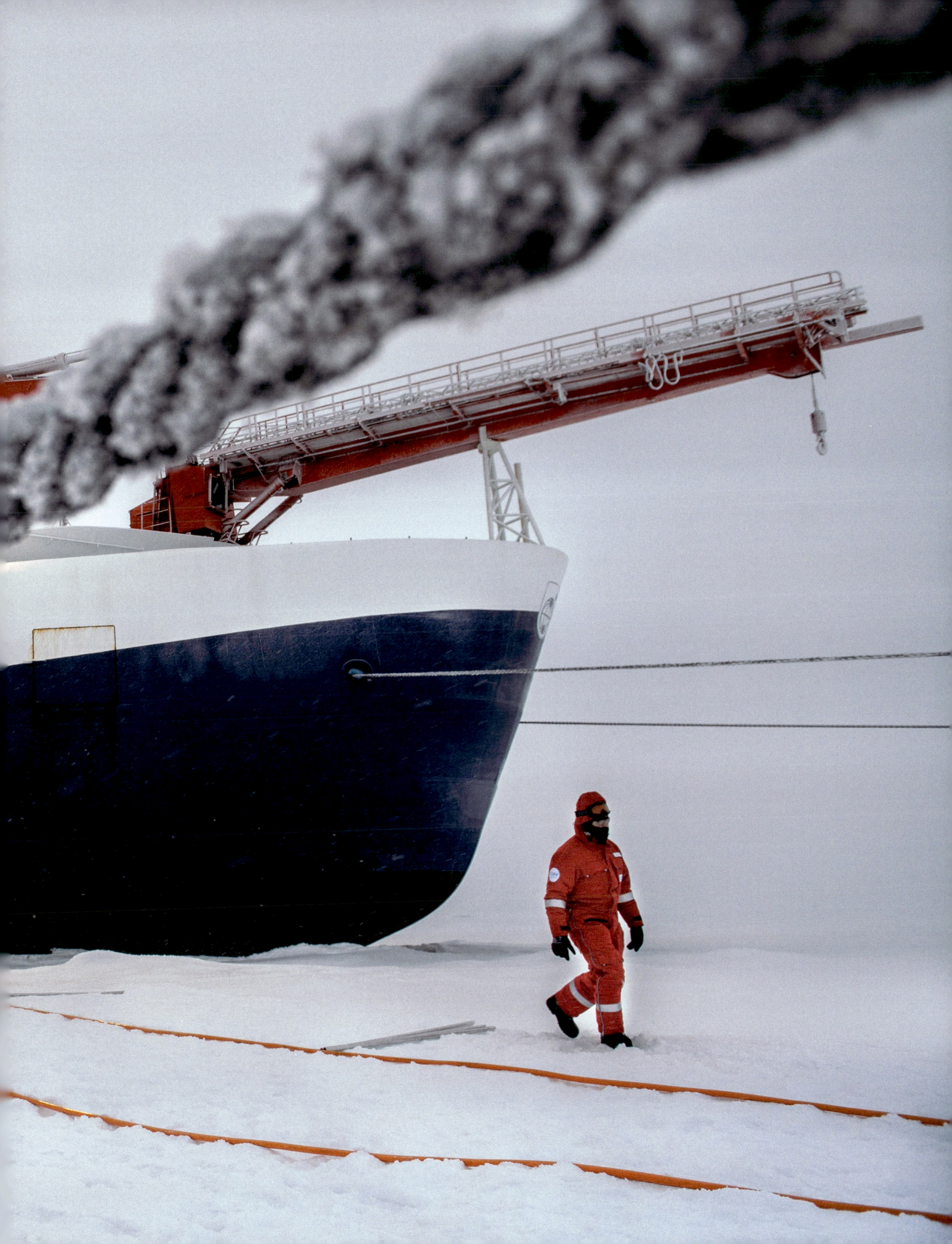

A new world, a new home, a new research subject: for the expedition, the ice floe is all of these things. Responsible research means always being considerate. But anyone who plans to spend an entire year on the fragile Arctic sea ice must take particular care. The first steps on the MOSAiC floe serve to orient oneself in the whiteness of the Arctic. With a neon-green flag, Captain Stefan Schwarze lays one of the first foundations for colonizing the ice landscape, and the researchers start to settle in on the MOSAiC floe. Although the ship will be frozen into the ice and use the natural motion of the ice drift, ice anchors have to be deployed. Otherwise winds and currents could pull at the hull of the *Polarstern* during the drift and tear it away from the ice floe.

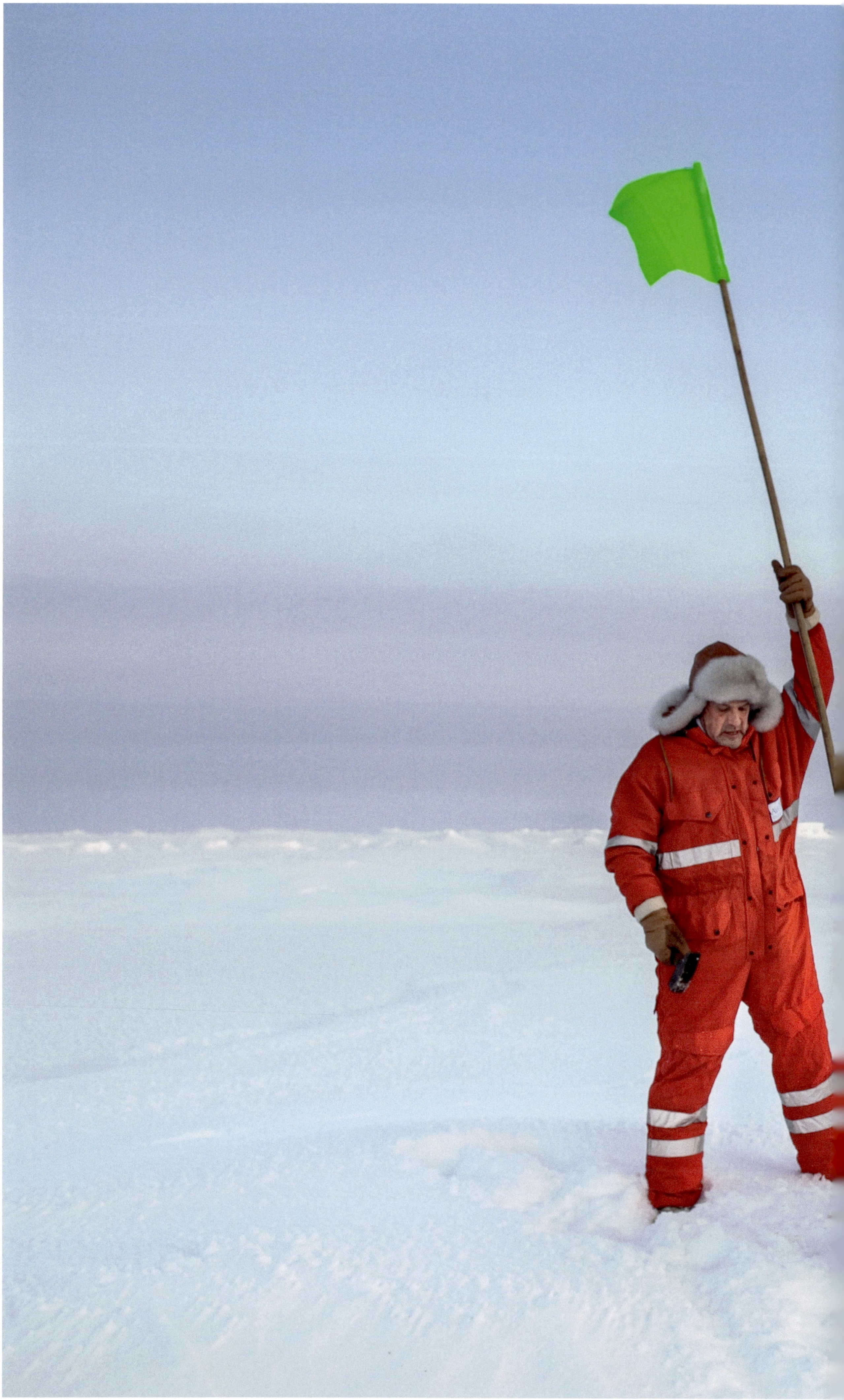

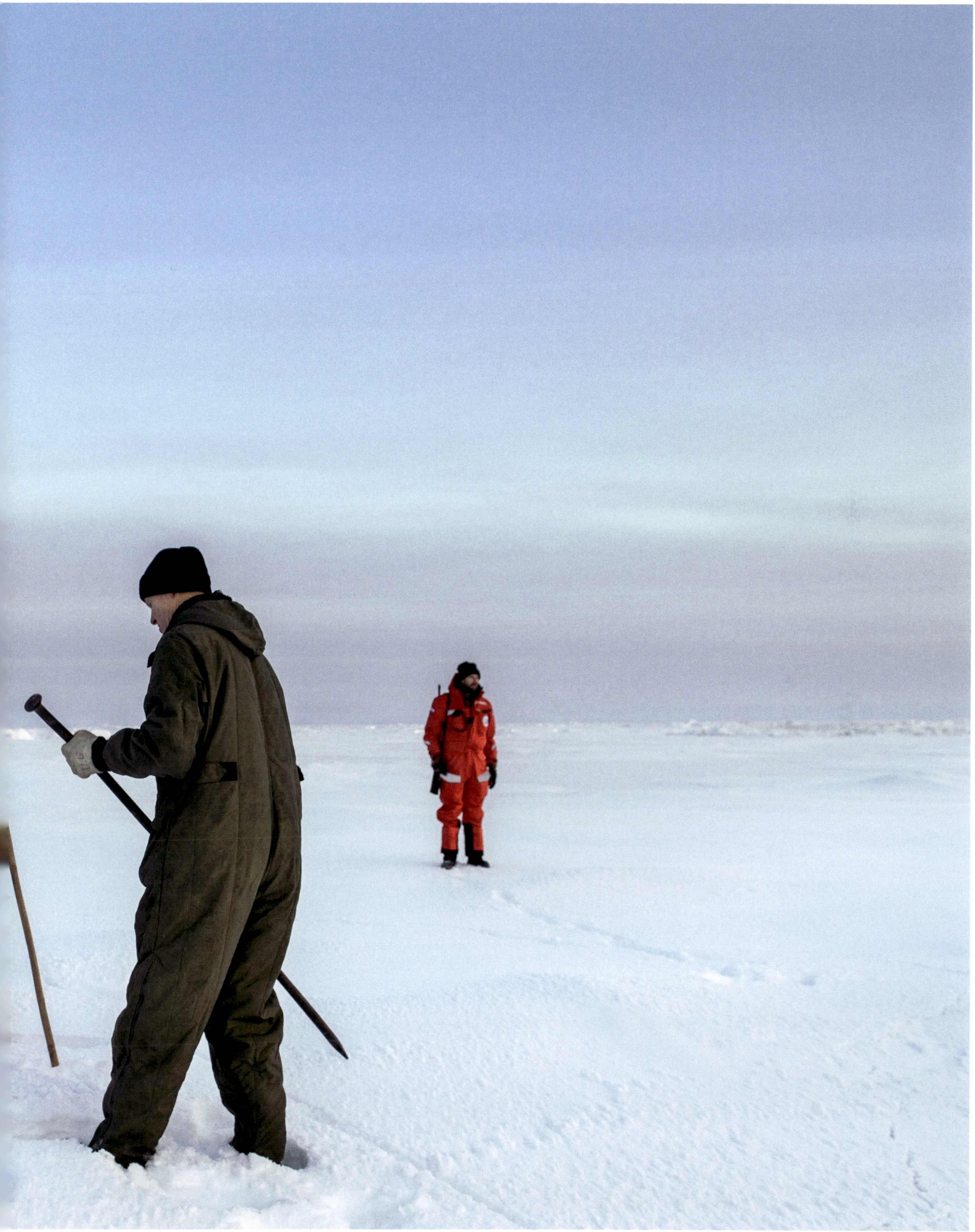

W

While in recent weeks the world has been dominated by white and grey, a strange glow now increasingly hangs over the ice, the sky and the *Polarstern*. It is mid-October and the polar night is coming ever closer.

The magic of the Arctic half-light creates seemingly endless shades of blue. It takes a good deal of discipline in this twilight to take your eyes off of your surroundings and return to work. But for the expedition these are once again days of a critical race against time. The team has to find its way around the floe and set it up as a research station and home base. Stable sites for scientific devices have to be found quickly. The Ice Camp must be set up before night and winter permanently close in on the ice floe.

The floe is therefore a hive of activity in these days of twilight. The set-up phase is one of the busiest of the whole mission. Every hand is needed, wrapped more and more thickly in protective clothing as the cold increases, and everyone joins in the work. On paper there is a clear division of roles among the scientific disciplines, logistics and the ship's crew, but now a maximum of teamwork is called for: polar bear guards become joiners, biologists build roads, doctoral students lay electric cables, and atmospheric researchers show their mettle as Arctic freight forwarders. Once routes have been marked out on the floe, the expedition members transport heavy instruments across the ice with the help of snowmobiles and archaic-looking but extremely practical wooden Nansen sledges. The on-board helicopters are also in operation. But it is above all the rolling out of the power supply with frost-proof electrical cable calling for a great deal of physical effort and sweat that represents a major logistical project. As the ice continues to grow, daylight diminishes. Artificial light sources are used more and more. The *Polarstern* with her three powerful front spotlights lives up to her name, "Polar Star", by showing the way to people moving on the ice, often returning to their secure home base after working at a distance of several hundred metres. At the same time the shadows that arise on the dark side of ice hummocks become deeper. The bluish glow of twilight increasingly gives way to hard contrasts.

Headlamps allow the polar researchers to keep their hands free for setting up their stations and instruments. At times the work demands simple physical strength, at times a sensitive touch. And thus one particular decision becomes increasingly tricky: to keep thick gloves on and protect your fingers or to expose your hands to the freezing cold in order to fine-tune the devices? Now the temperatures are often far below −20 °C (−4 °F). When wind hits bare hands it is biting cold, and it cuts through one's threshold for pain. Frostbite is a constant risk in polar research.

Nevertheless, thanks to this collaboration across professions, disciplines and languages, the world's most northernmost settlement comes into existence at lightning speed: a small research town, criss-crossed by a network of tracks and power lines, with shelters, hangars and districts for the various fields of research. These small districts even have their own names: Ocean City, Balloon Town, the Met City, the Remote Sensing Site and the ROV Oasis, the home base of the diving robots. In short: orderly civilization comes to the ice floe. At least temporarily. Because from the very start, this short-term colonization of the Arctic ice by a group of humans attracts the attention of the true inhabitants of the region: polar bears. With their curiosity, which can be as destructive for sensitive scientific instruments as an Arctic storm, every now and then they disrupt the setting-up of the Ice Camp. All the more urgent is the erection of tripwires and deterrents to stop the creatures – for their own safety as well – from making unobserved incursions into the research camp. Flare guns with their hissing, glowing projectiles fired into the sky prove to be a good means of deterring the highly inquisitive polar bears.

During the construction phase, the Russian support vessel *Akademik Fedorov* circles the central research camp at a distance of up to 50 kilometres. The crew's mission is to take a distributed network of automatic measuring stations out on to the ice. With robust

Polar research is often a genuinely tough job, but sometimes it is also a task for subtler senses. Searching for a suitable spot for Ocean City, the biologist Allison Fong carries out taste tests of the properties of the ice. The human tongue is used as a measuring instrument and provides reliable information: if the ice is hardly or not at all salty, there might have been a melt pond here in the warmer season. At such sites the ice could still be too unstable to bear the weight of a research station.

Russian Mi-8 helicopters they place buoys in the ocean and on the ice. This task involves risks, but the weather favours them. Fog and wind rarely hold up the work, and thanks to their vast experience, the crew and pilots complete their mission right on schedule. As soon as the distributed network reaches completion, the darkness of Arctic night finally falls on the expedition. The *Akademik Fedorov* moors on the MOSAiC floe for a last exchange of several scientists and instruments and for a last farewell. Then the *Polarstern* is left alone with some one hundred people on board to face the almost unexplored Arctic winter. The collective drift of the sea ice, *Polarstern* and research stations has begun.

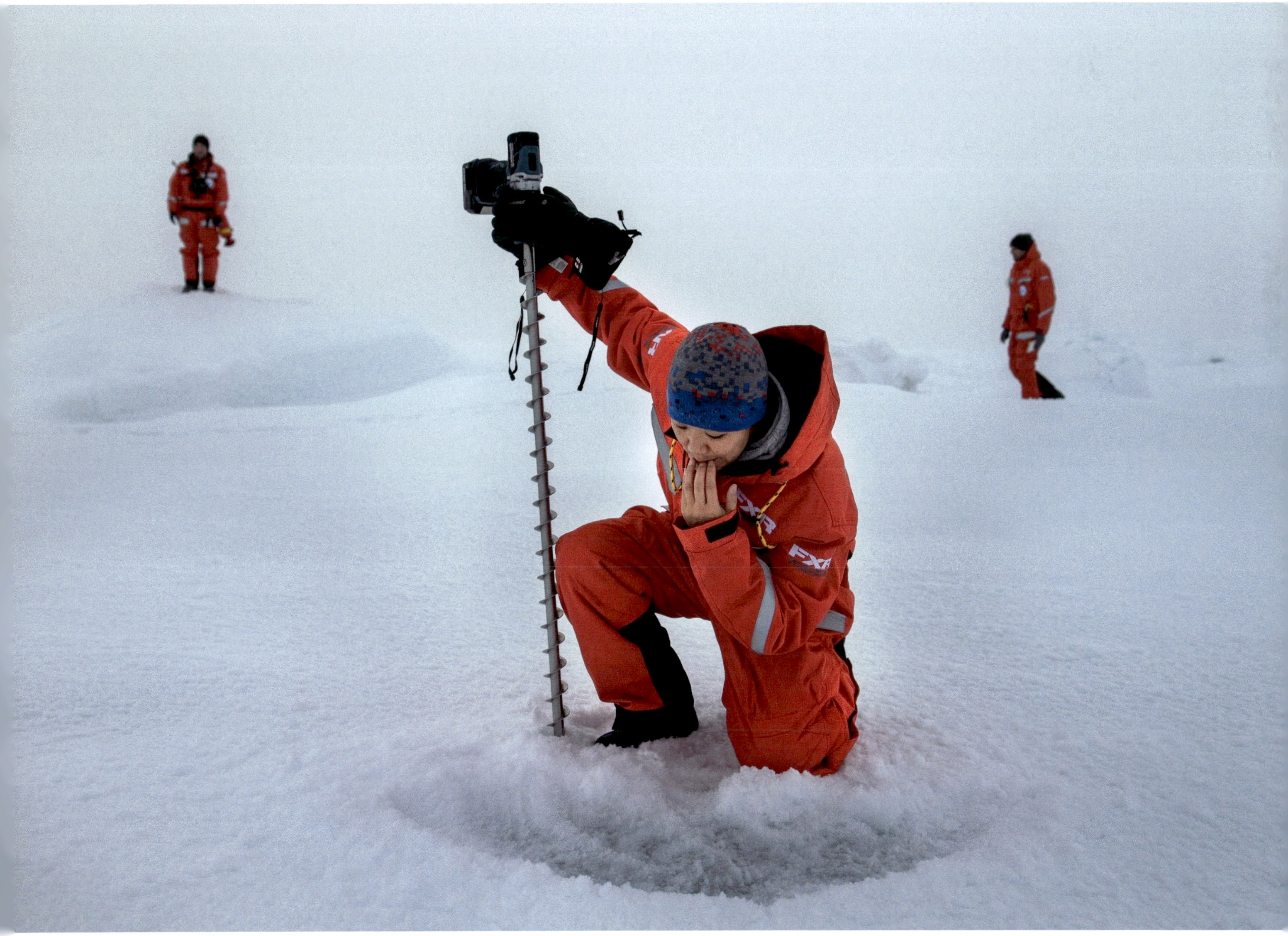

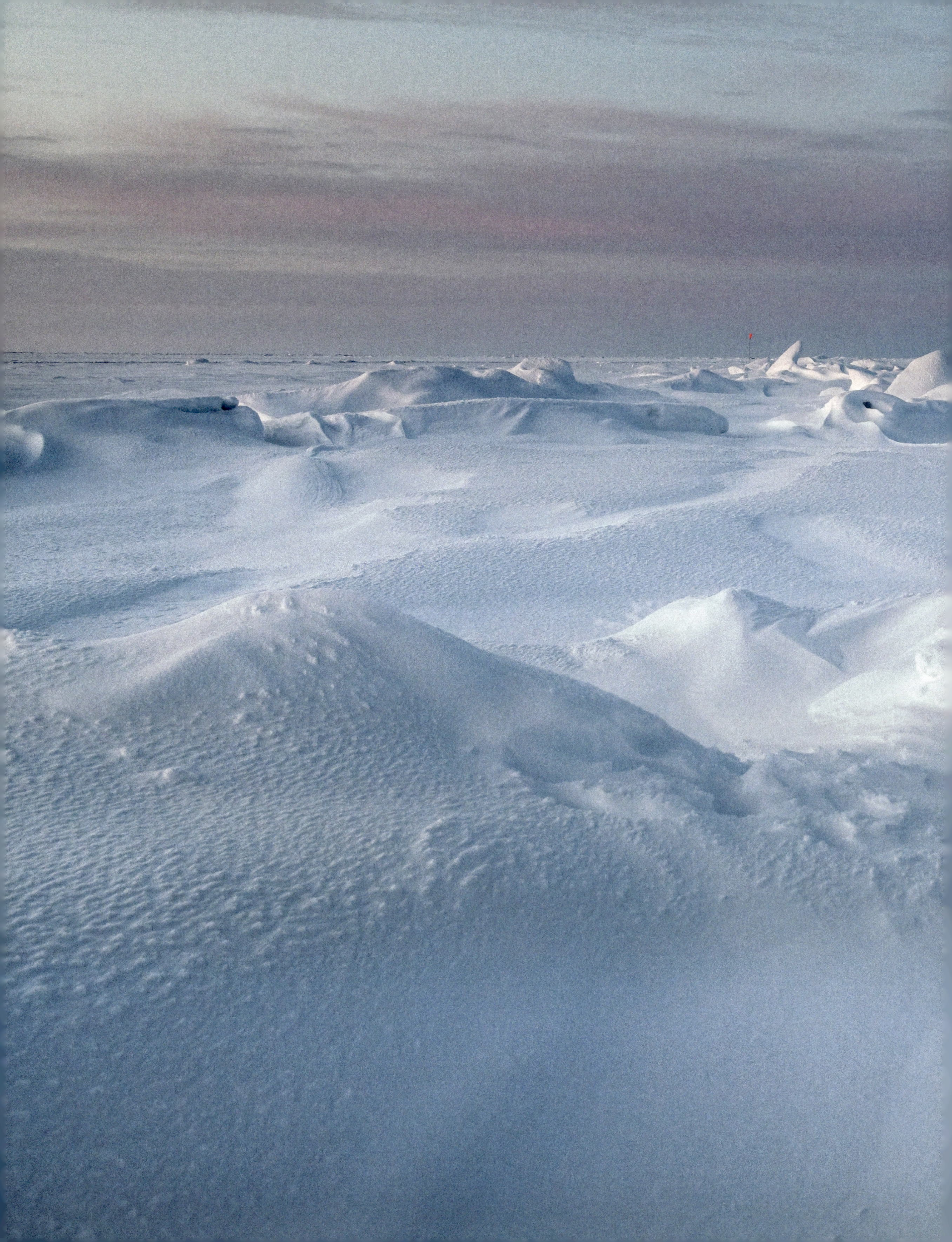

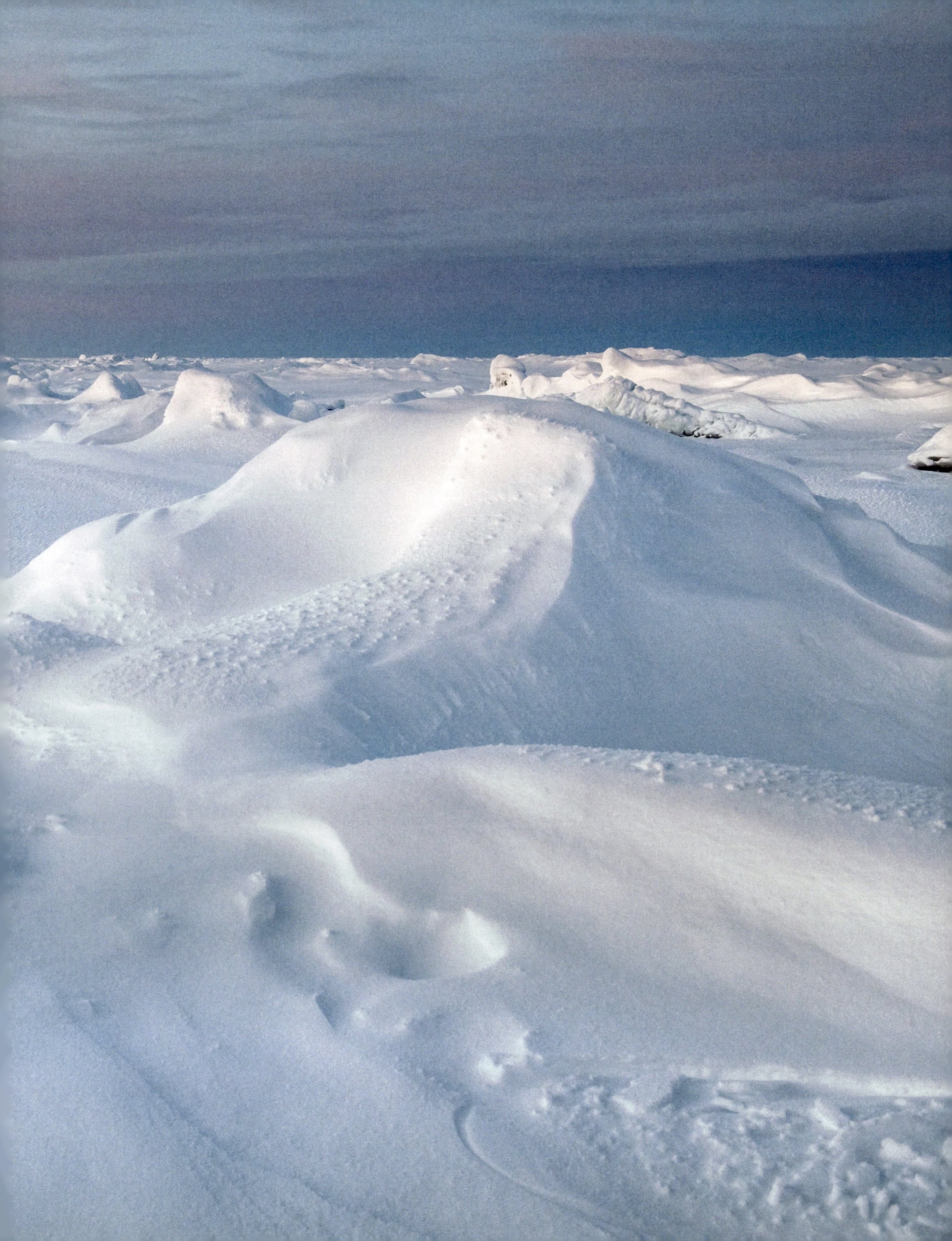

Is the ice stable enough? Its thickness must be checked meticulously. With a drill and ice-thickness probe, Julia Regnery and Rainer Graupner are searching for a site for Ocean City. The ice should be at least 60 centimetres thick here in order to support the station.

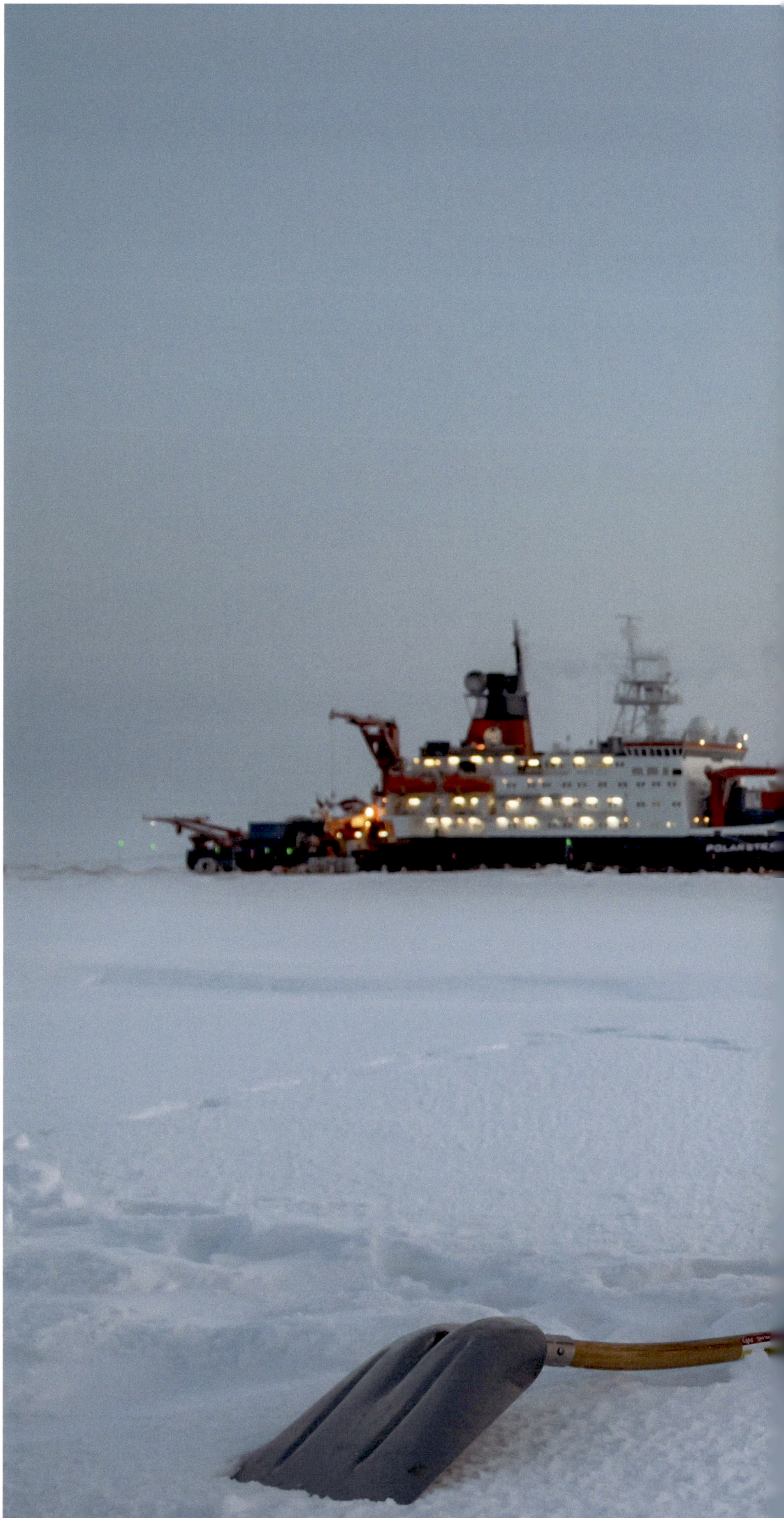

Drift route on the first leg of the expedition:

200 km

Actual distance travelled due to the floe's zigzag course:

720 km

Drift speed record on leg one of the journey (on 16 November 2019):

1.4 km per hour

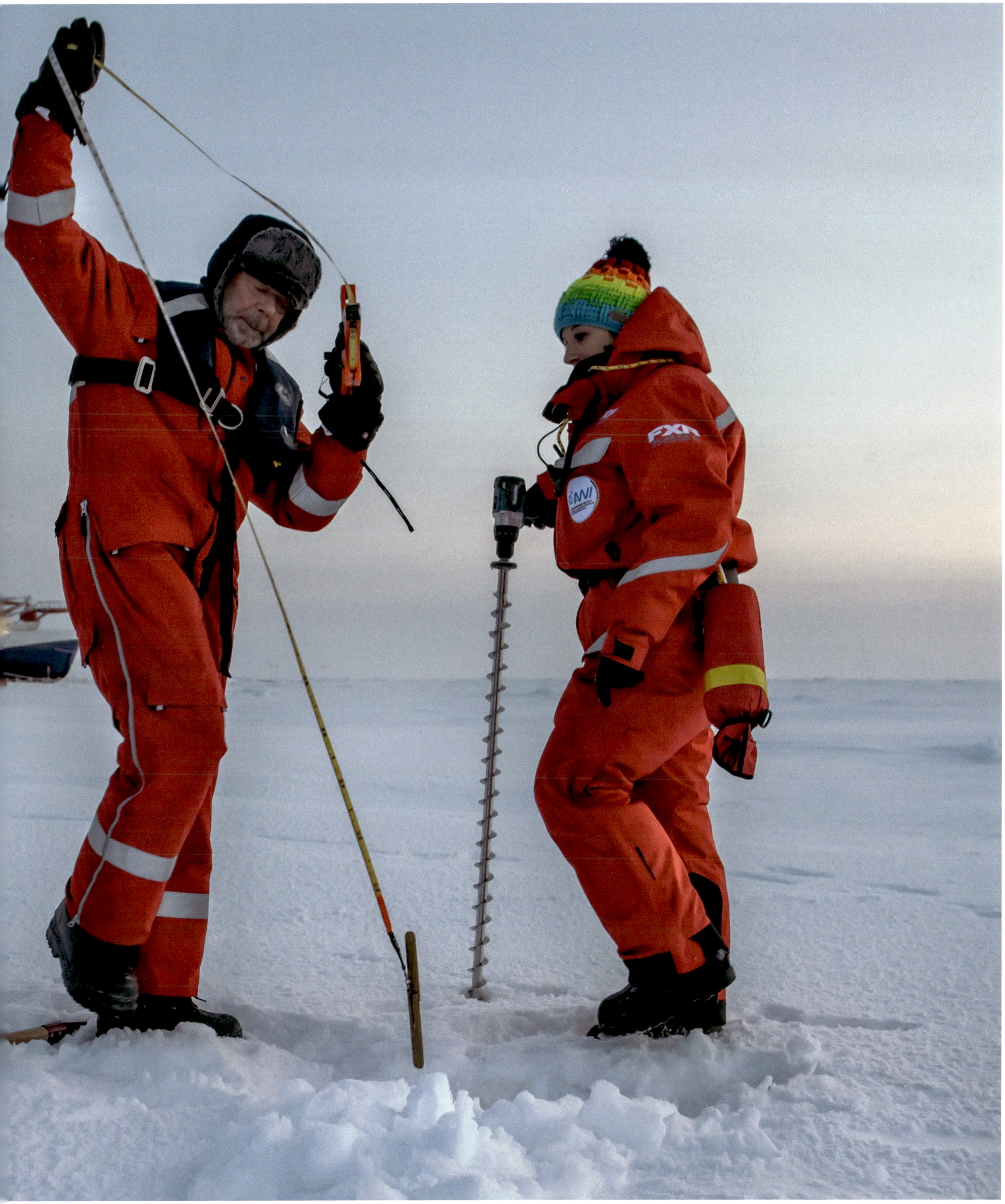
FXR

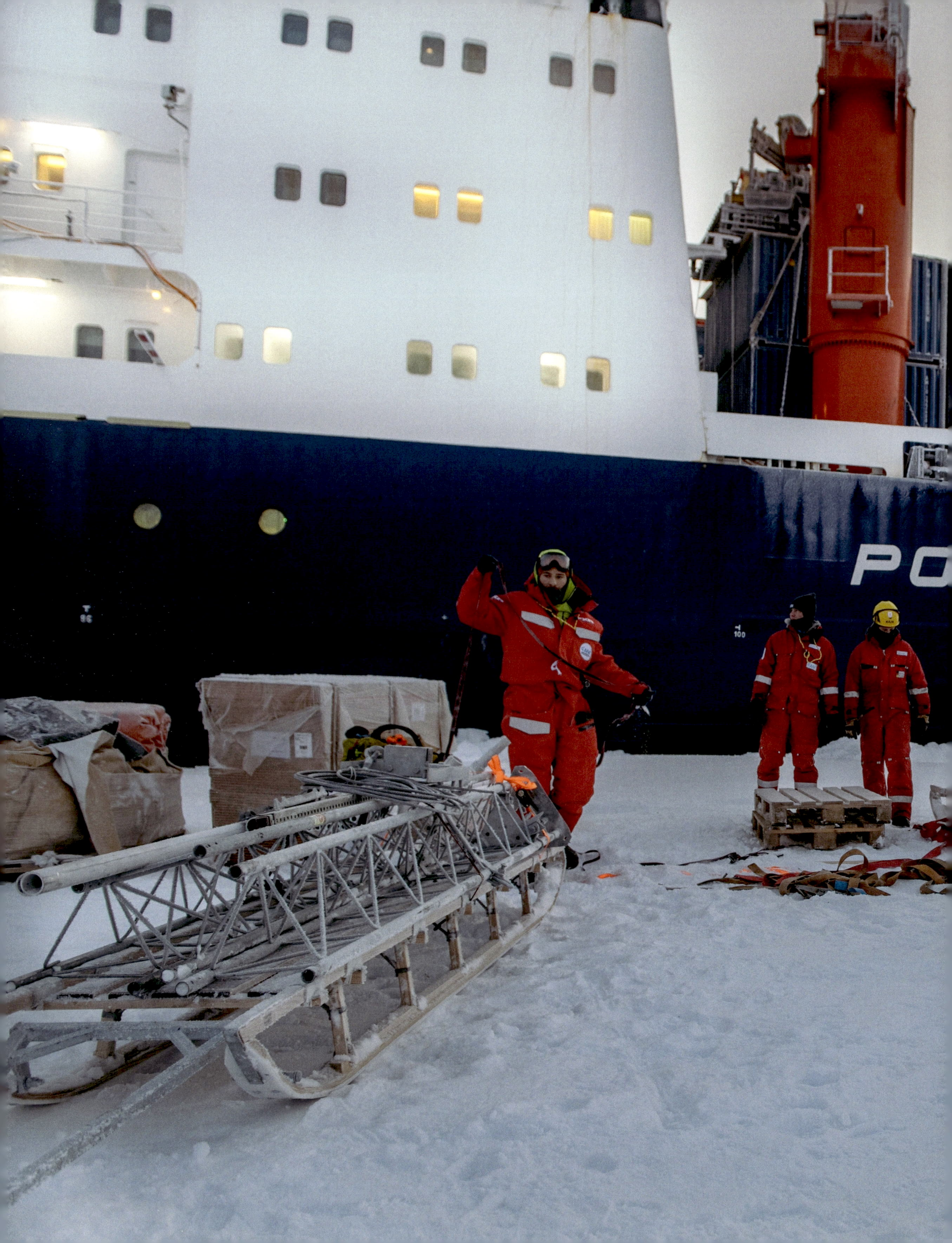
PO

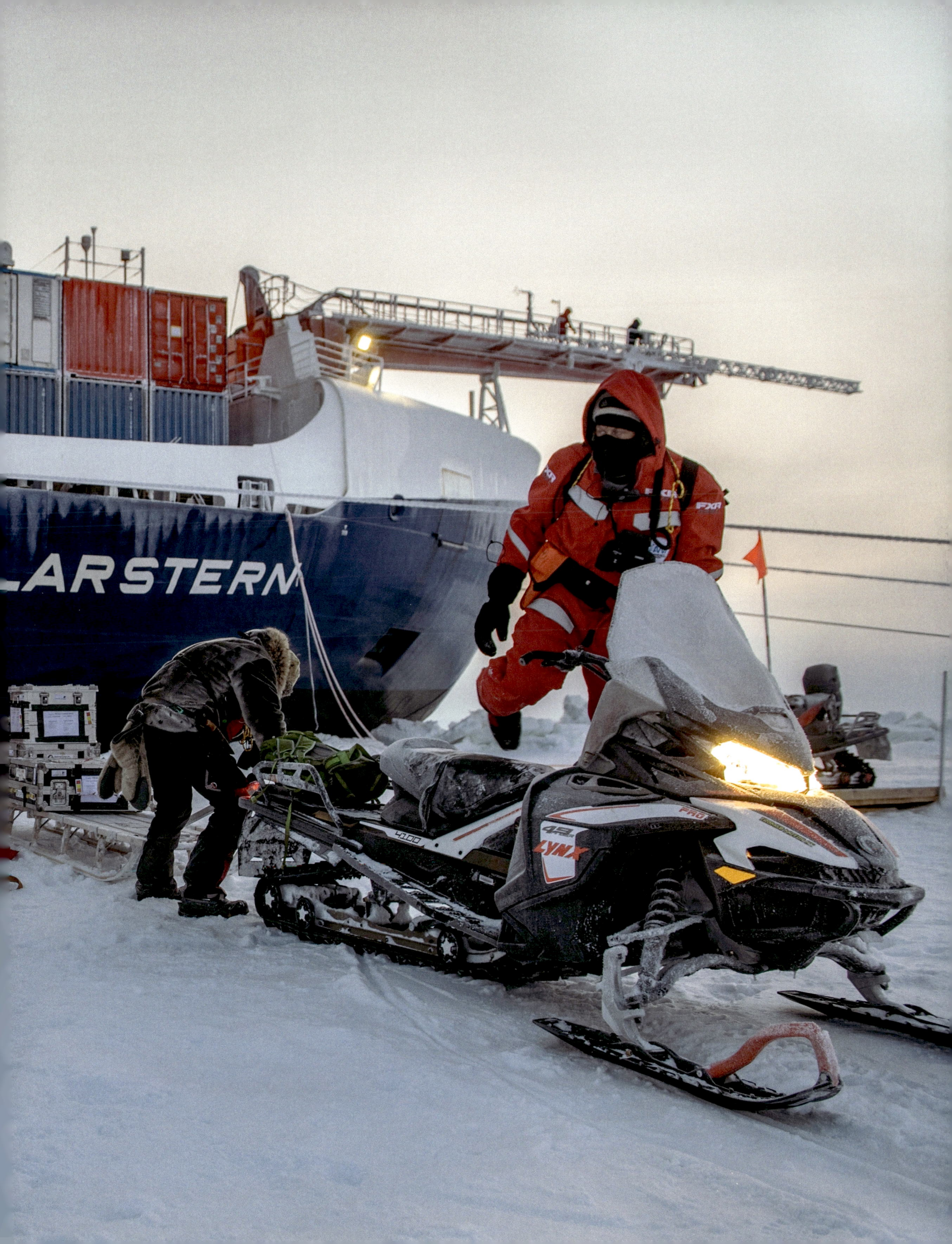
LARSTERN
LYNX

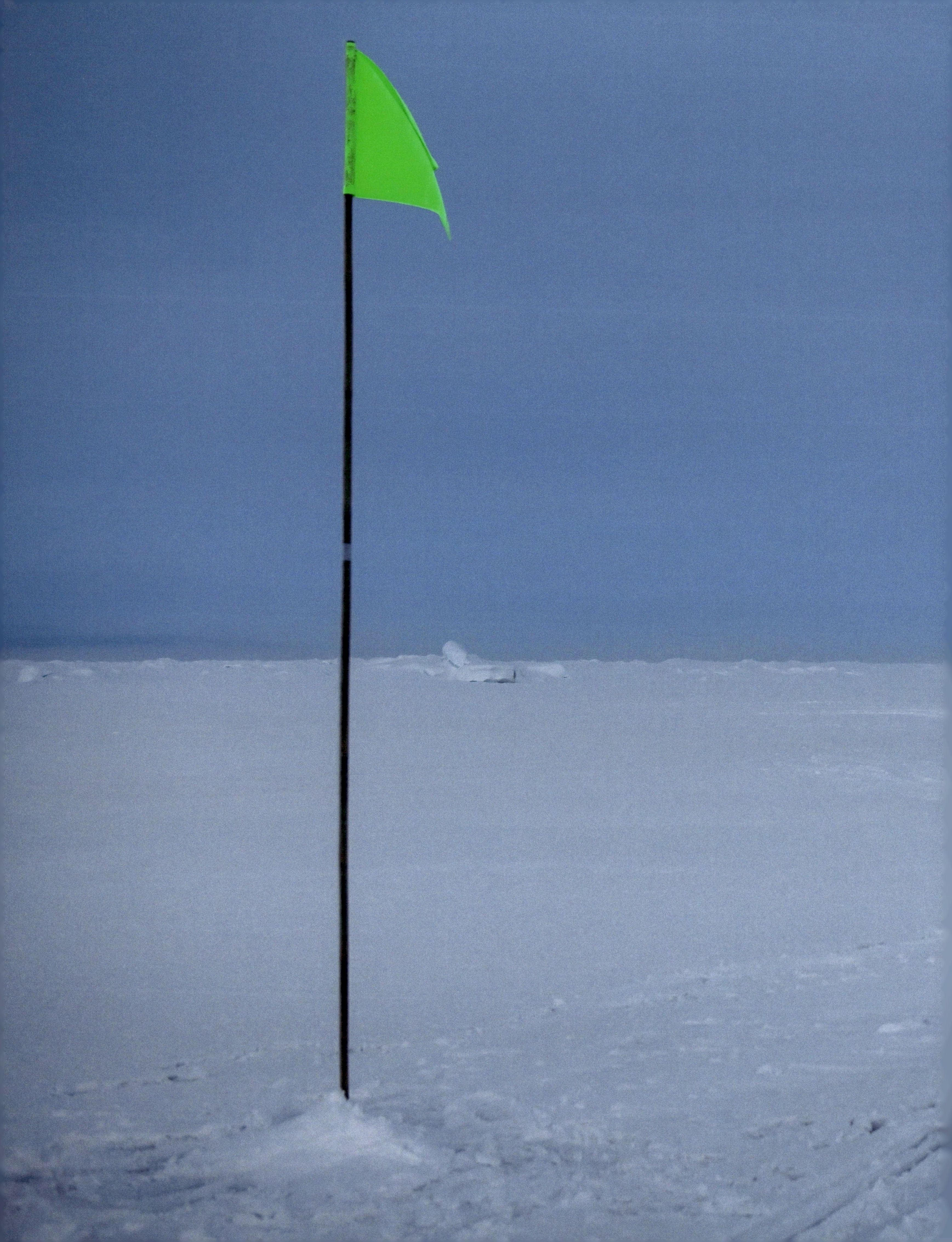

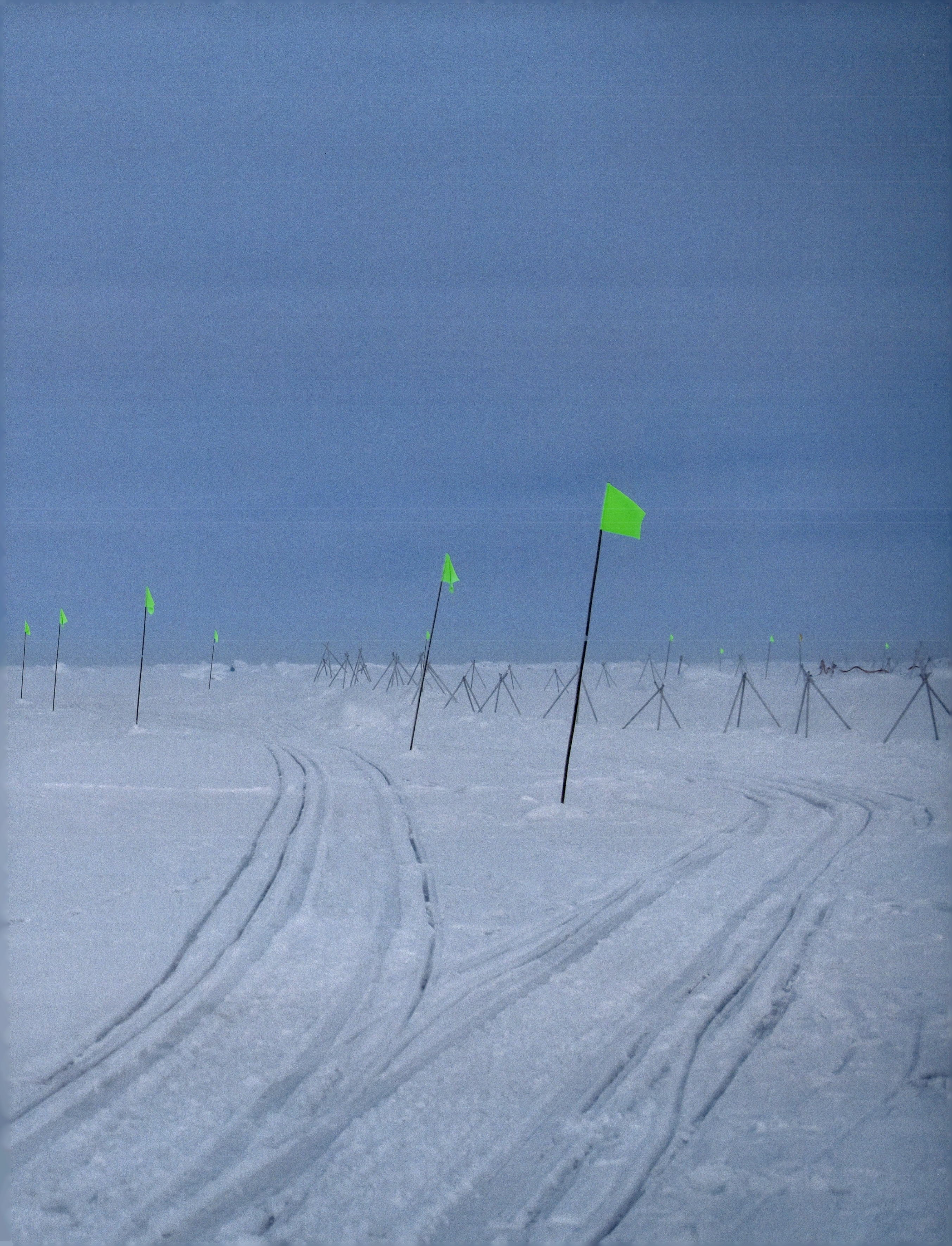

Some 600 metres separate the *Polarstern* and Met City, the most distant research station and a centre for atmospheric research. The great distance to the ship makes the presence of a shelter essential: a protective hut at the outermost edge of the research camp, right next to the icy, heavily compressed ice mounds of the so-called Fortress, the stable heart of the MOSAiC floe. On a windy day in late afternoon, the team begins to erect the wooden hut, energetically assisted by Gaute Hermansen (on the ladder, bottom right), the Norwegian polar bear guard who is a carpenter by trade. When some of the team return from their short lunch break on the *Polarstern*, they are amazed: the hut is ready and finally provides shelter from the icy wind for those who have remained in Met City to continue working.

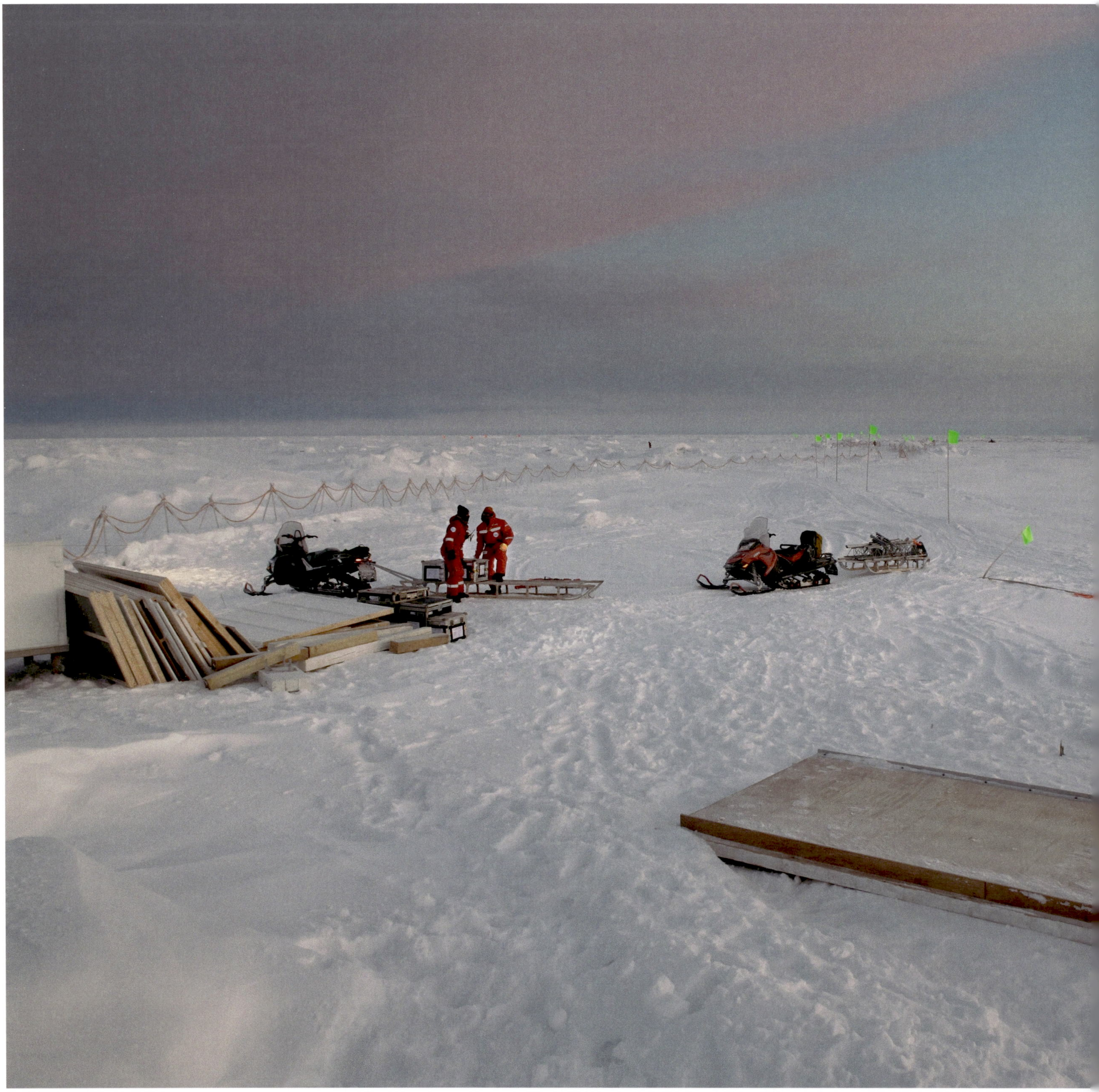

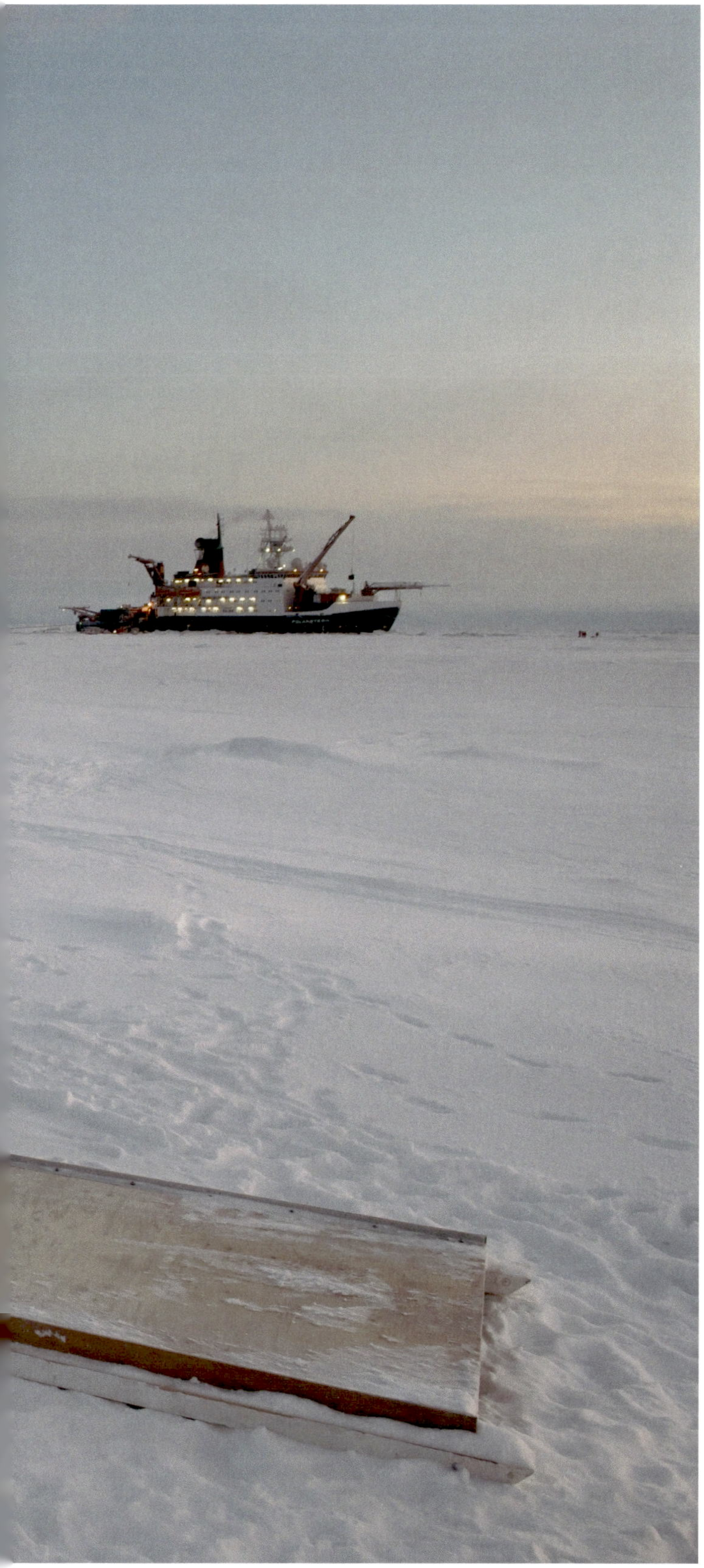

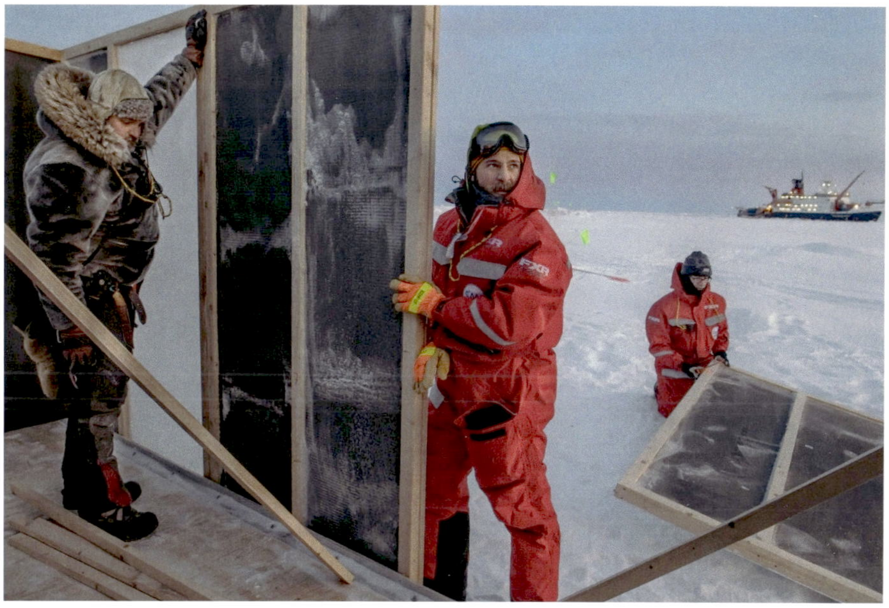

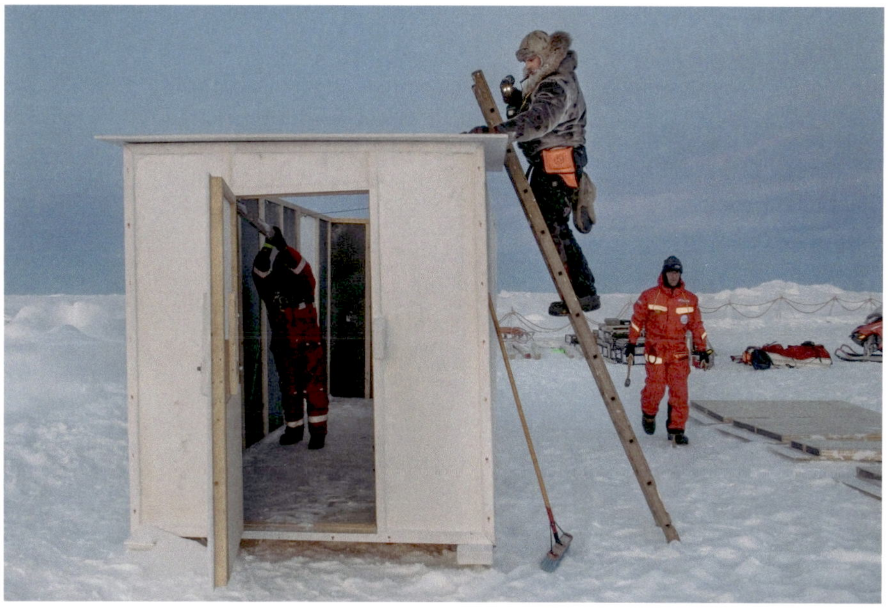

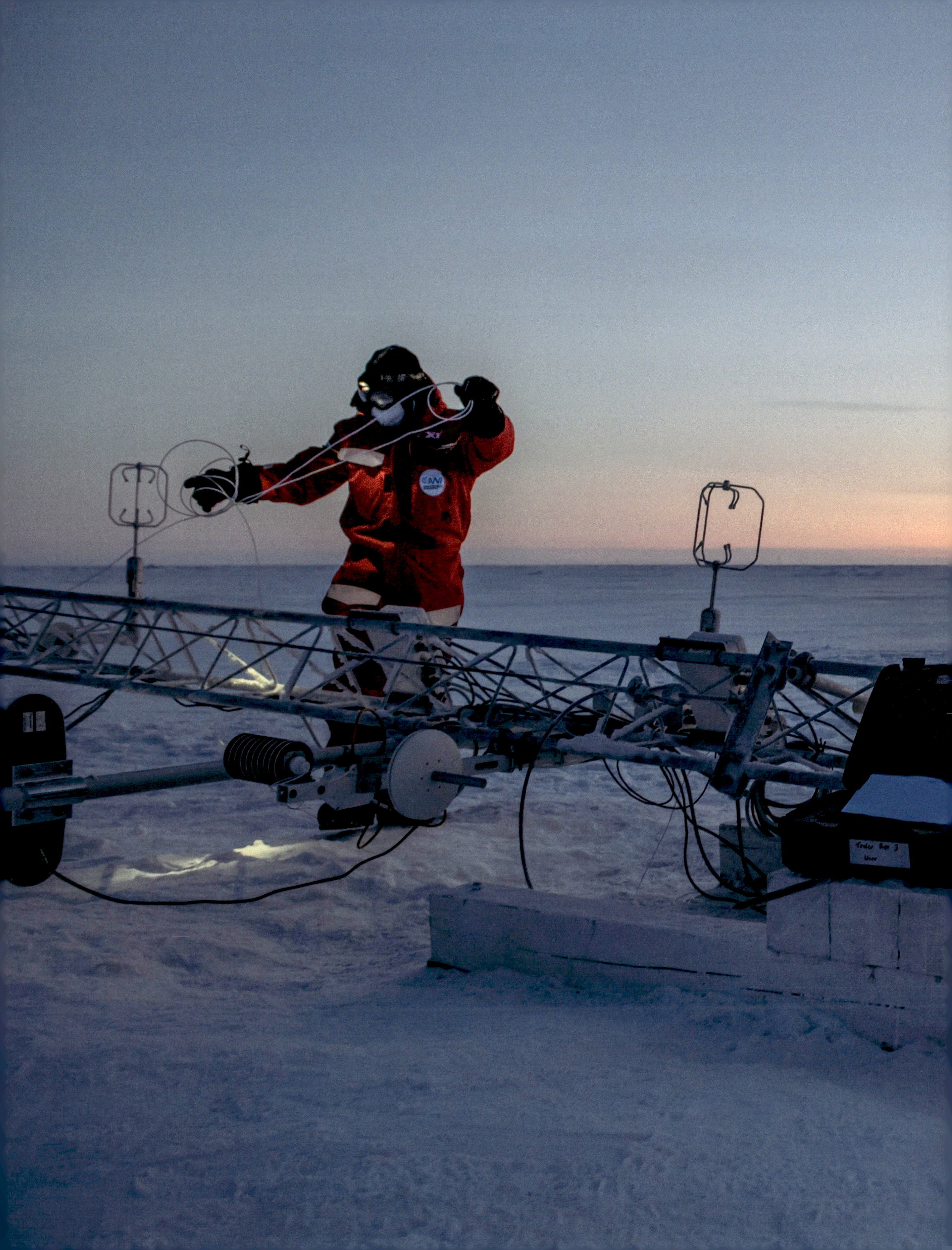

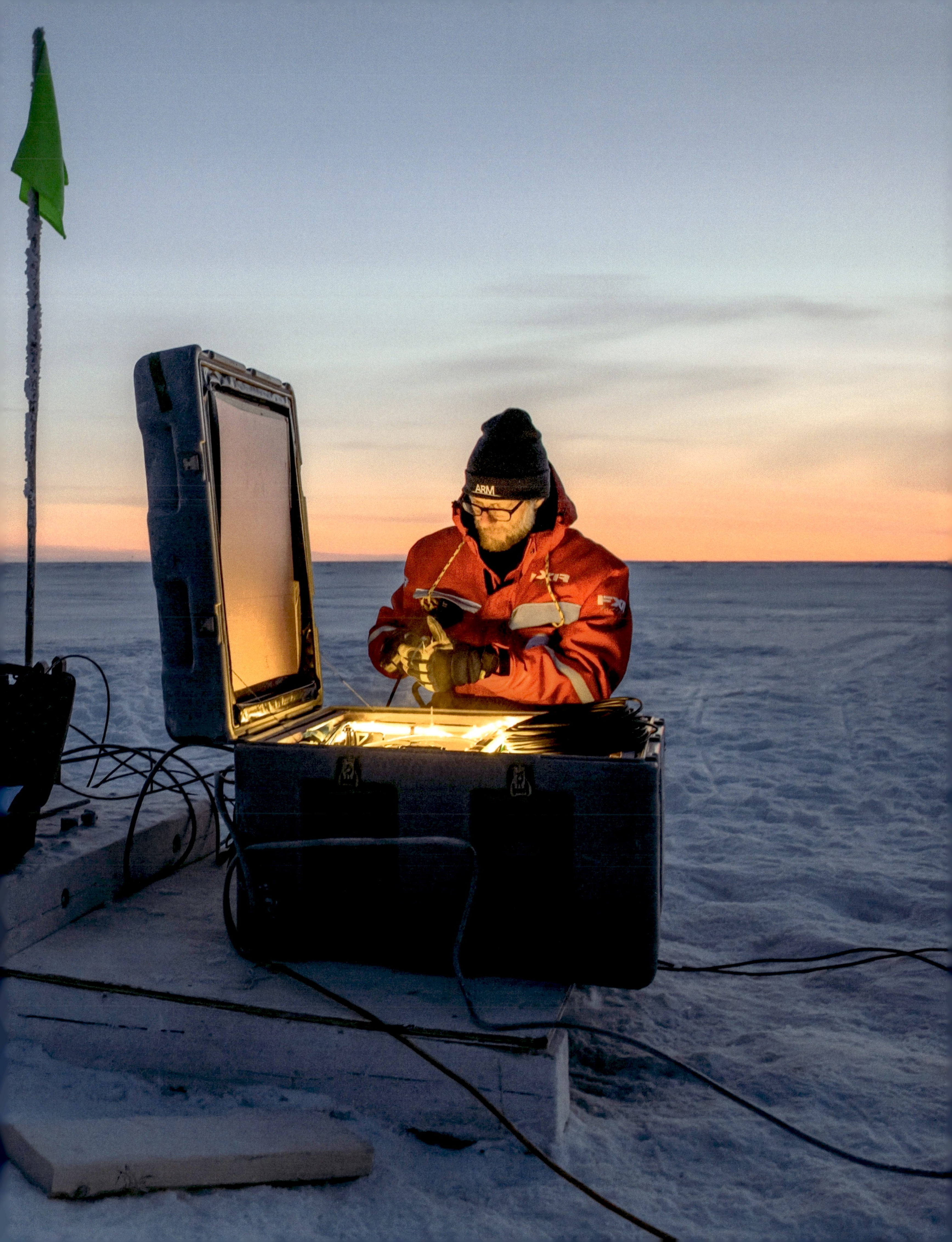
ARM
FXR

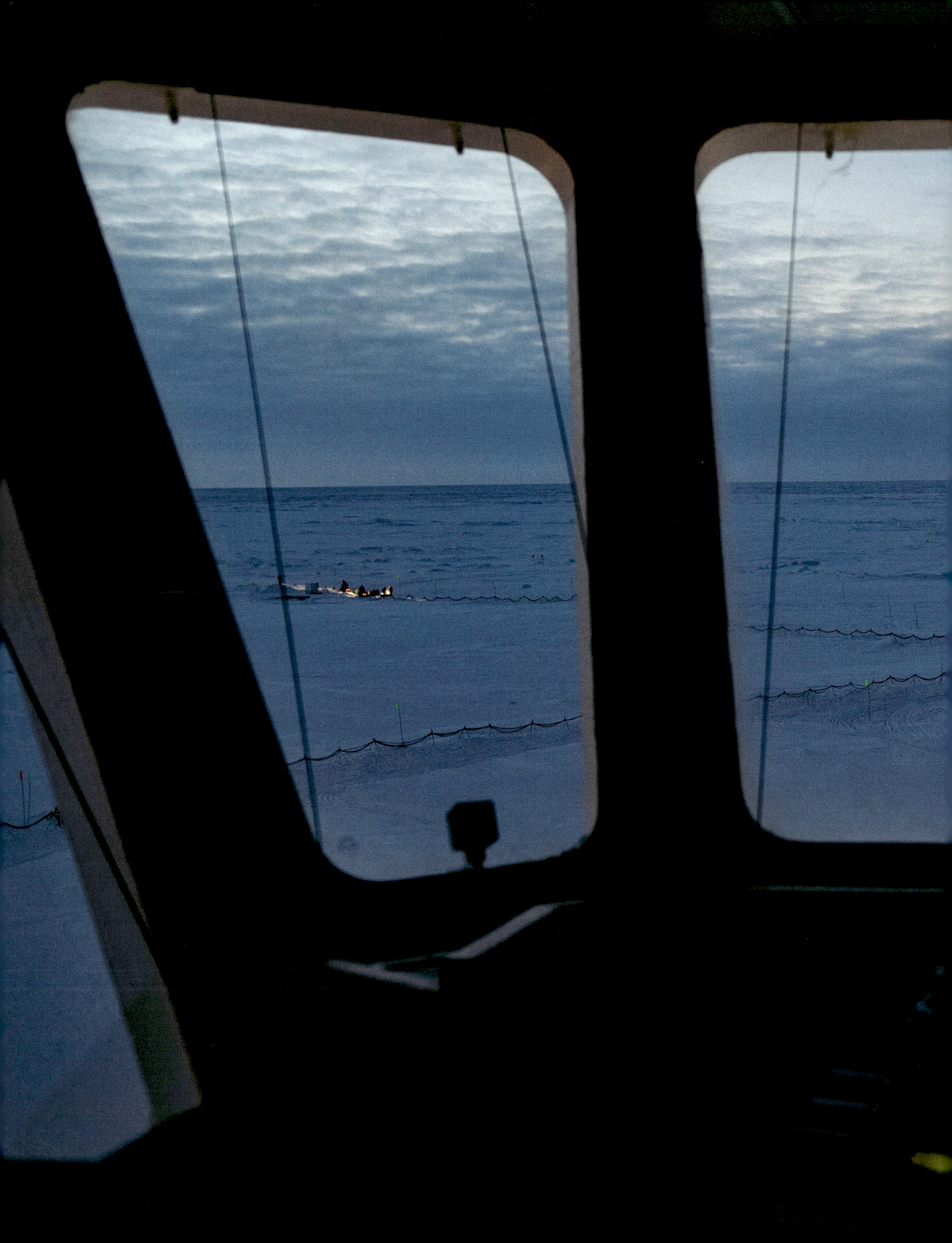

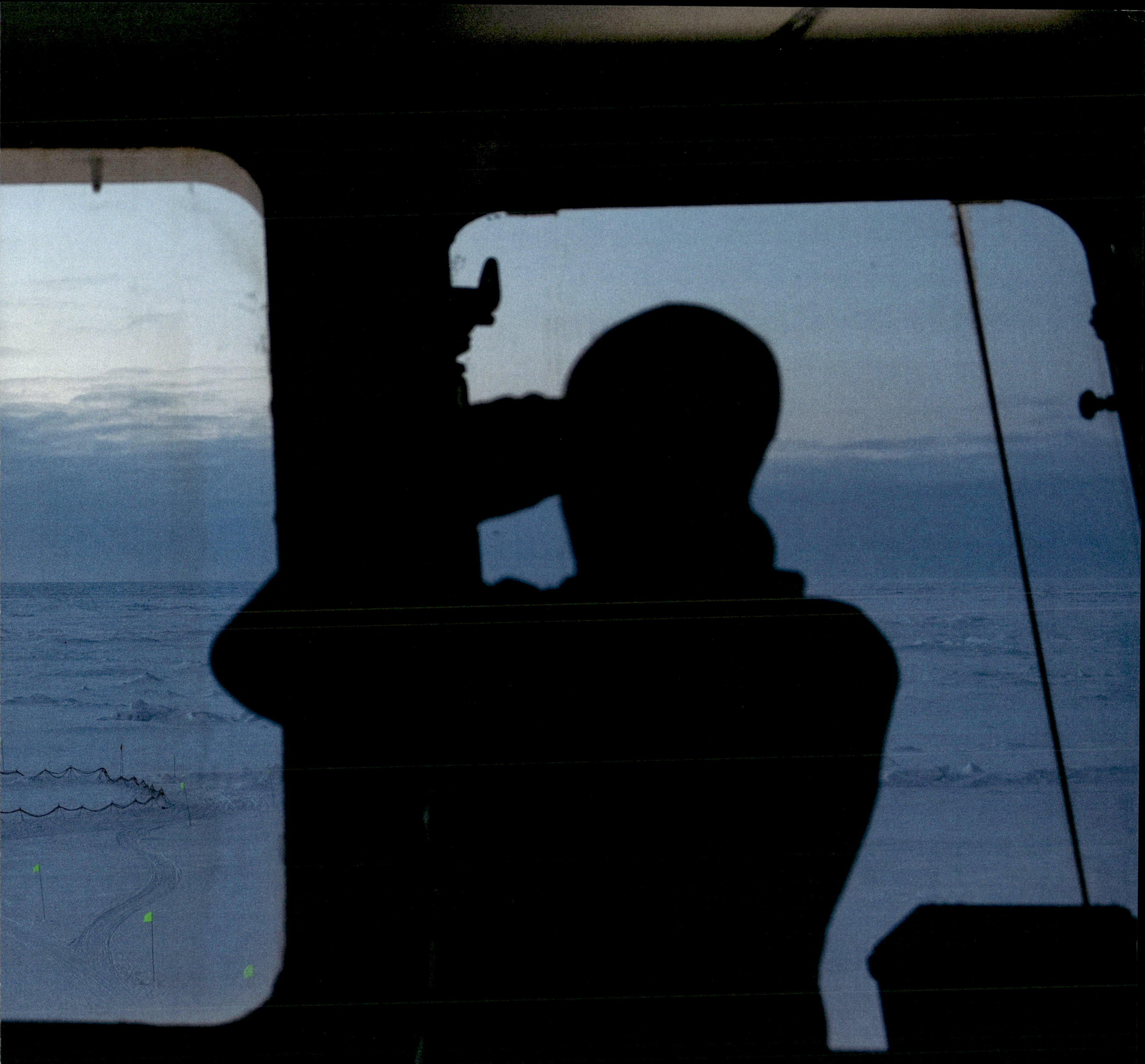

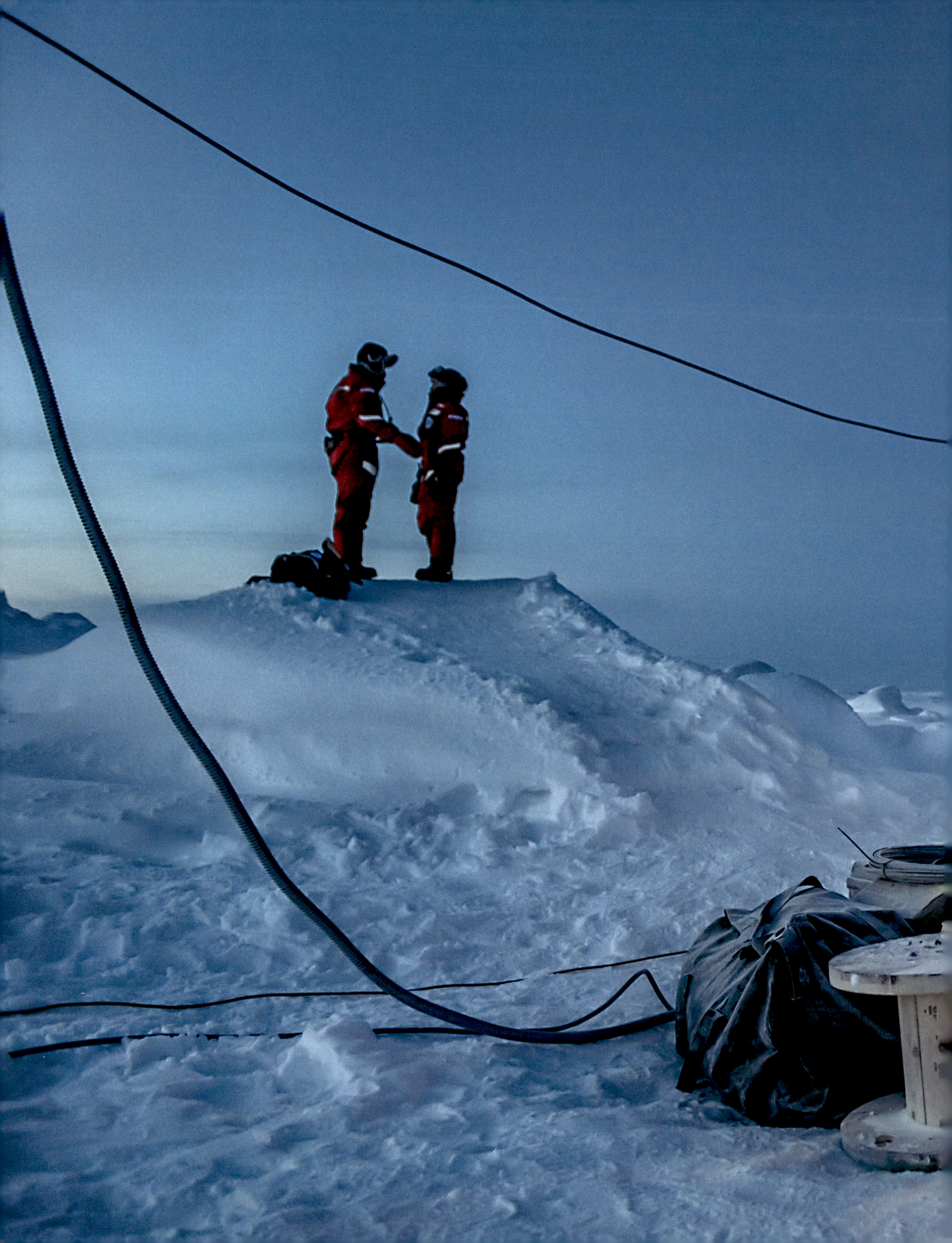

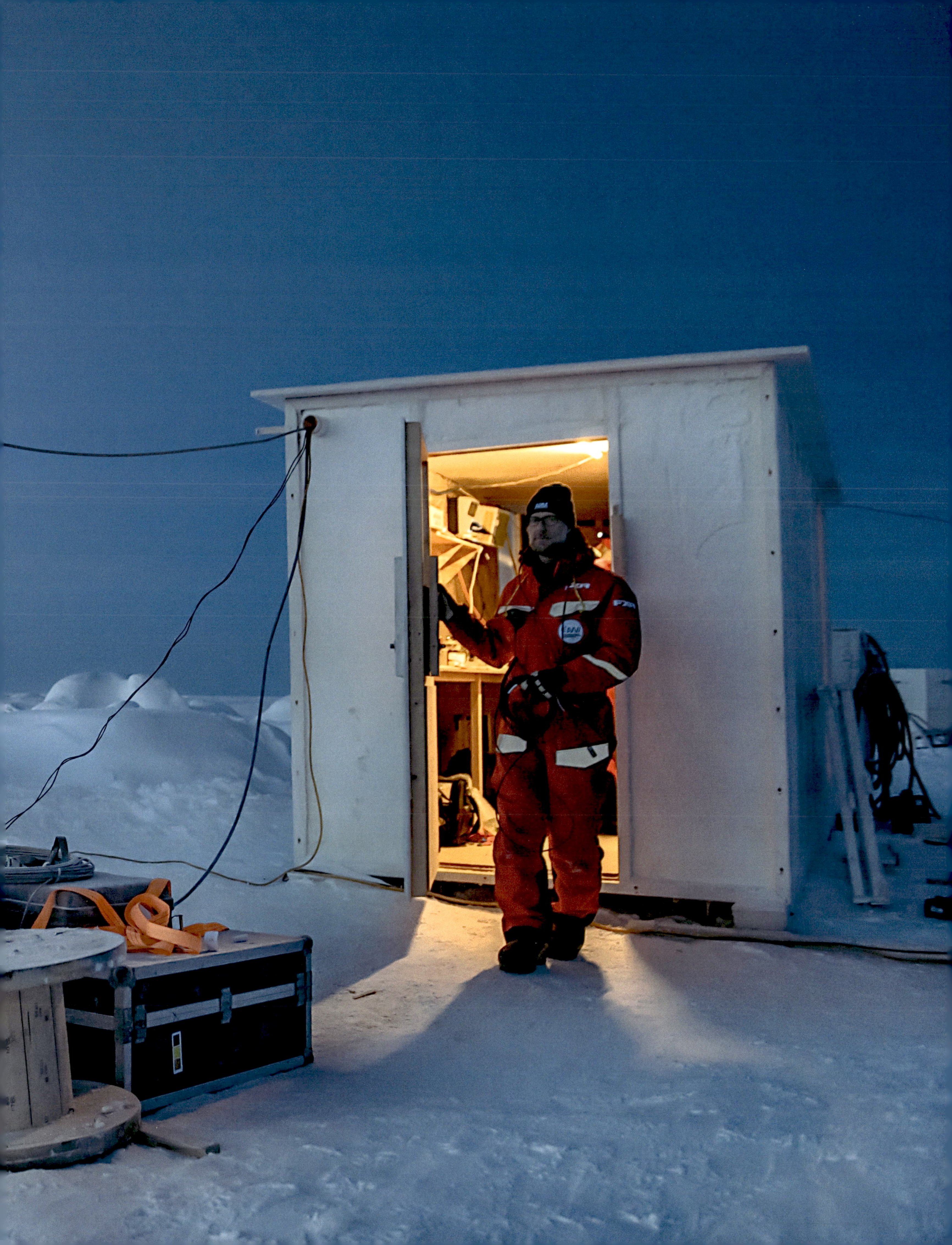

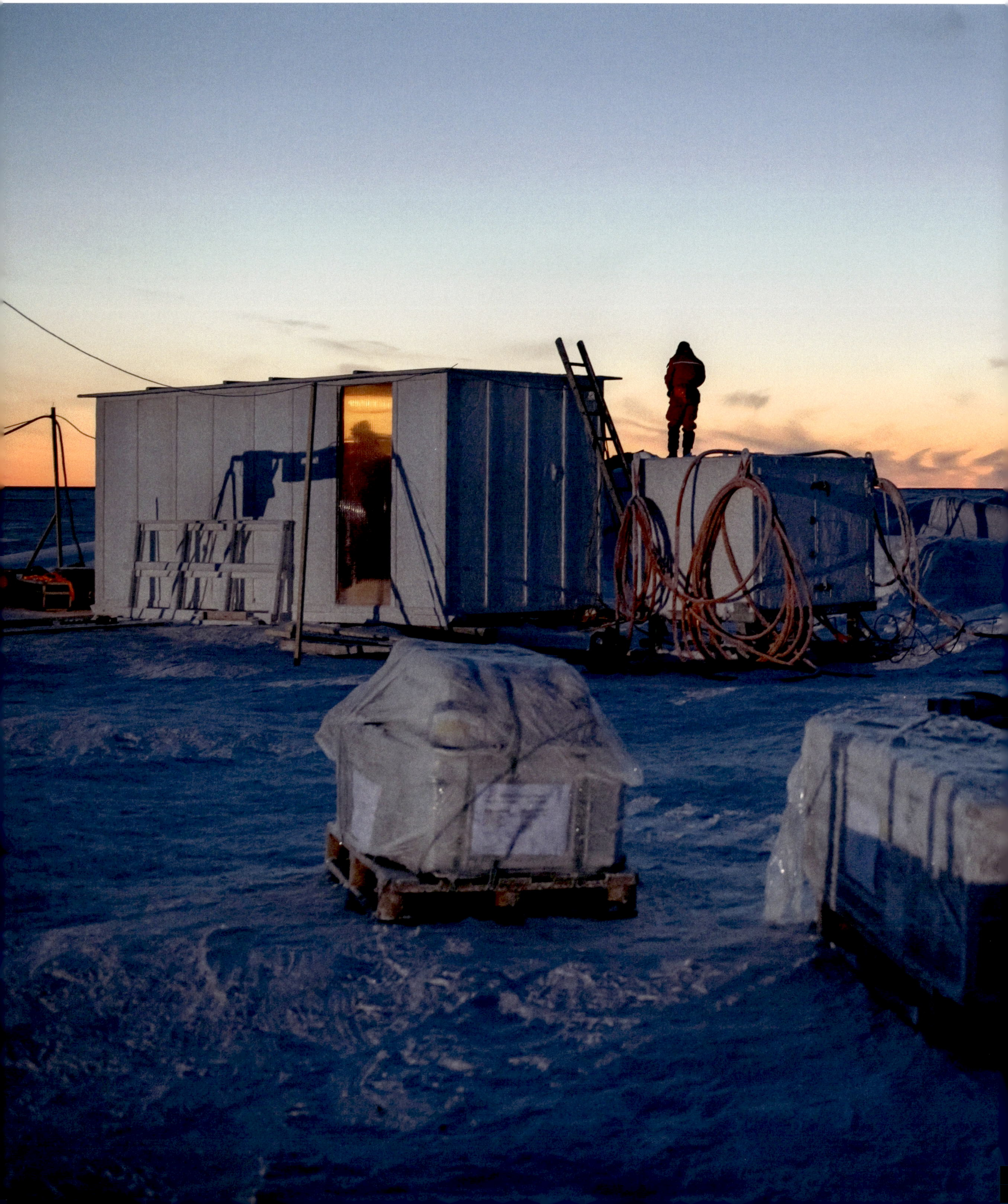

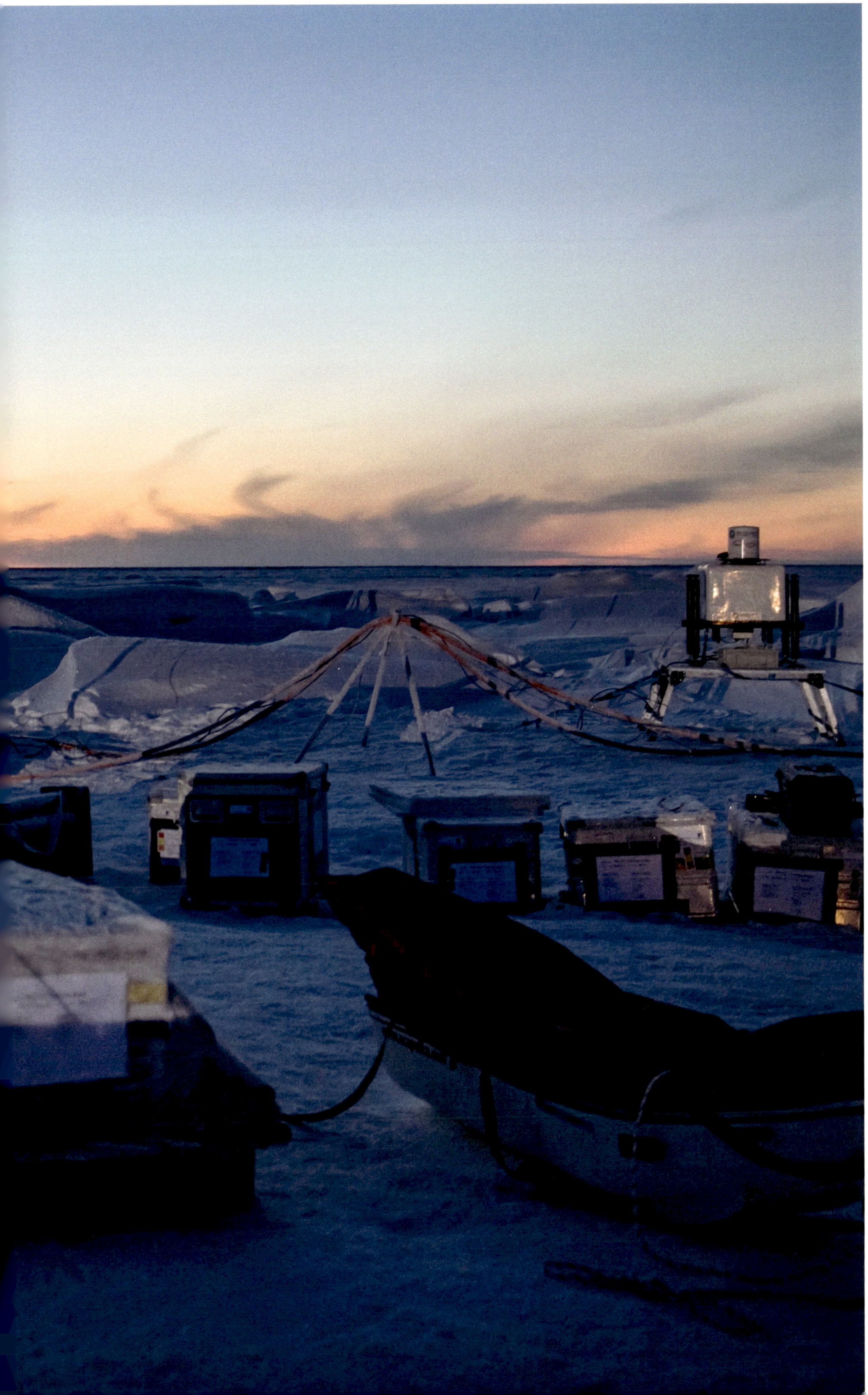

Both lonely and inviting, the Met City shelter marks the outermost edge of the Ice Camp. It can be heated to a little above freezing – an almost cosy refuge in Arctic conditions. Just the sight of it can make you forget that there is no solid ground beneath, only a layer of sea ice measuring 60 to 70 centimetres. On the neighbouring hummock of ice and snow, the bear guards keep an attentive watch over the surrounding area.

Following pages: Researchers including Matthew Shupe, co-coordinator of the expedition, have already rehearsed the complex installation several times on land. The procedure must be well thought out and every hand movement must be in just the right order to successfully set up the sensitive instruments under the difficult conditions on the ice.

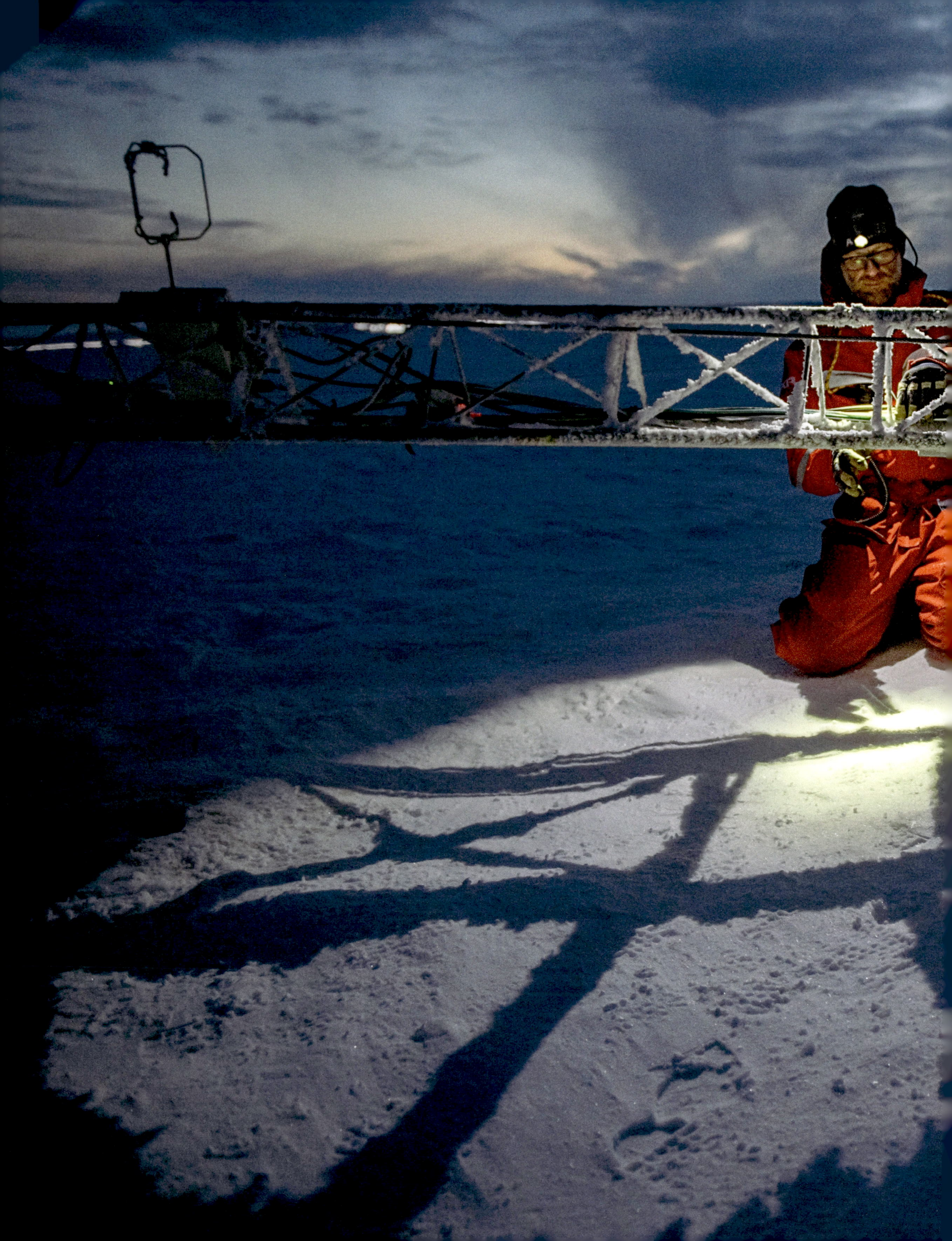

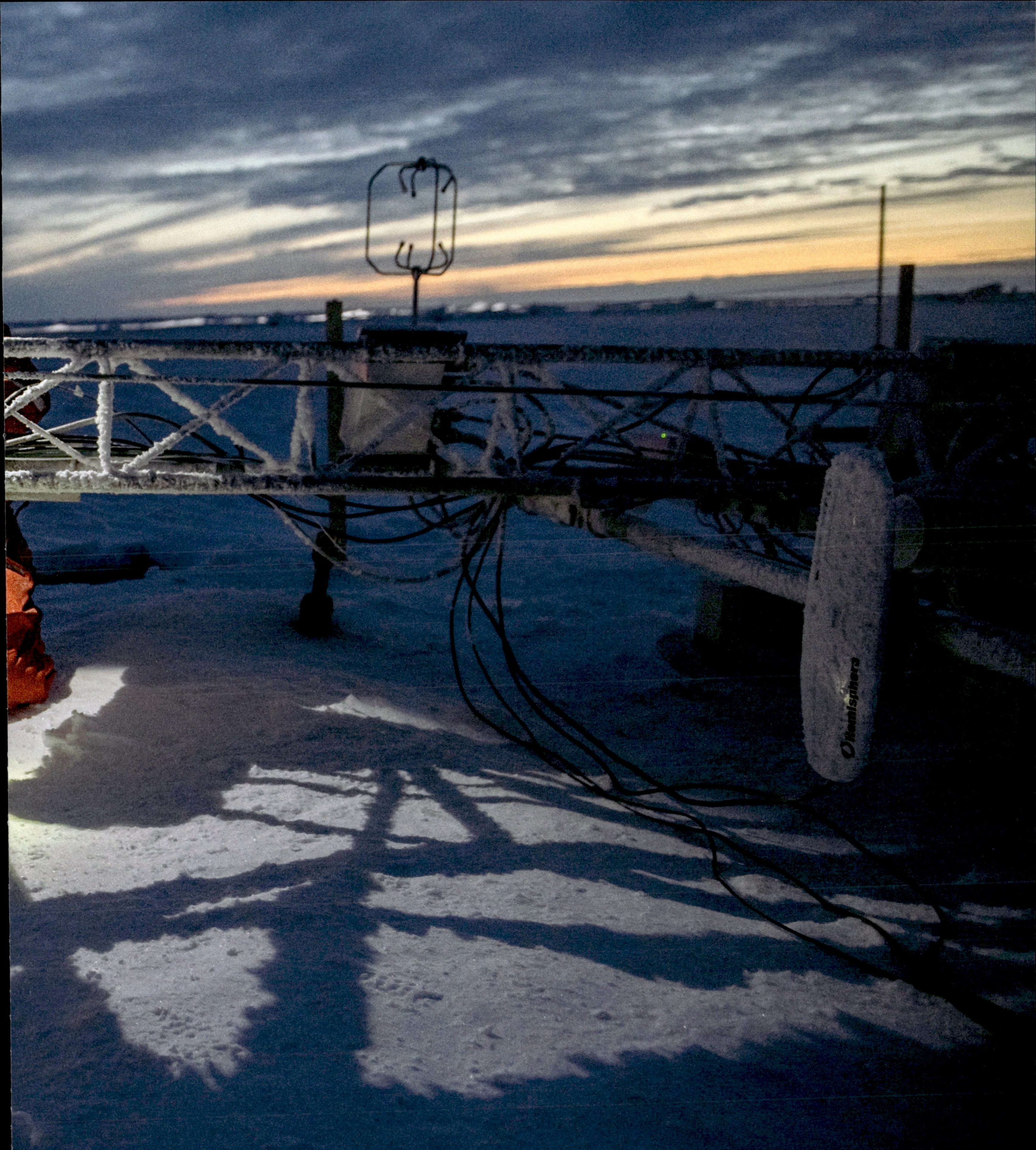
Hemisphere

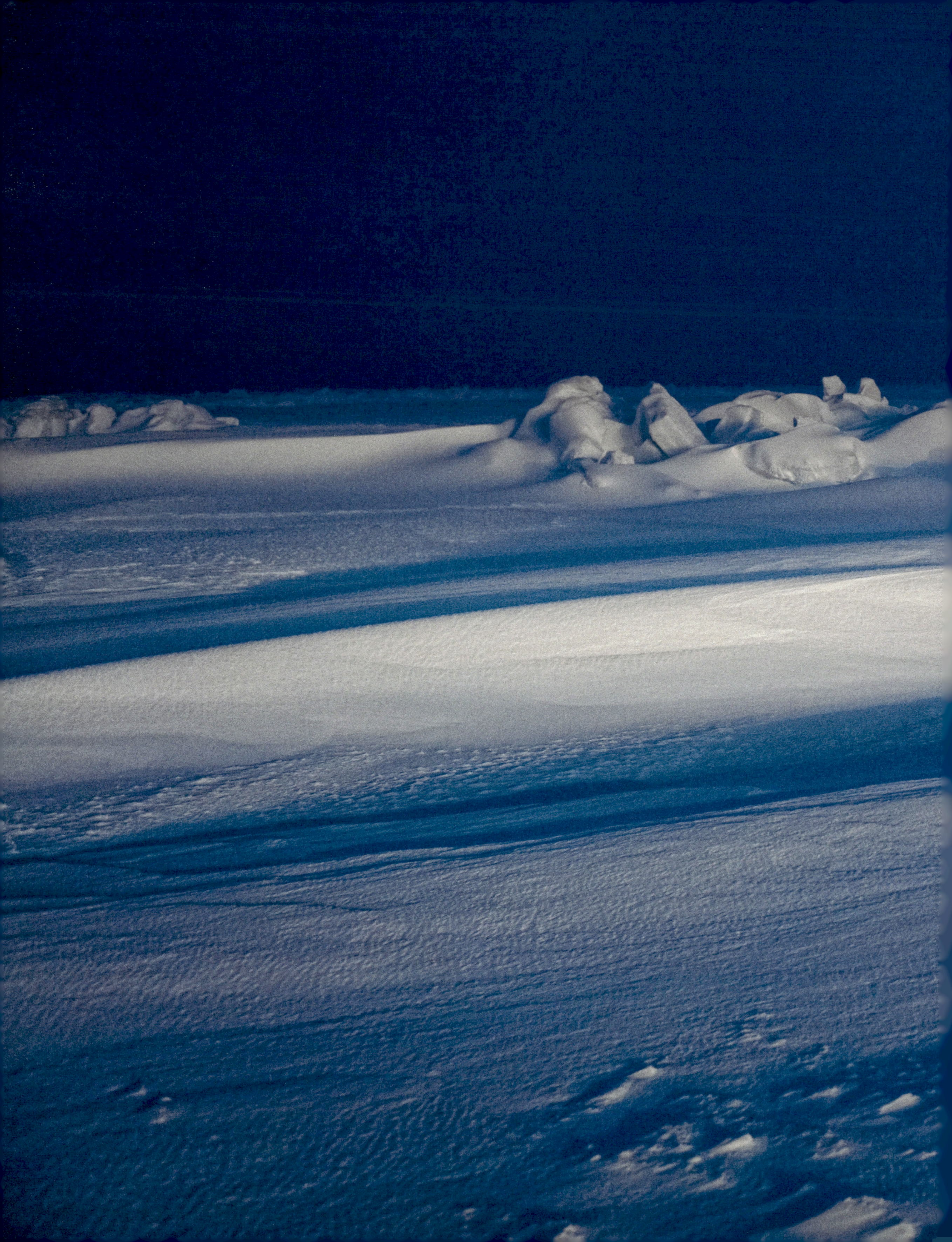

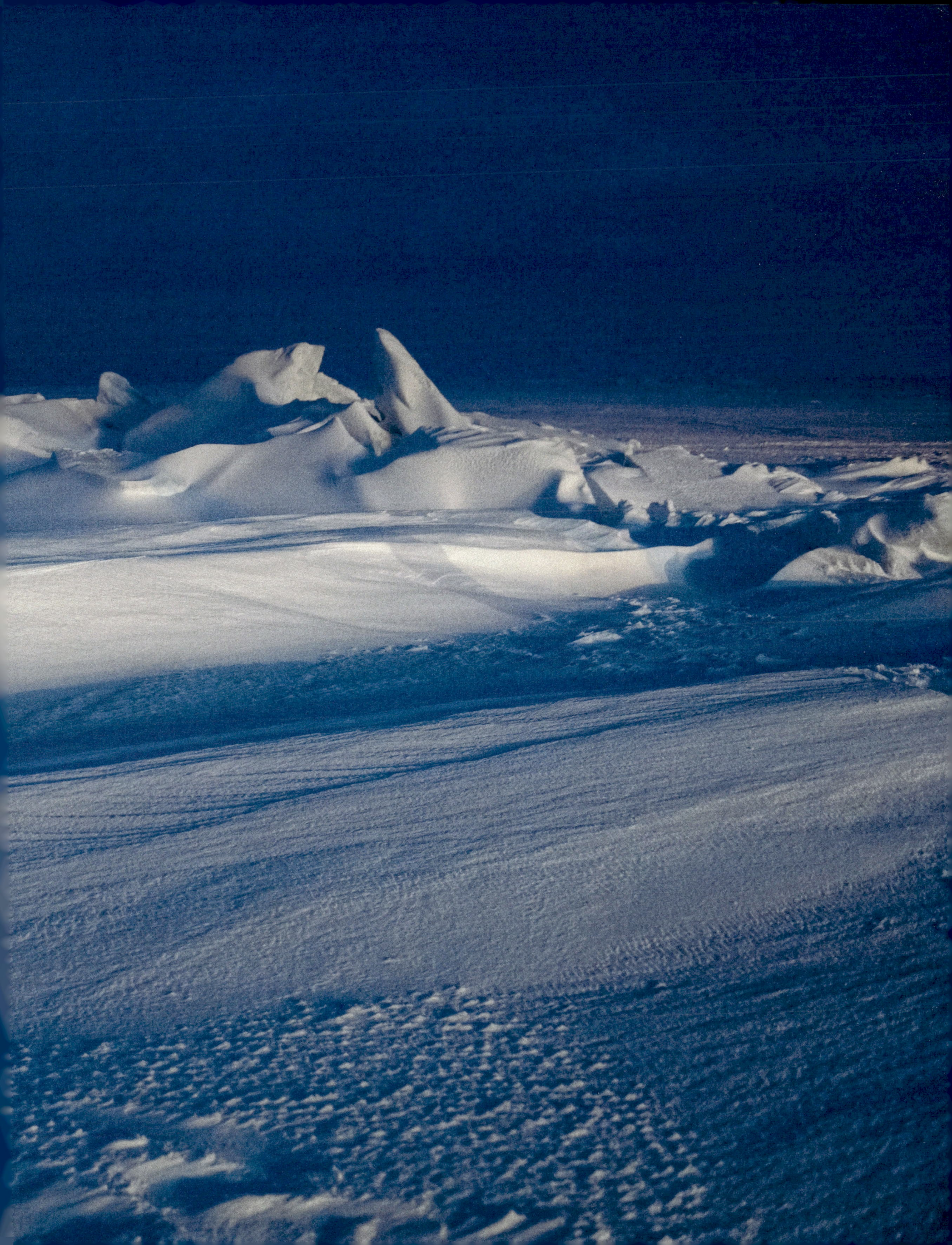

A talent for improvisation is invaluable on an expedition; after all, the weather and the sea ice are a constant source of uncertainty. At the same time, the construction of the research stations calls for excellent coordination and planning. The team around Matthew Shupe and David Costa, both from the United States, discusses the next steps in setting up Met City.

Every move is carried out carefully so as not to damage the precious instruments. Errors can be devastating, delaying the work by hours or even days. Thanks to the precise collaboration, the team succeeds in raising two sensor towers on the fragile sea ice – 11 and 30 metres high.

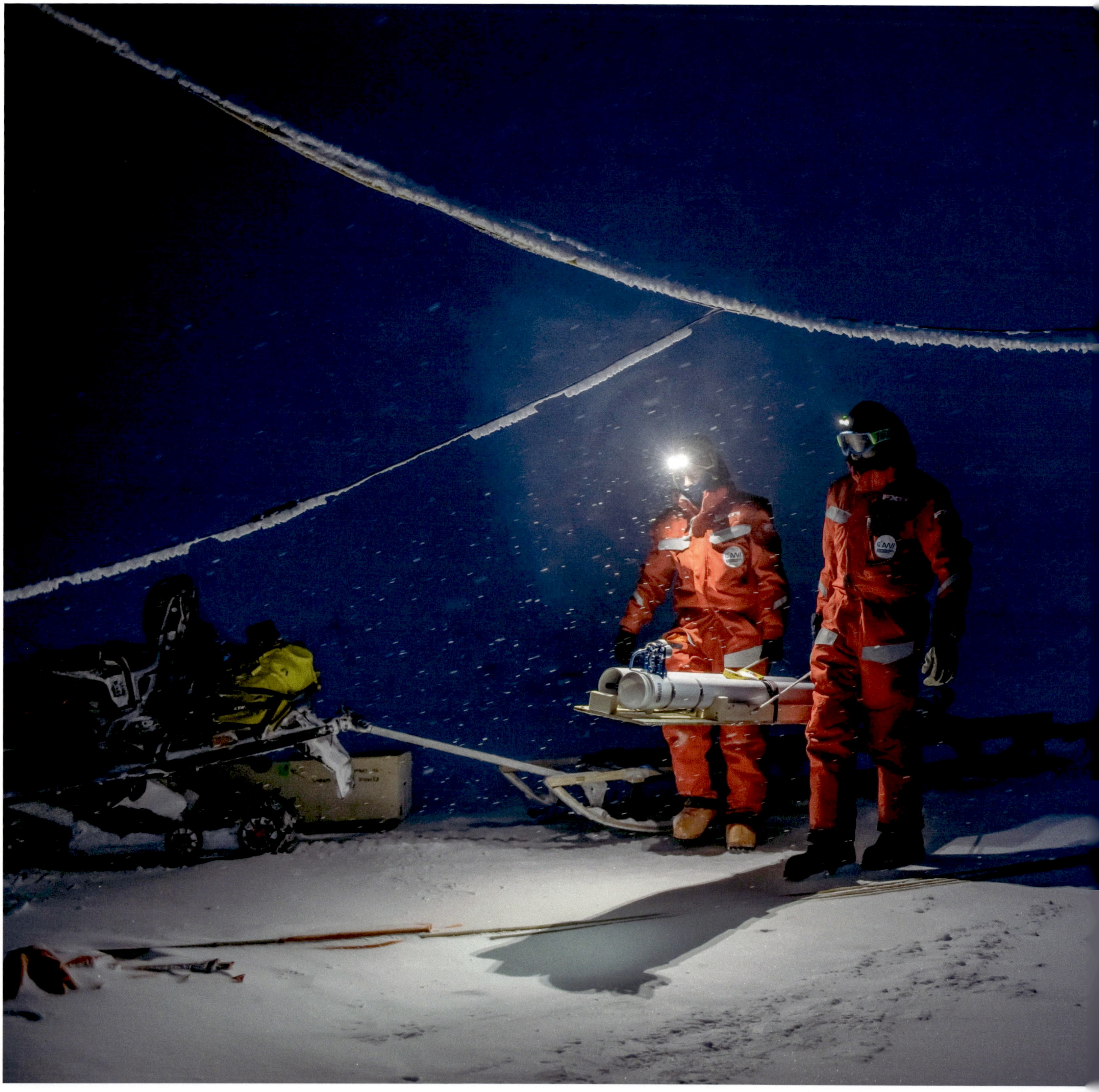

"Teamwork, across projects, across disciplines, across nations, is so essential for the success of MOSAiC."

Matthew Shupe, co-coordinator of the expedition

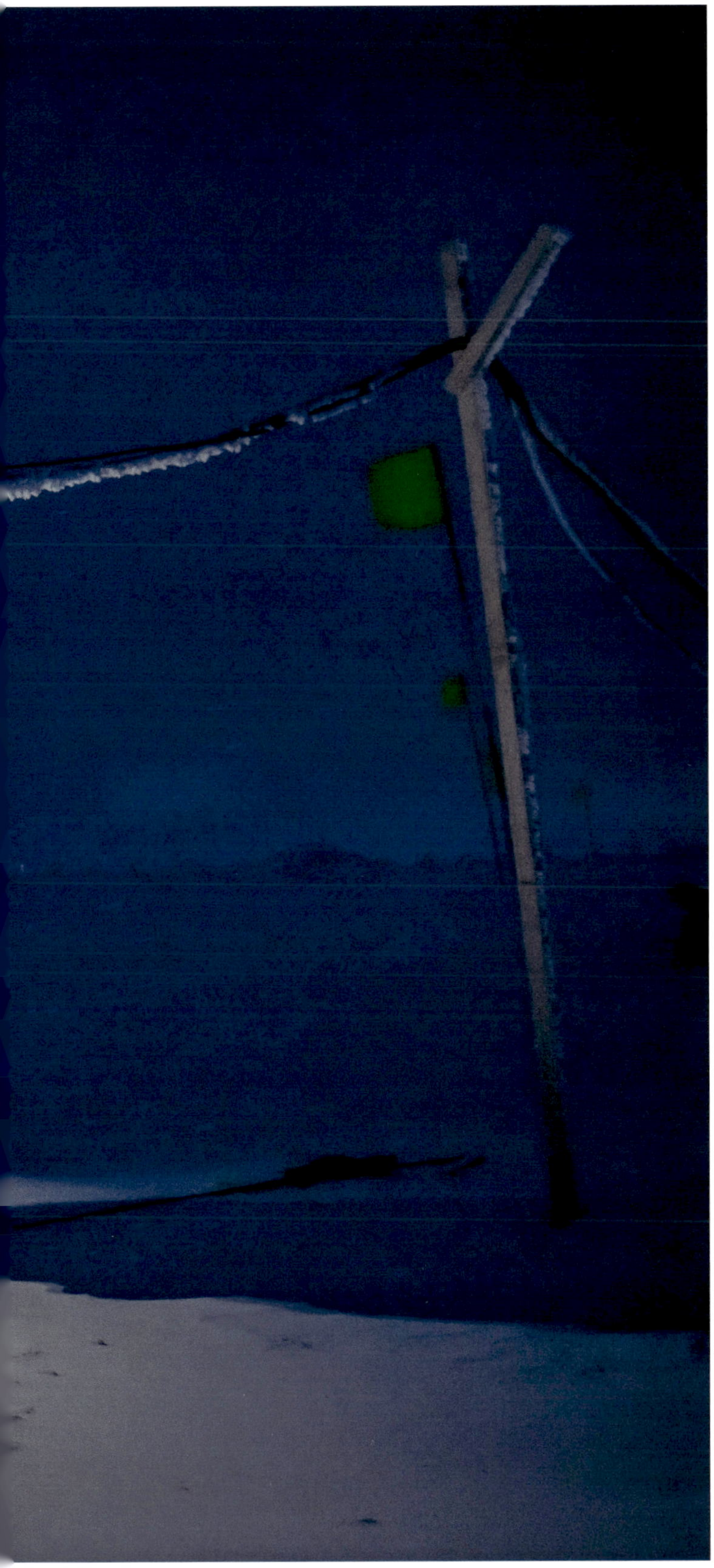

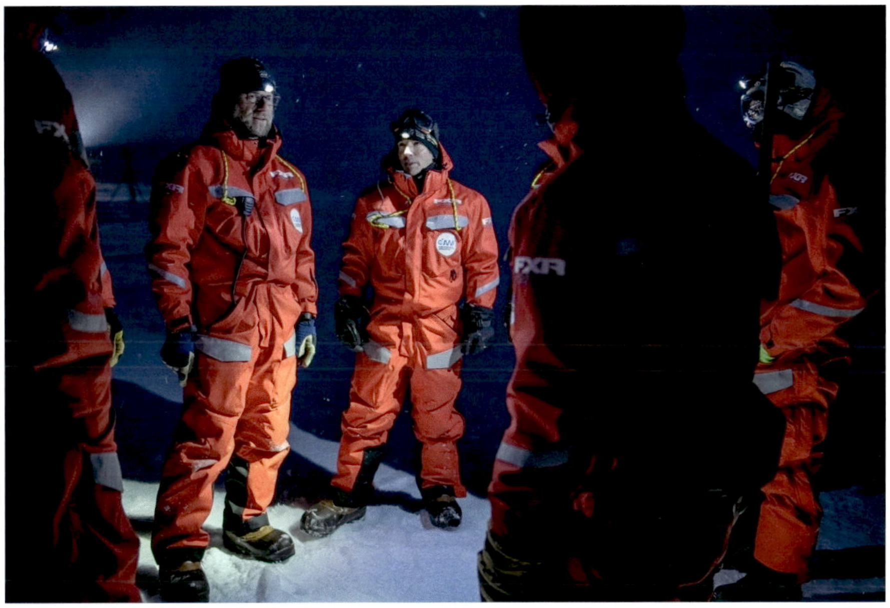

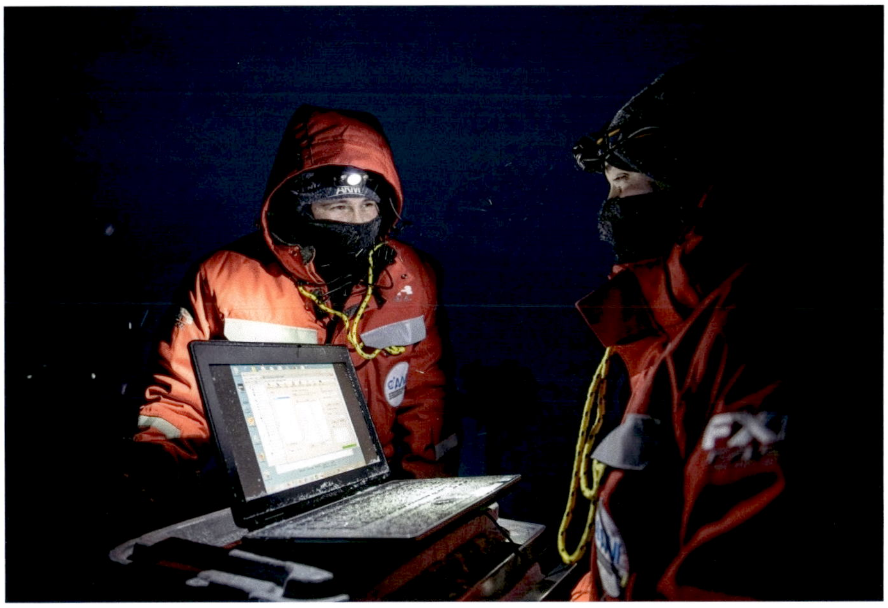

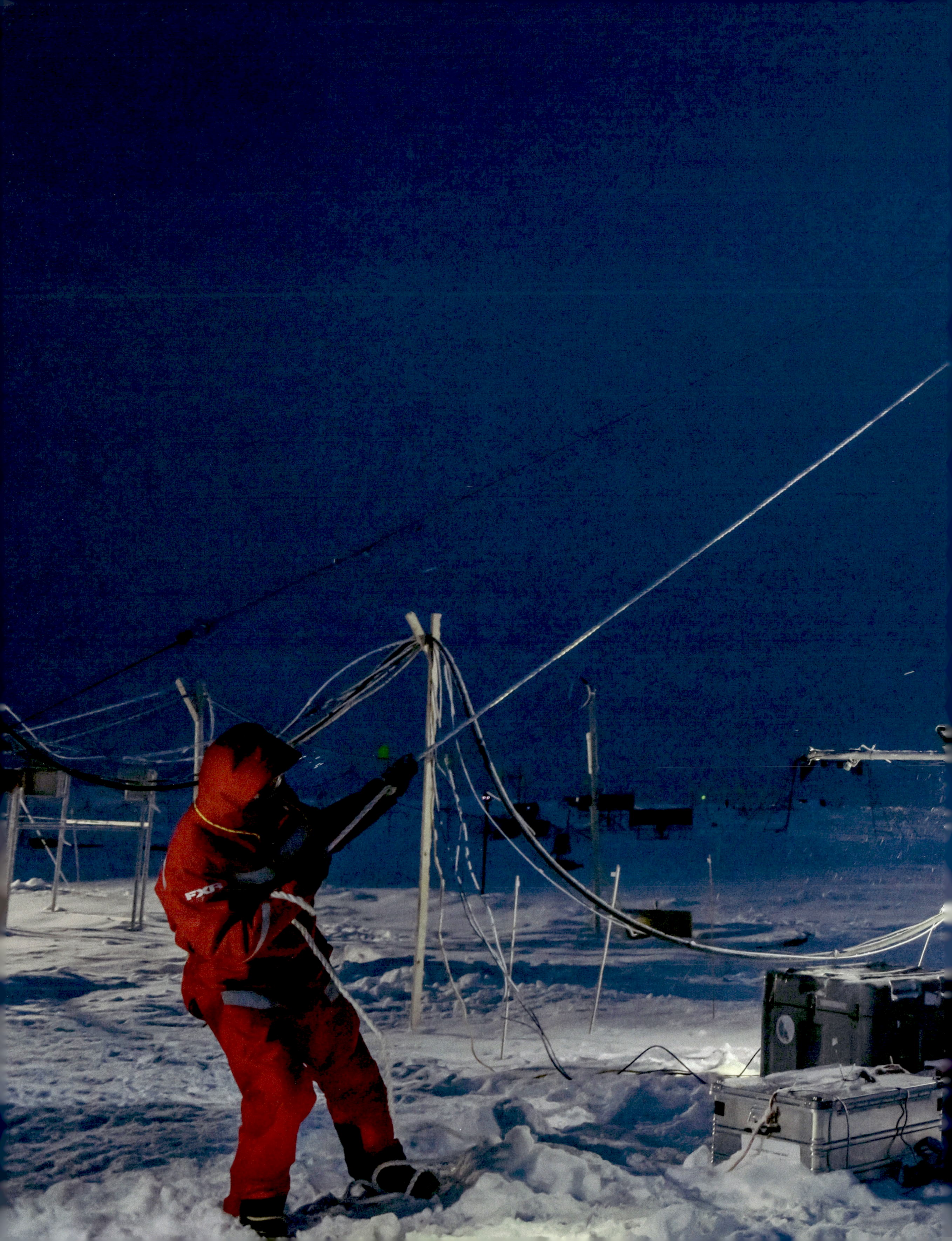
FXR

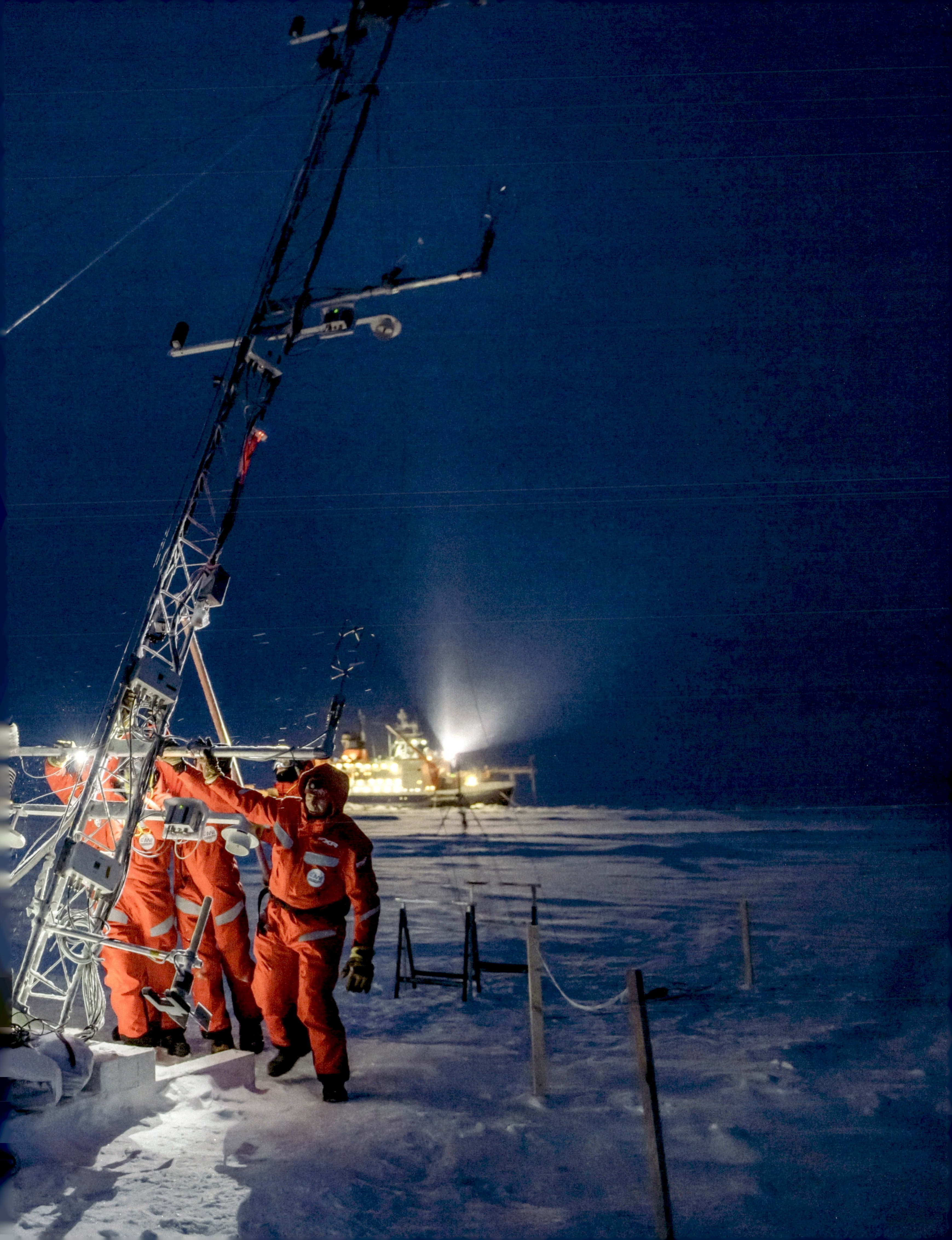

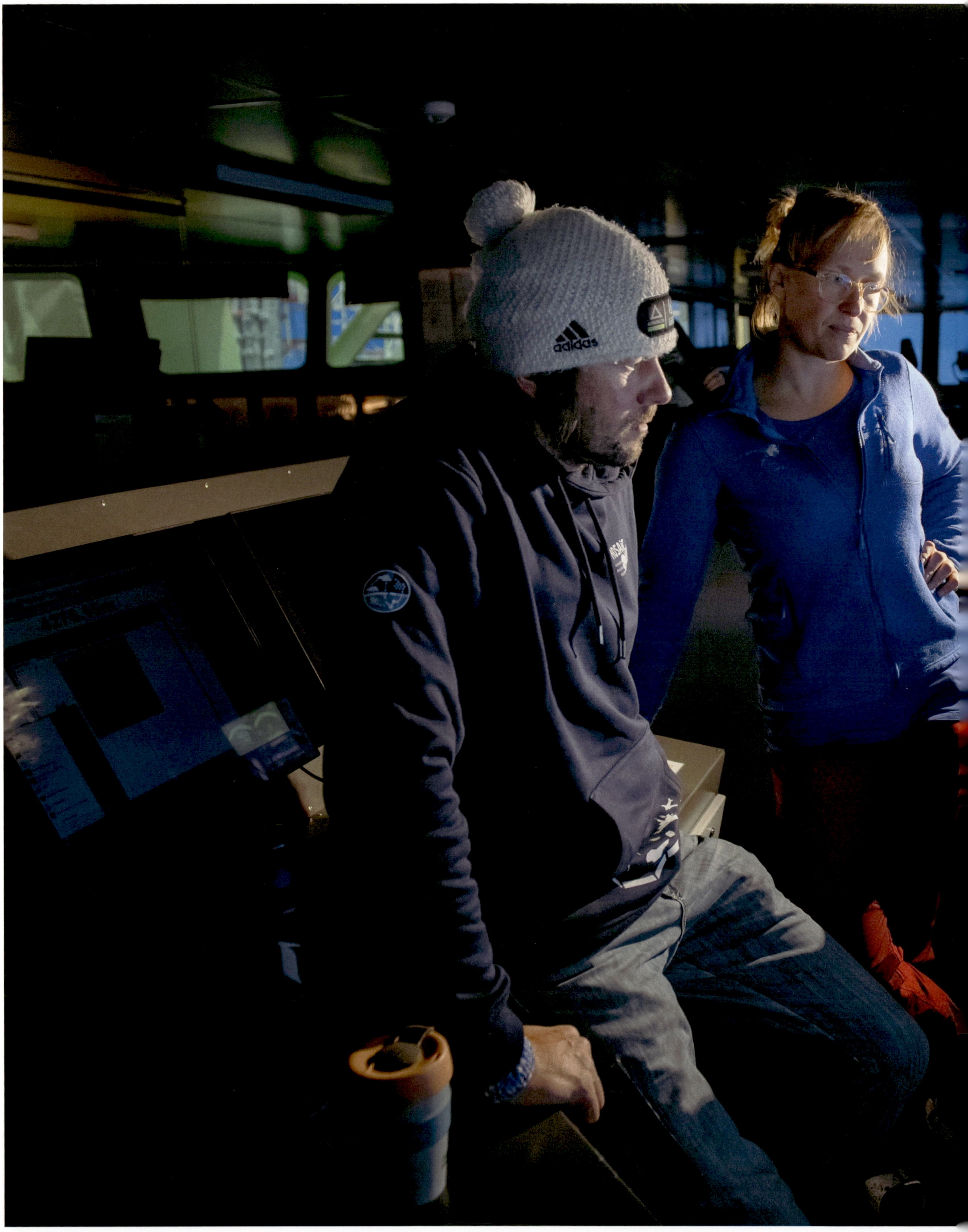
adidas

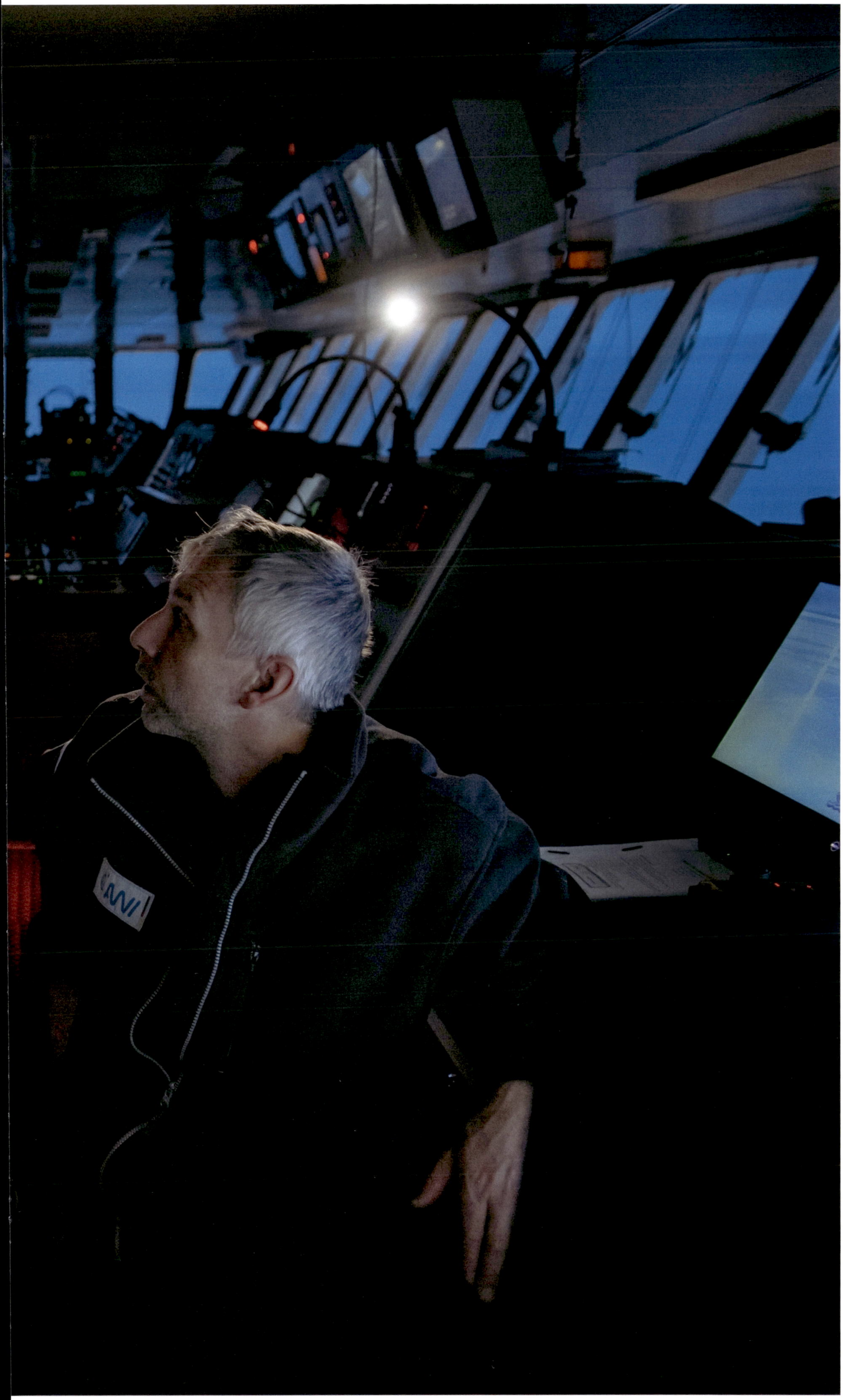

Spontaneous visits from polar bears are not the only potential danger. The sea ice, too, which is fragile, dynamic and oftentimes slippery, can be treacherous. While the camp is being set up, the safety team keep an eye on events on the ice at all times. Safety expert Bjela König is having a discussion on the bridge with her colleagues Hans Honold (left) and Henning Kirk in front of the infrared camera. The camera allows them to monitor their surroundings even in poor visibility. The safety personnel on the bridge are also in constant radio contact with the bear guards who accompany the team on to the ice. In case of danger or emergency they would send help immediately or evacuate the floe.

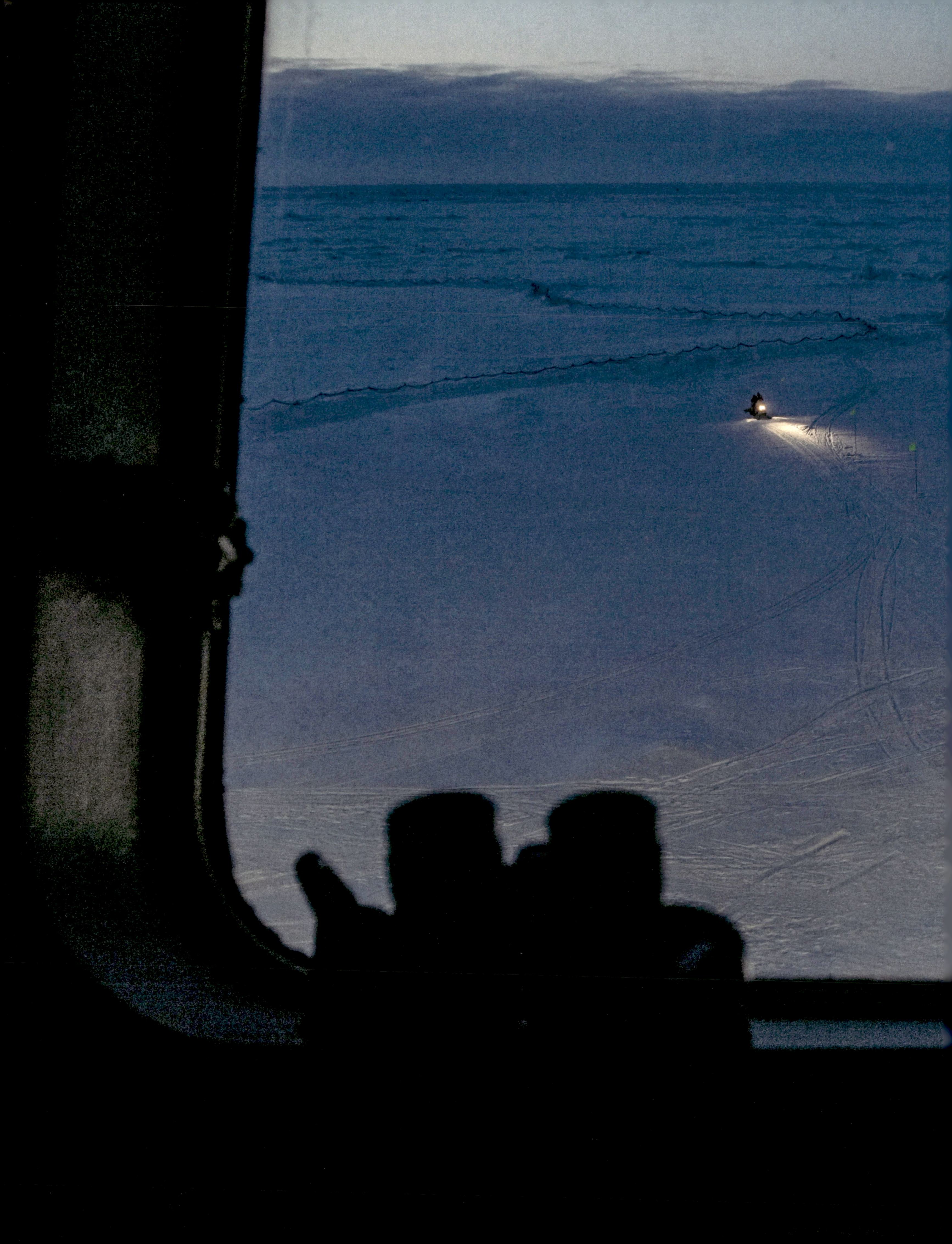

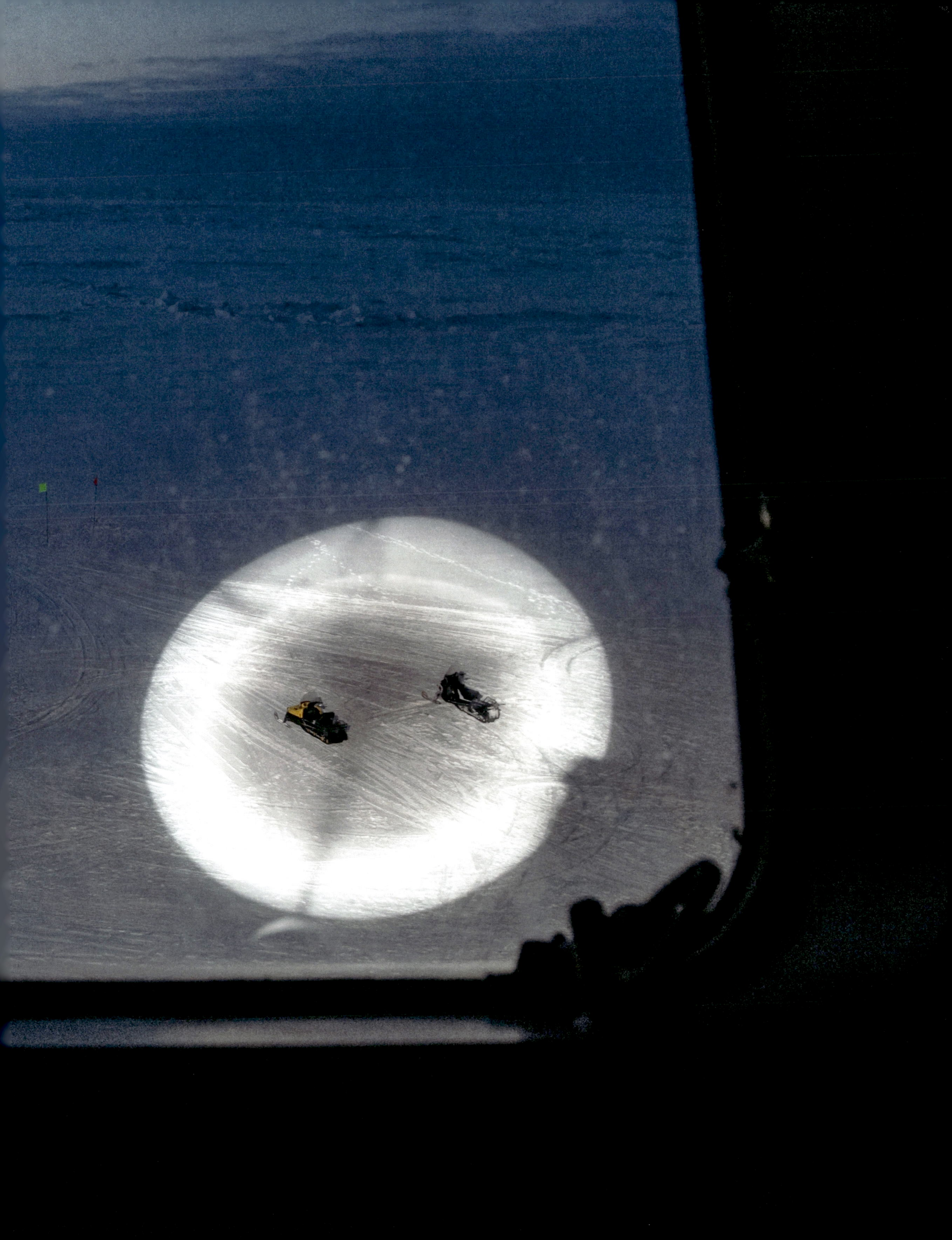

Atmospheric Research

The quest for winds, clouds and particles

The skyline of Met City, the district of the atmospheric researchers, rises conspicuously from the icy landscape. One's gaze is drawn especially to the two measuring towers, 11 and 30 metres high, which with their metallic structures look both fragile and alien – resting as they do during the drift on a thin, dynamic and unreliable layer of sea ice rather than on solid ground. Despite the constant threat of falling over in strong winds, the instruments on these towers provide a wealth of valuable data about the Arctic climatic system during the expedition. When weather permits, the researchers' airborne apparatuses climb to dizzying heights. Research balloons in different colours and shapes, drones, research aircraft and helicopters as well as measuring instruments such as radars and laser scanners, microwave and infrared radiometers – the arsenal of devices is as diverse as it is spectacular. The measurements extend from the surface of the ice to a height of 35 kilometres, about three times higher than passenger planes typically fly. Up there, the researchers are looking into the ozone layer deep in the so-called stratosphere. This is the second layer of the earth's atmosphere – the one below it, the troposphere, is where we live and where our weather happens.

However, the Arctic itself is an even greater spectacle: when winter storms rage through the polar night and the drift pushes the sea ice, the ice breaks up with the sounds of cracking and crashing. Suddenly the air and the ocean, otherwise separated by the ice, come into contact. Then, in the last place on planet Earth where one would expect it, water vapour rises in this cold region – and with it large amounts of heat. Although the ocean is cold (at about −1.5 to −1.8°C [29.3 to 28.6°F], the temperature of this marine saltwater is well below the freezing point of freshwater), in comparison to the air of polar winter at temperatures as low as −45°C (−49°F), the Arctic Ocean seems almost invitingly warm. Like a gigantic underfloor heating system that draws its energy from more southern regions of the earth, the sea warms the polar night through the cracks in the ice.

As spectacular as the sight of these cracks allowing massive amounts of heat and moisture to escape from the ocean might be, these events received little consideration in previous climate models. Given the massive Arctic sea ice layer, their influence seemed to be too little. But climate change is making such cracks increasingly common. The thinner, generally younger ice of recent times breaks up much more easily than the thick ice growing over several years and winters that was characteristic of the central Arctic before global warming. What influence do these gigantic heat fluxes from seawater have as they warm the air with ever greater frequency? Finding an answer to this question is one of the main tasks of the researchers on the MOSAiC expedition.

Clouds are born from the water vapour escaping from the ocean through the cracks in the sea ice. And, like everywhere else on our planet, in this region the heat balance – both in terms of cooling and warming – depends on whether and which clouds appear in the sky. A cloudy sky is generally associated with little sunshine, with precipitation and with cool weather. But on winter nights, for example, clouds in our regions also demonstrate the ability to ensure mild temperatures: they trap the heat energy of the day in the atmosphere. In contrast, a cloudless night sky allows the heat of the day to escape into the expanses of space. Similarly, greater cloud formation in the Arctic can also lead to milder temperatures.

Whether a cloud has a cooling or warming effect depends on the type of cloud and its altitude. And it also depends on how often a certain type of cloud forms in order to have a lasting influence on the Arctic climate system. For example, cirrus clouds, with their delicate appearance, form at lower temperatures. They carry crystals of ice and snow within them and float high in the atmosphere. Cirrus clouds are so fine that they barely trap any heat in the atmosphere. Wetter clouds, by contrast, hang lower in the atmosphere like thick banks of droplets. The air beneath them becomes milder. Thus, with global warming, the types of clouds

The MOSAiC expedition is devoted to the Arctic climate system in all its fascinating complexity. During their year-long drift mission, the international team of experts has the opportunity to closely observe the interplay of the different parts of the climate system. The individual teams with their scientific focuses on atmosphere (1), sea ice (2), ocean (3), ecosystem (4) and biogeochemistry (5) work closely together in order to fully understand the Artic. The data gleaned will make it possible to produce the urgently needed more accurate forecasts for how climate will develop – not only in the Arctic, but globally as well. This complex and diverse collaborative approach to research is reflected in the name of the expedition: MOSAiC – Multidisciplinary drifting Observatory for the Study of Arctic Climate.

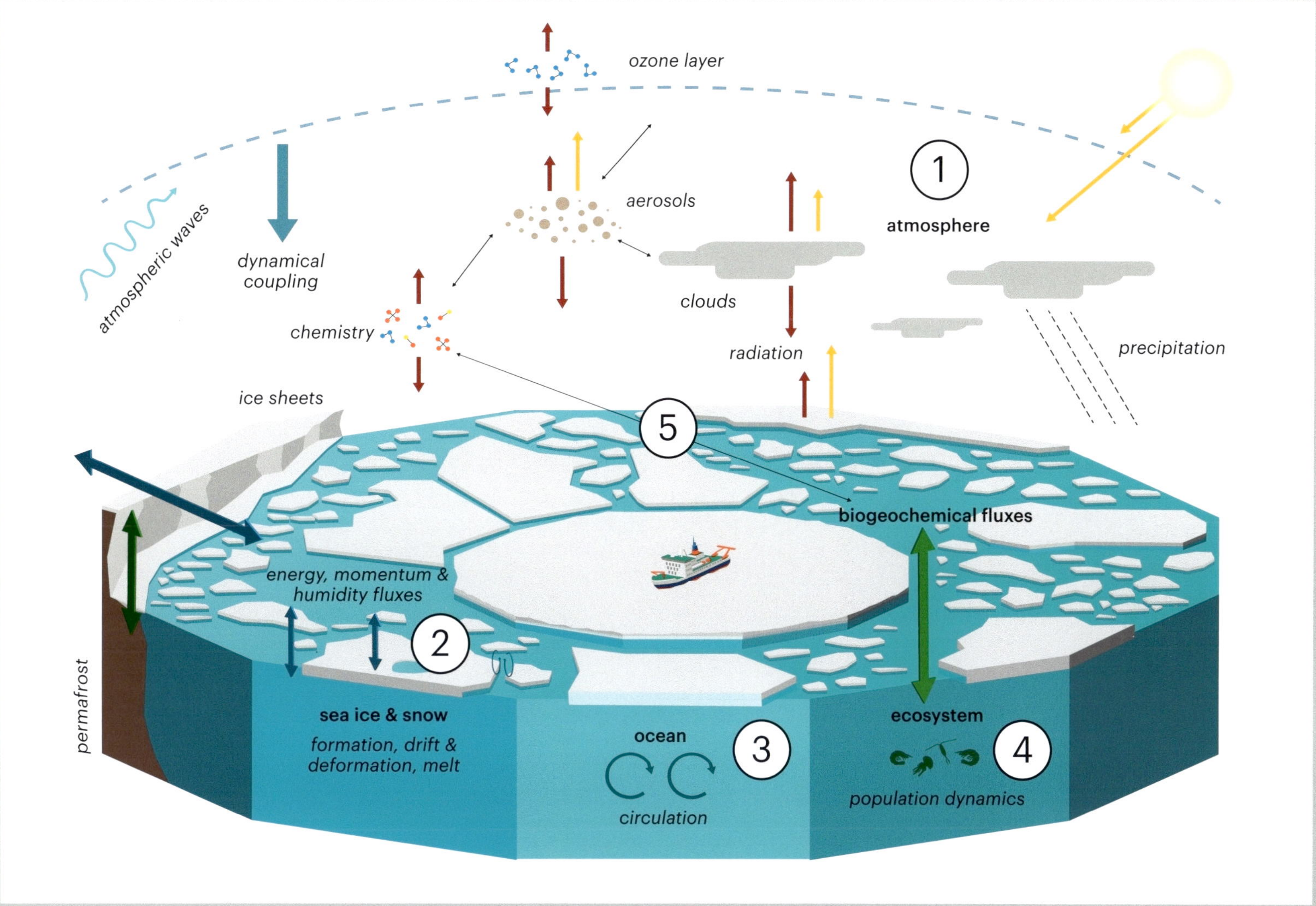

that are becoming more likely are precisely the wetter ones, which can result in higher temperatures. Climate change then drives itself forwards.

This is why the scientists of the MOSAiC expedition are chasing clouds. They aim into the sky their radars and the so-called lidar, a laser that shoots a beam of light straight as an arrow to great heights in true science-fiction fashion. Once a cloud is reached, the reflections of radio waves and laser light permit the analysis of its composition. When the wind is not too strong, "Miss Piggy", a plump red tethered balloon equipped with several instruments rises up to 2 kilometres into the sky on its rope in order to collect data in the lower troposphere and transmit them to ground level. For the researchers, the most entertaining measurements include scenes that for onlookers would seem like a pleasant afternoon with flying toys: standing on the ice, remote control in their hands, the researchers send nimble drones chasing after clouds and winds in order to broaden their knowledge of the atmosphere.

On the *Polarstern*'s helicopter pad, Jürgen Graeser, who is experienced in the Arctic and in expeditions, releases a research balloon into the Arctic atmosphere.

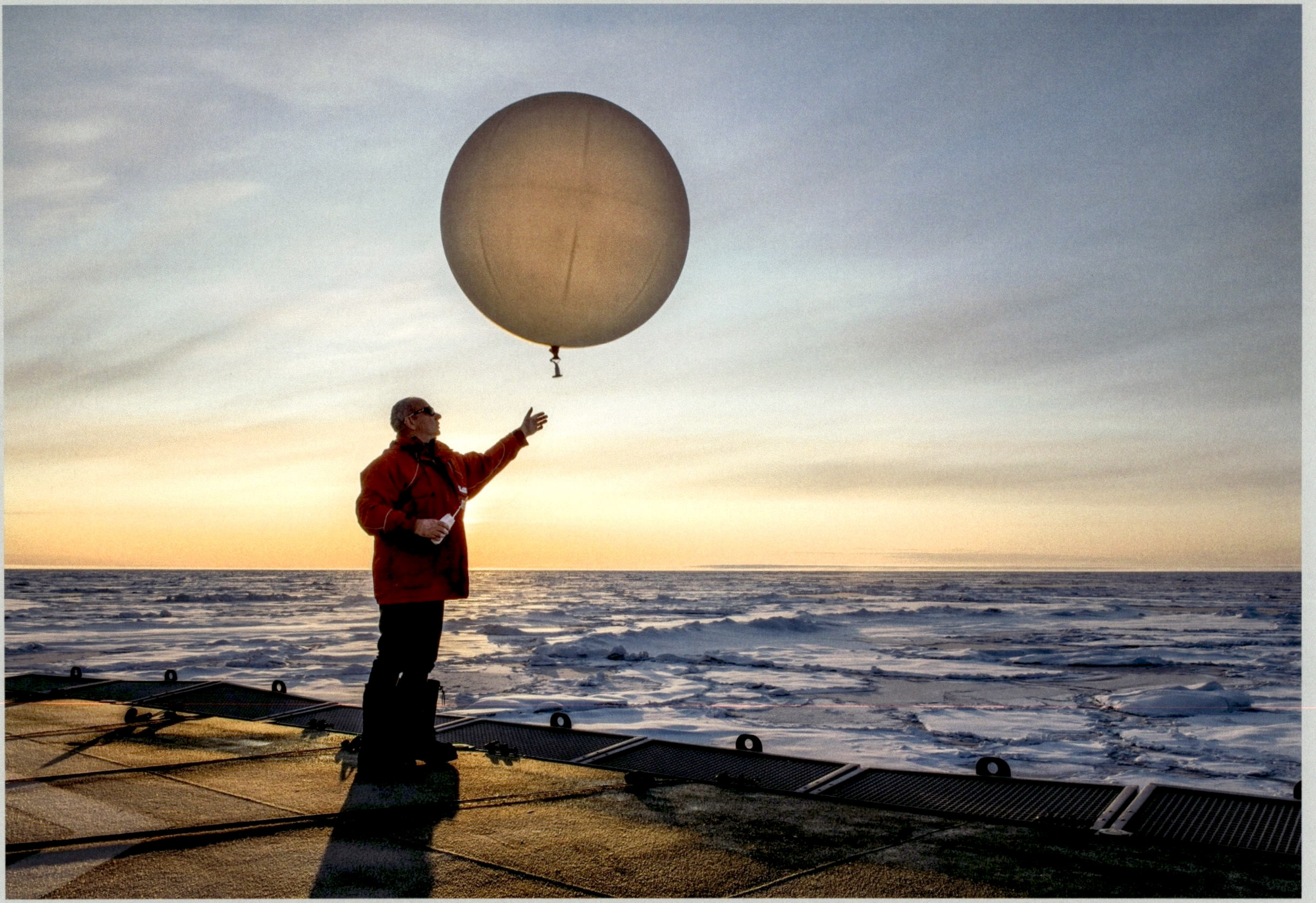

In order to understand the climate, one must understand clouds. But clouds are transient, and thus the processes associated with them hold many unknowns for science. In climate models, clouds are among the greatest sources of uncertainty. One type of cloud is of particular interest to the researchers on the expedition, the so-called mixed-phase clouds. As the name suggests, this type of cloud combines different properties, and it is composed of both ice crystals and of supercooled liquid water droplets. They float at mid-altitudes and it is these clouds that precipitate snow. When blusters of Arctic snowflakes rage over the sea ice, the flakes, so apparently delicate, become painful projectiles for unprotected skin. For the scientists on the ice, a gathering storm means beating an immediate retreat to the shelter of the ship before whirling snow and ice crystals reduce visibility to zero. The frozen precipitation also has an enormous effect on the Arctic climate system: the layer of snow lies on the ice like a protective cover, insulating floes against the cold Arctic air. The snow thus has the effect of preventing more ocean water from freezing. During the expedition scientists were amazed at the readings: the thermometer registered a difference of as much as 7°C (44.6°F) between the top and the bottom of the snow layer. The result is thus hardly surprising: where there is snow, the sea ice remains thinner.

When Miss Piggy's red silhouette sways in the atmosphere, however, this bulging blimp is searching for something else: her sensors are looking for aerosols, tiny particles that float in the atmosphere. These suspended particles form the nuclei of clouds as rising water vapour condenses around them. Droplets that form around aerosols grow into clouds, and the type of aerosol determines what kind of cloud can be formed. When one says that the smallest particles have the greatest impact, it is aerosols about which they might be speaking: a human hair is about 100 to 1,000 times thicker than an average aerosol. Still, the human eye can discern these minuscule floating particles: on the horizon they regularly provide us with colourful moments, such as when they scatter the visible blue/red part of the spectrum of sunlight and the sky glows red at sunrise and sunset. On the other hand, a layer of aerosol above heavily congested and industrialized regions is also a common sight, and it negatively affects visibility as well as prospects for human health: suspended particles concentrate in urban smog and raise alarms about fine particulate matter. Aerosols can be of natural origin, in the form of sea salt particles scattered by ocean spray, blown desert dust, volcanic ash or flower pollen. However, many particles suspended in the atmosphere are man-made, coming from industrial emissions, factory chimneys and car exhaust, often the same source as the greenhouse gas carbon dioxide. Forest fires also put aerosols into the atmosphere. And these too are largely caused by human activities such as slash-and-burn or arson for the purpose of gaining pasture or arable land for soy plantations, which in turn go to feed the vast numbers of livestock in industrialized nations. Resulting particles such as soot are small enough to get through the defences of the human respiratory system, penetrating deep into the lungs and attacking respiratory organs and the cardiovascular system.

While it has long been known that these suspended particles can cause illness in humans, still lacking is adequate research into how they influence the climate. Little was known about the origin and composition of aerosols in the Arctic. However, the first results of the MOSAiC expedition have shown that fine dust from human sources and forest fires, carried by winds from North America and Siberia, also pollutes the Arctic atmosphere. What effect do such particles have on cloud formation in the Arctic – and, in turn, on temperature? But there is indeed more, as the influence of aerosols on climate is even more complex. Even without cloud formation, anthropogenic air pollution can, depending on the type of particle, contribute to the cooling or further warming of the earth. Soot particles, for example, absorb sunlight and thus heat up the surrounding atmosphere. In contrast, particles of sulphur compounds that enter the atmosphere through a complicated process as a result of the burning of coal or oil reflect sunlight. Thus smog, of all things, has so far been responsible for moderating anthropogenic climate change – possibly by as much as a half-degree Celsius (0.9°F). But this is only a temporary shroud over the true extent of climate change: this cooling effect would probably end if humanity succeeded in taking the ultimately unavoidable step of reducing the burning of fossil fuels in order to decrease the emission of greenhouse gases, or succeeded in improving the air quality in Asia by treating the flue gas of power stations, as has largely been achieved in Europe.

Even in the Arctic, where neither pollen nor forest fires are considered "native" sources, not all aerosols derive from human sources. Even in this region there are plants that emit a substance into the Arctic air from which aerosols are formed. The name of this substance is dimethyl sulphide, a gas created when the metabolism of algae and other microorganisms produces a sulphur compound that is further processed by bacteria. Anyone who has vacationed at the beach knows the unique smell of dimethyl sulphide: this gas is whirled into the air with saltwater spray and enters the nose as the smell of the sea. But does it ultimately produce clouds – and if so, what kind? Does the sea's microalgae – phytoplankton – not only produce 50 per cent of the oxygen that we breathe but also play a part in our weather and climate through their excretions? Answering these fundamental questions is an important part of the joint work of the atmospheric and biogeochemistry researchers in the Arctic. Ultimately, the complex code of the climate system can only be deciphered through a team effort.

The Observatory

Notes from the ice

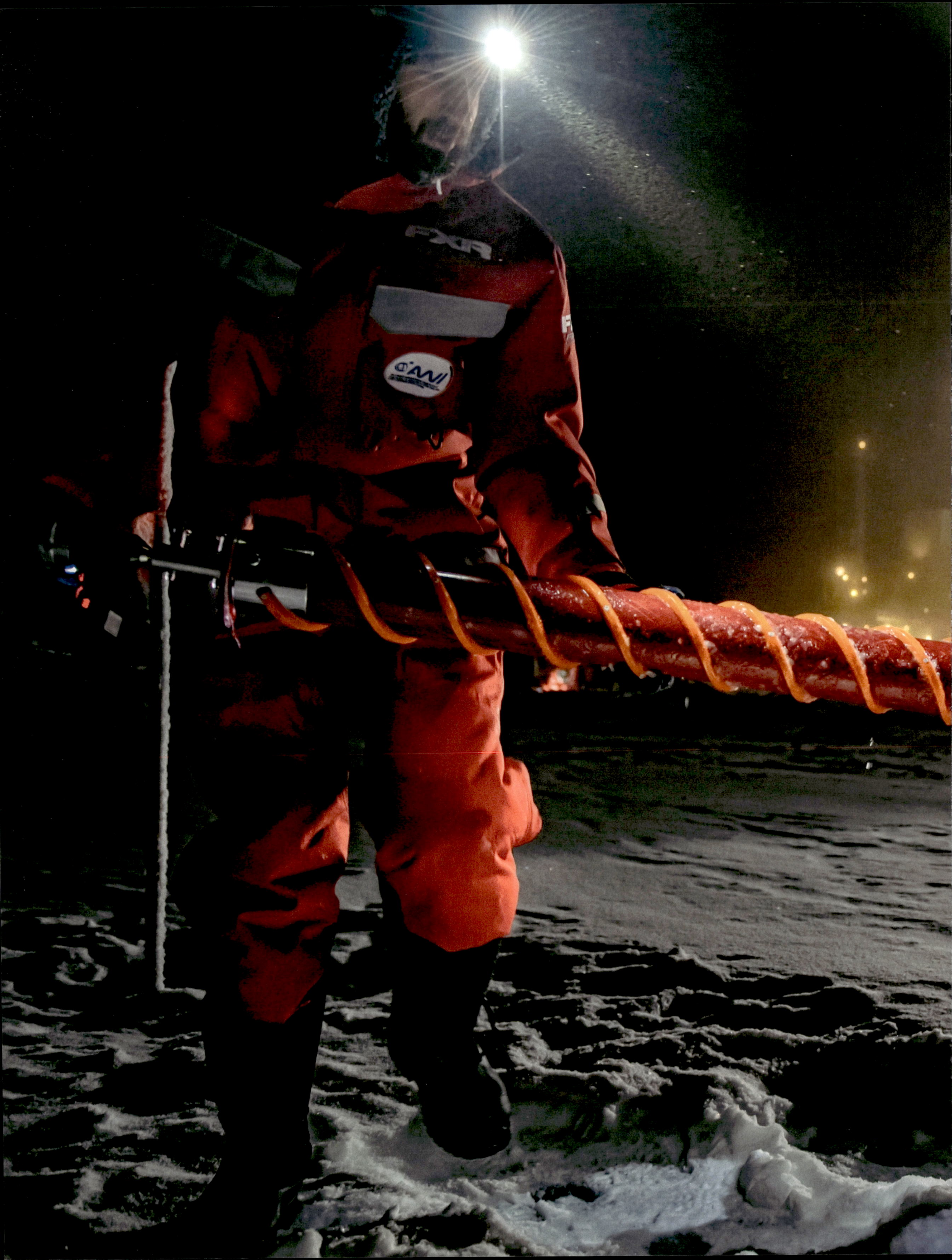
FXR

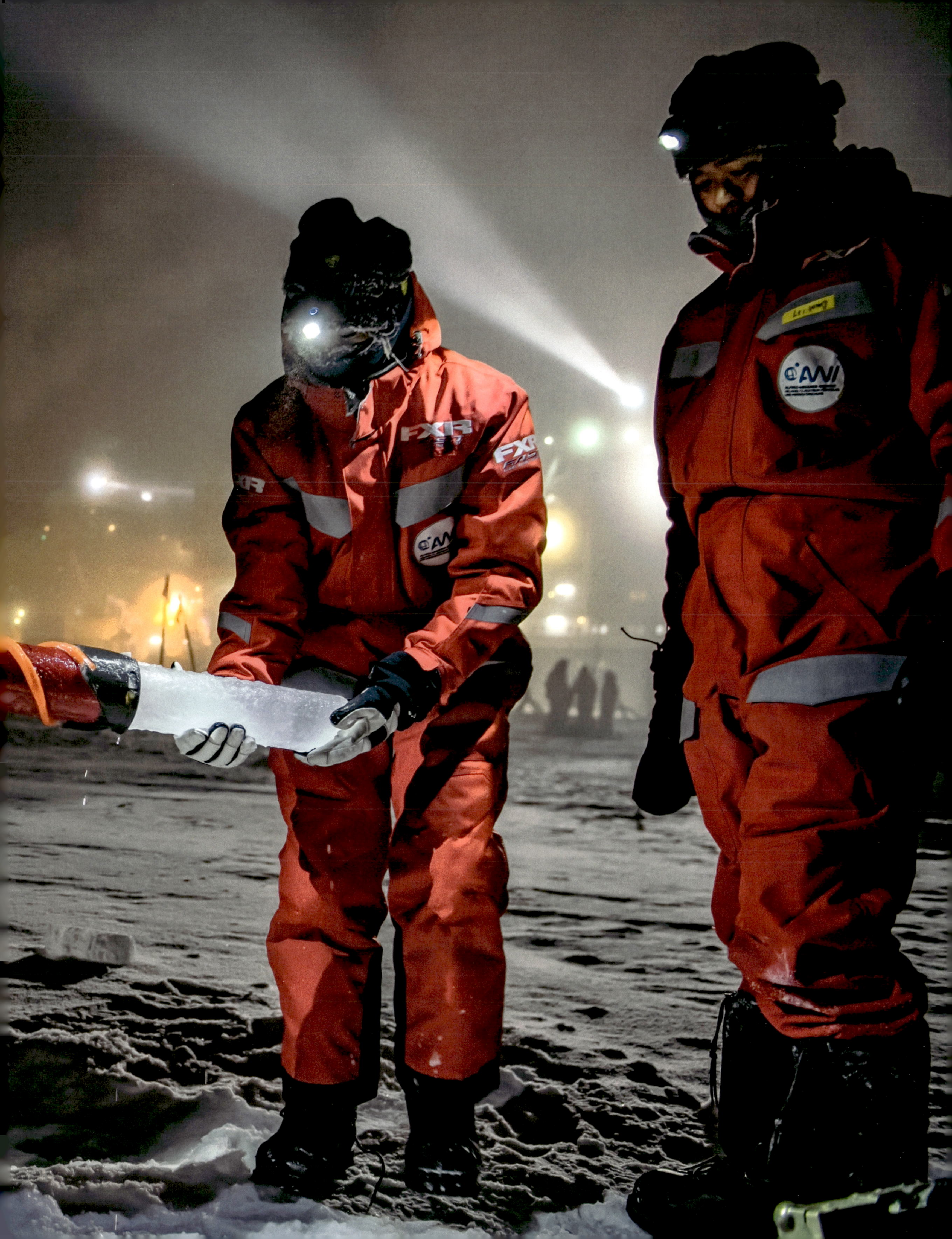
FXR
FXR
FXR
AWI
AWI

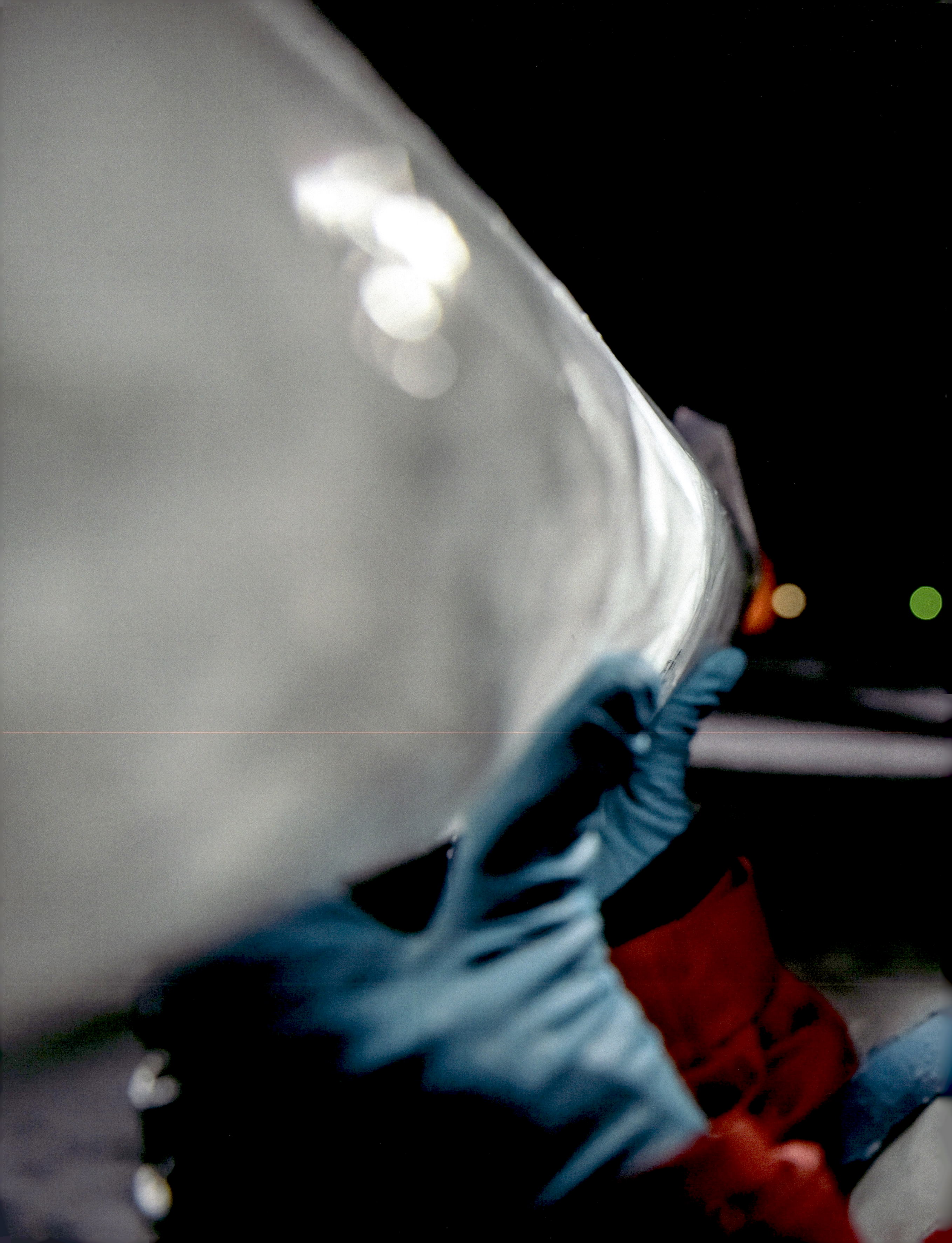

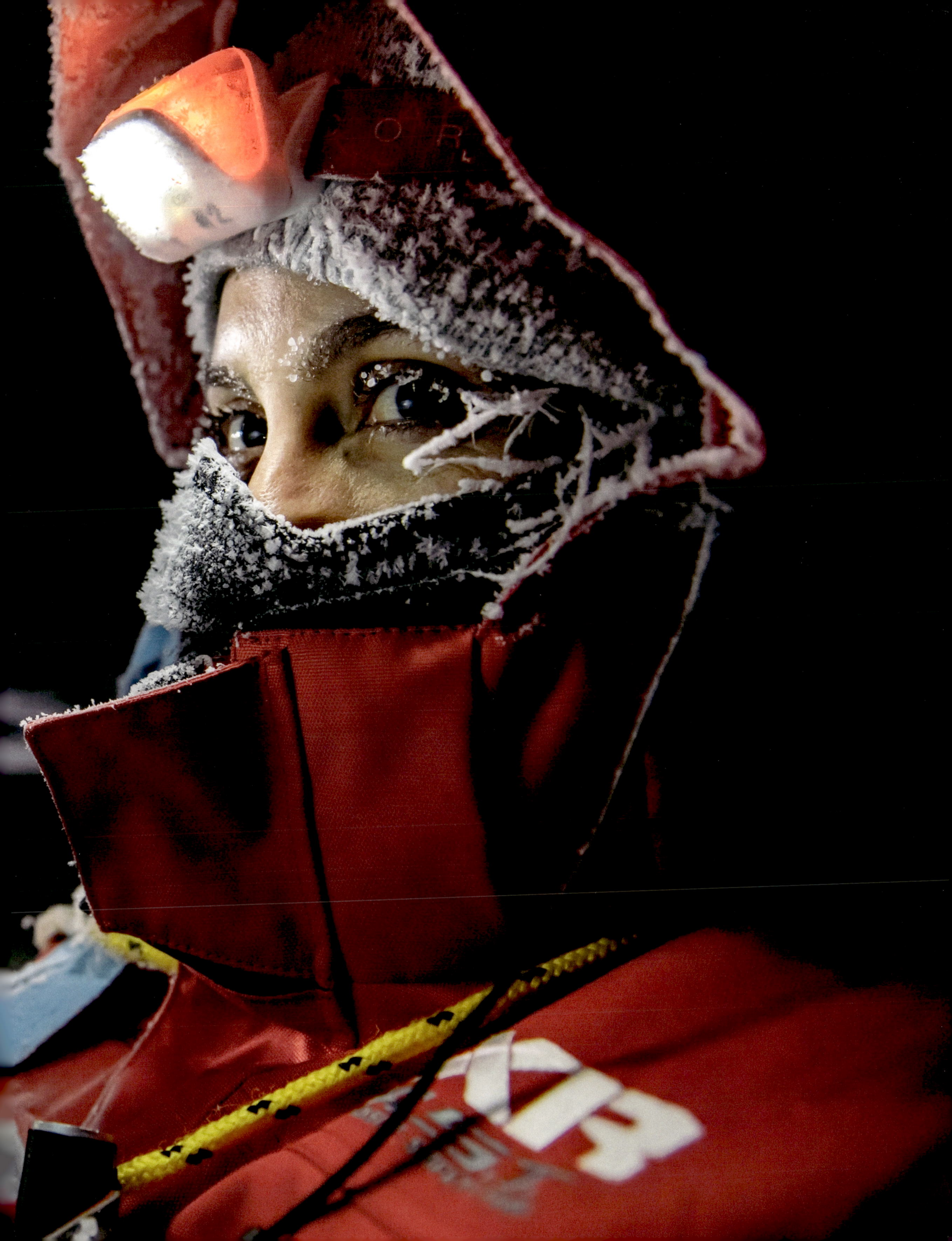

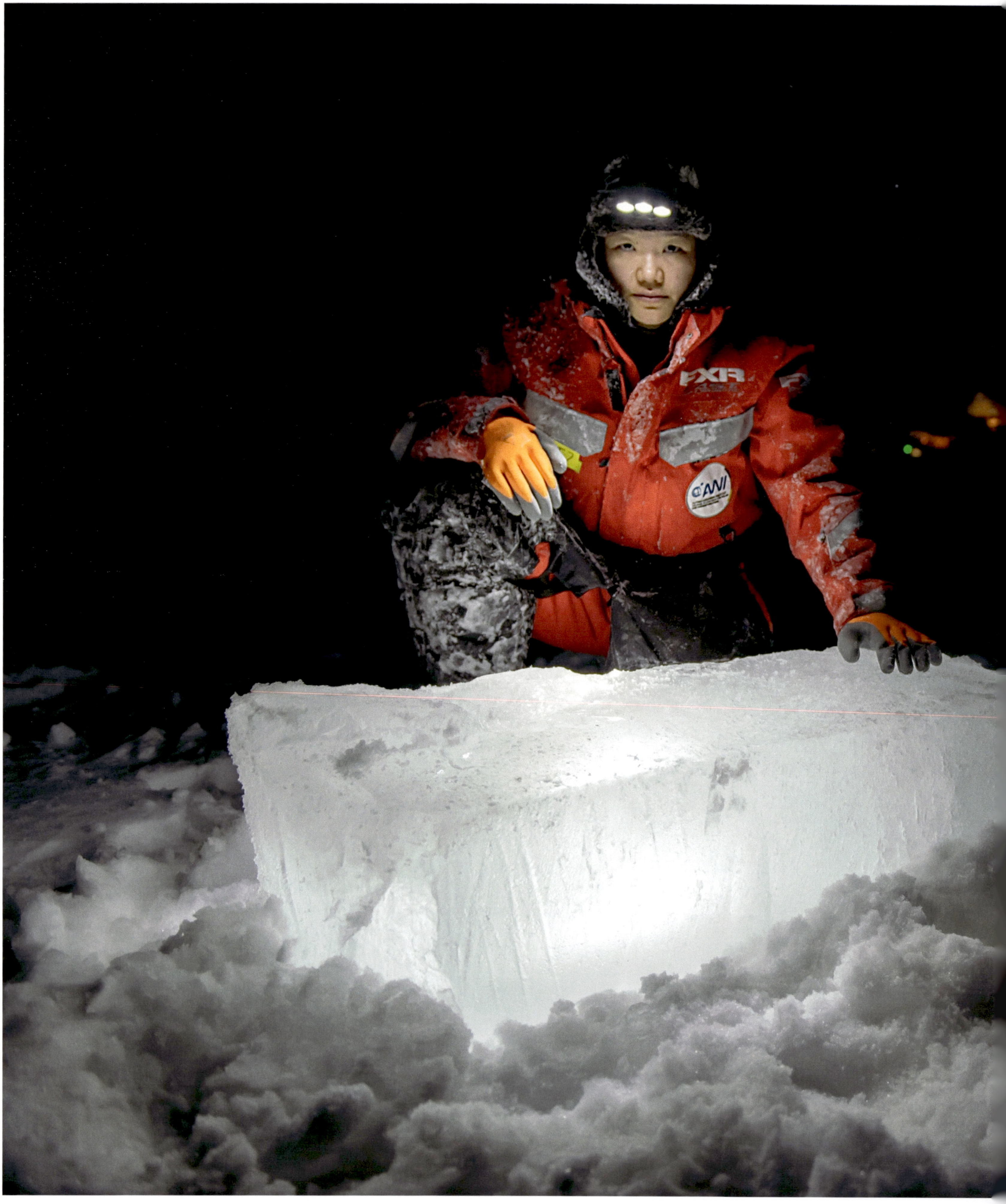
FXR

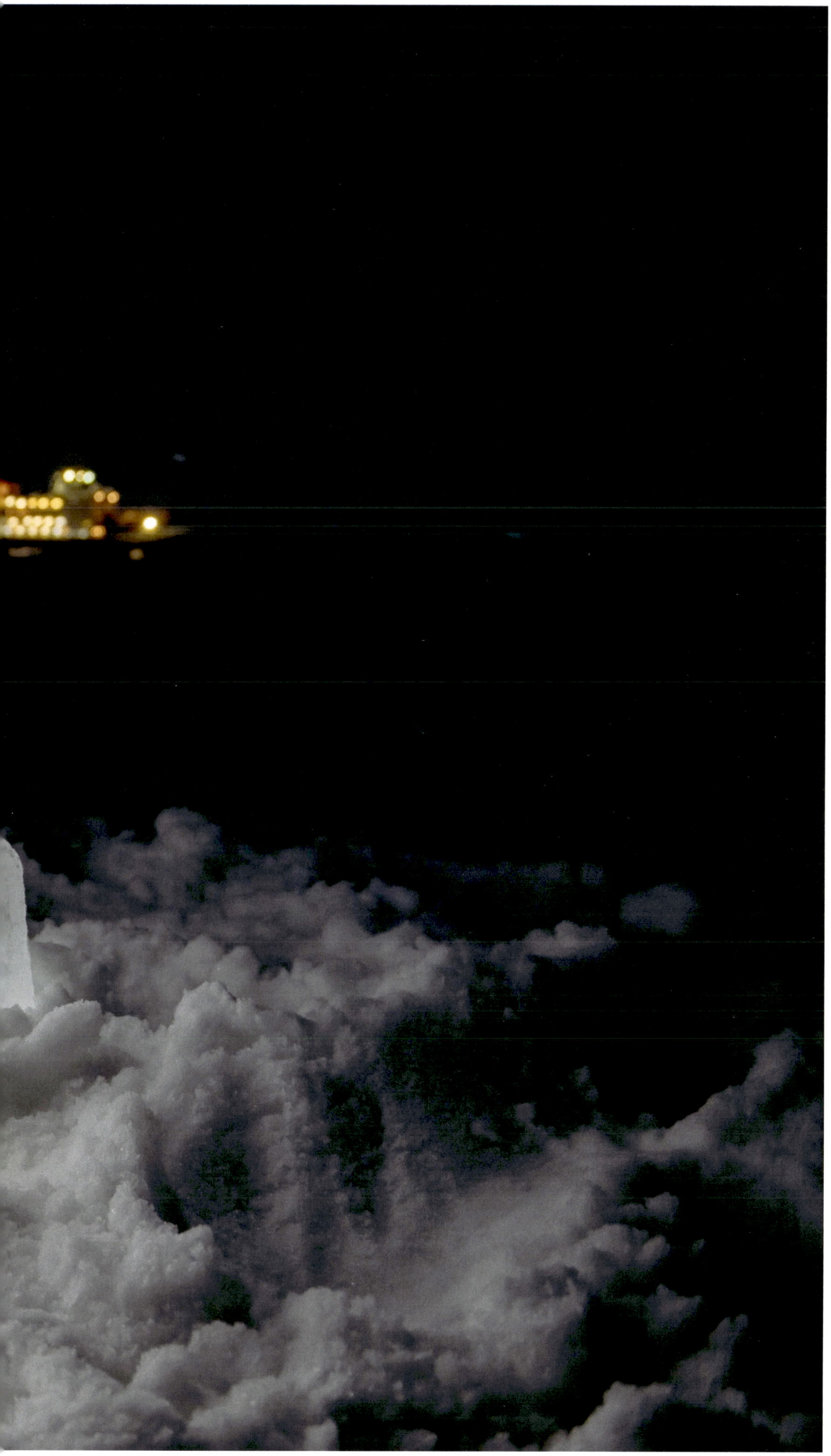

The scientists remove ice from the floe in different ways and in various forms, for example as ice cores for samples, or to install scientific instruments. Allison Fong, a biologist and head of the Ecosystem team, has sawn a large block out of the floe in order to launch the so-called FishCam in the water. At a depth of 250 to 350 metres the FishCam records images of the world of Arctic fish – and occasionally of other animals. During the expedition, the FishCam supplies images of marine creatures, but at times the team has to rescue it, along with other devices, from cracks in the ice so as not to be swallowed up by the ocean.

The expedition's research teams:

- Atmosphere
- Sea Ice
- Ocean
- Ecosystem
- Biogeochemistry

The Ice Camp is divided into different districts, including:

- Met City
- Balloon Town
- ROV Oasis
- Ocean City
- Remote Sensing Site
- Fort Ridge
- Dark Sector

On windy days like this, when scientist Michael Angelopoulos reaches for his pen on the ice, merely writing notes becomes a challenge. Although the temperature is "only" about −17 °C (1.4 °F), wind gusts sweep across the ice at more than 15 metres per second.

The team nevertheless devotedly carries out its work. There are few days when the Arctic weather rages so severely on the ice that nobody can leave the ship. On other days the women and men of the expedition venture out even in the harshest conditions.

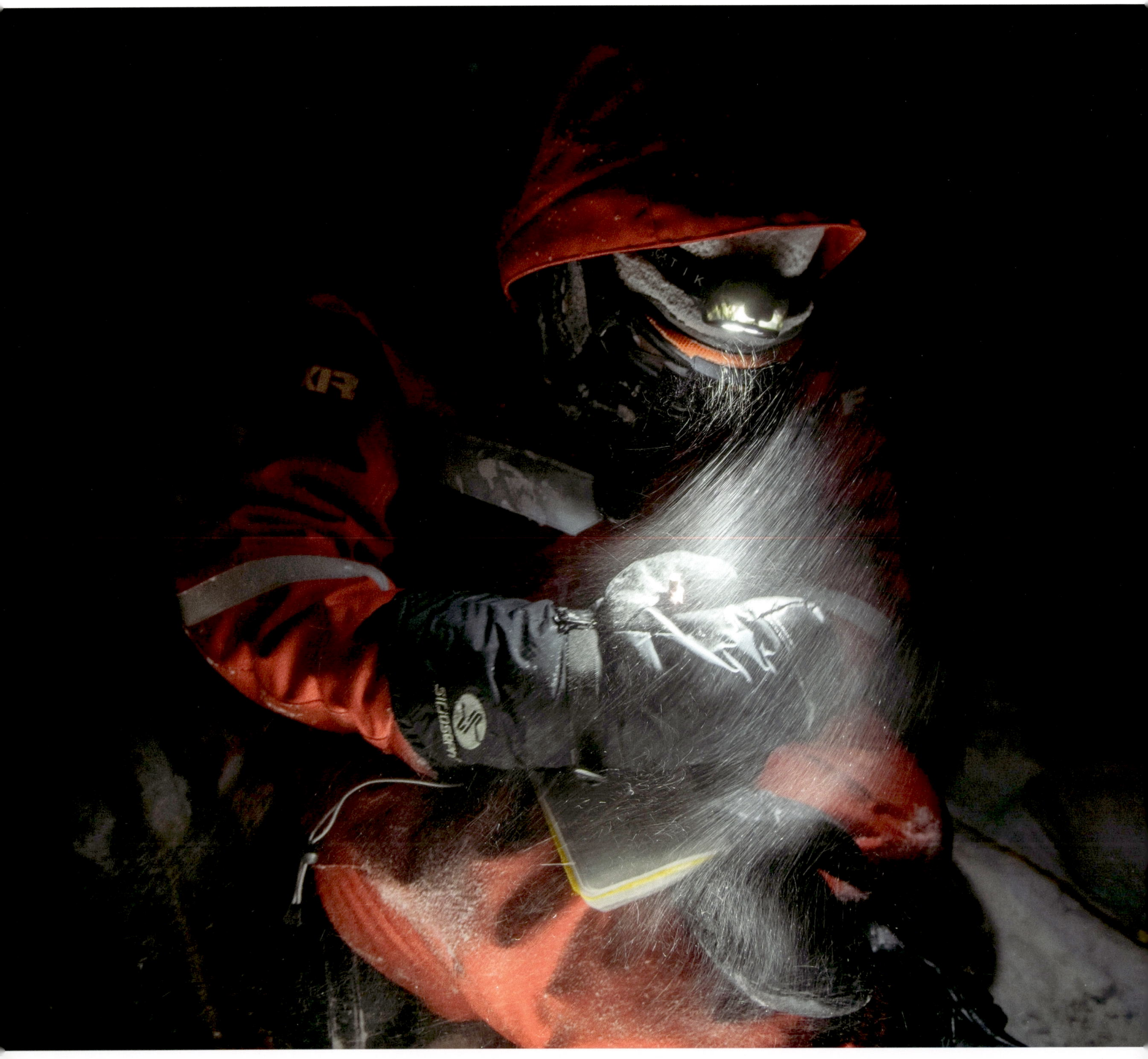

T

The Arctic climate puzzle is complex – and how it fits together in the Arctic winter was largely an enigma before the MOSAiC expedition.

What exactly happens when the Arctic sea ice suddenly cracks and the insulating layer that separates the relatively warm ocean from the bitterly cold atmosphere is removed? There has hardly been any previous research on the exchange of heat and energy that ensues from encounters involving extreme temperature differences of up to 40 °C (72 °F). The whole subject of exchange between the ocean and the atmosphere, for example the exchange of gases such as carbon dioxide, is one of the major questions of the expedition. How does this affect our climate? When do clouds form, what properties do they have, and how do they influence the Arctic climate system? How do life forms spend the polar night in months of darkness beneath a closed sheet of ice? And what happens in spring, when the ice gradually breaks up and life bursts explosively forth again within a few weeks?

Data from the central Arctic are mainly available from the summer months, but hardly any exists for a full annual cycle, as the North Pole region in winter has to date been a blind spot. The MOSAiC expedition should provide answers to such questions during its year-long drift. The scientists on board and in the Ice Camp will measure more than one hundred different parameters. While many of the instruments have long been tried and tested in polar research, some of the measuring devices have been under development for years and made specially for this expedition. Expedition members go on to the ice in cold and windy weather to gain long-awaited data. Some instruments have to be rescued during the months of operation – rescued from the crushing mass of a ridge of compressed ice or from the depths of the ocean under the opening ice cover. The daily research routine is recorded in a separate logbook available to the general public – daily snapshots full of hope, exhaustion and enterprise, desperation and curiosity, hardship, courage and ingenuity.

2 November 2019 · At Met City the second tower was raised. The 30-metre-high telescoping mast was installed by a team of six, winching up twenty-seven sections of the mast one by one. This was one of the coldest days so far, with wind chill of −40 °C (−40 °F). The two towers measure, among others, ground and air temperature, relative air humidity, wind speed, water-vapour pressure, air pressure and carbon dioxide.

4 November 2019 · The big CTD hole next to the ship is almost ready and scientists can soon start their first oceanographic measurements. This operation needed manpower and coordinated work between crew and scientists. After the hole was cut at the edges, big pieces of sea ice were removed by the crane. Through this hole, our biggest CTD rosette will be deployed down to the sea floor, currently at about 4,200 metres, to measure seawater properties and collect water samples.

20 December 2019 · When Nansen tried to sound the water depth 126 years ago, he did not have enough line to reach the seafloor as he never expected the Arctic Ocean to be deeper than 2,100 metres. Today we just need to check the various monitors on board *Polarstern* to see the depth, which is automatically logged by sonar. On our current position, for example, we are 4,415.43 metres above the sea floor. These data are obtained by the so-called multibeam echo sounder. It sends out sound, which is reflected by the sea floor and received by the device again. The depth is calculated from the travel time between transmission and reception: the longer the travel time, the deeper is the water. With different sound frequencies, the multibeam sonar can also register what happens within the water column, because everything swimming around also reflects the sound.

16 January 2020 · In the past few days we have set up a station that has been missing so far from Leg 1: the ridge observatory, which we call "Fort Ridge", as it includes parts of the area initially called the Fortress. It is approximately 100 metres long and 15 to 20 metres wide and located 400 to 500 metres north-west of *Polarstern* and can be reached by the ROV. Drilling and coring indicates that the ridge is not fully consolidated as it has several soft layers and wet cavities instead of only compact ice. In a common effort, teams Ice, Ocean and Ecosystem installed several instruments to study the processes, fluxes and biota on, in and under the sea ice, which we are going to present in the next days.

The young scientist Oguz Demir looks out from a shelter at the Remote Sensing Site. This area is home to an ensemble of sensitive instruments such as microwave radiometers and radar units. They are fitted with the same sensors as satellites that monitor the Arctic sea ice from space. Thanks to the survey of large areas by satellite measurements, we know about the huge retreat of sea ice – one of the most important and at the same time most alarming findings on climate change. However, until now the precision of satellite measurements had reached its limits, partly because of the snow cover in the Arctic: the properties of newly fallen snow differ from those of old, lumpy snow, and so the snow-covered surface reflects back the microwaves transmitted by the satellites in various ways, which means that, depending on their properties, they appear differently on the satellite images. A large-scale measurement programme at ground level that is being carried out during the MOSAiC expedition at the Remote Sensing Site will ensure greater precision. The aim is to compensate for the weakness of the satellite measurements and to include the snow cover, properties and depth more accurately in the observations.

20 January 2020 · On Sunday morning the remote sensing team noticed that the radiometer feed horns of an instrument deployed on the ice were looking in different directions than before. They immediately checked the photos from a nearby camera, which takes photos automatically every five minutes. To their great surprise they discovered the reason for the disorder: one and only one photo showed a polar bear checking out the instruments at the Remote Sensing Site shortly after midnight. Luckily, it approached the site quite carefully and only ripped the covering of the antenna tracker but did not damage any other installations, and cautiously stepped over all cables without ripping them.

24 January 2020 · Yesterday morning we recognized subtle deformations of the ice floe in the vicinity of the *Polarstern* on the ship's radar. A logistics team went out for scouting and came back with the information that a big lead 50 metres wide and 1 kilometre long had formed along a shear zone. This was the long awaited "event": the chance to measure energy and gas fluxes between the water and air, through the newly formed and older ice. It also allowed to study the initial population of newly forming ice-inhabiting biota. This was a great interdisciplinary effort to unravel the secrets of the central Arctic in winter.

30 January 2020 · We have a new buoy collecting data on fish and zooplankton. It detects these animals in the water column by transmitting out acoustic signals and receiving acoustic reflections from them, just like an echo sounder does. We deployed it some hundred metres away from the ship and from Fort Ridge to avoid interference with other instruments there. The buoy's total weight of 200 kilograms and its height of 2.2 metres made the naming: the Monster!

8 February 2020 · The Atmosphere team just managed to bring an automated surface flux system (ASFS) back to life at the end of January. It is part of one of our distributed network stations that are as far as 40 kilometres away from *Polarstern*. Only a few days later, the system did not communicate anymore and we went back for troubleshooting. Amazingly and disappointingly we found that the whole system was partly eaten by a newly formed pressure ridge and badly damaged. As flight conditions deteriorated during the visit, we had to leave the site before we were able to disassemble the sensors or other parts of the station. Hopefully the ridge does not bury these parts before we are able to revisit and rescue them.

13 February 2020 · For years, we have hoped to collect two types of coincident data of the same ice: a precise map of surface typography from the helicopter-borne airborne laser scanner (ALS) and a detailed map of the ice underside which is invisible from above but can be mapped with the multibeam sonar (MBS) of the remotely operated vehicle (ROV). We have finally succeeded to obtain this unique data set: for the first time we have gathered a full, three-dimensional image of both sides of our floe. Apart from the nice animation, the data are essential to understand how the ice grows under different conditions of snow and thickness conditions, and how surface and bottom roughness correlate.

17 February 2020 · We are obviously surrounded by ice and snow as *Polarstern* is frozen into the pack ice. But did you know that ice crystals are also omnipresent in the air around us. They have different forms depending on the atmospheric temperature: solid plates, prisms or columns, stellar plates, needles and dendrites. Their sizes and appearances also vary with humidity. To observe these tiny crystals, they need to be collected and preserved in a synthetic substance to produce an imprint before they can be photographed under the microscope as they would otherwise melt away. We are interested in their forms, as the various shapes scatter light and radiation differently.

21 February 2020 · In today's diary entry in 1894 Nansen reported on the zooplankton sampling during the Fram expedition: in a net "... there was a variety of small crustacea and other small animals shining with such a strong phosphorescence that the contents of the net looked like glowing embers as I emptied them out." Team Ecosystem has deployed nets today and we found the amazing phenomenon, copepods of the genus *Metridia* – zooplankton that is sparkling bluish when the samples from the plankton net are filtered in the lab.

26 February 2020 · Last Saturday was a first in nine years of Alfred Wegener Institute's under-ice ROV diving: for the first time ever, we got to see a seal in the ROV video stream! It was final proof that at least one seal is roaming the MOSAiC area in the midst of winter! Previously a seal had been spotted breathing in the ROV access hole by ROV operators on 4 February, but there remained some doubt among some colleagues because they were not able to catch it on camera before it had disappeared.

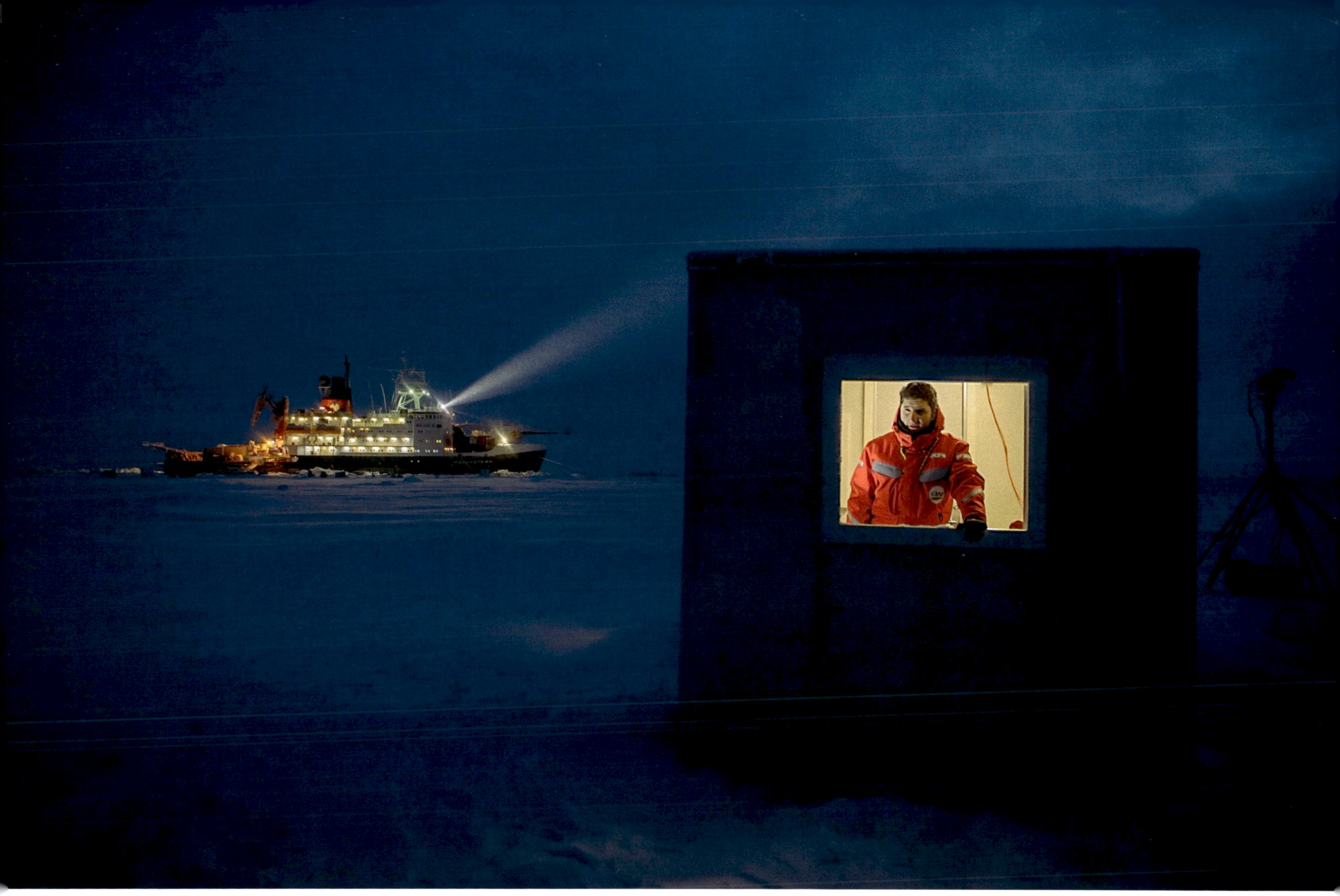

9 March 2020 · A large lead of more than 5 kilometres long and 500 metres wide opened north-east of *Polarstern* in a distance of about 1 nautical mile. Leads are very important for the winter Arctic energy balance and biogeochemistry. Therefore six scientists from the Atmosphere and Biogeochemistry teams set out to scout the way for future instrument deployment and take some samples on the spot. A couple of hours later, we saw on radar images from the ship that the extremely impressive lead is closing again.

16 April 2020 · In order to be able to predict the possible course of the *Polarstern*'s drift in preparation for the campaign, satellite data were used to reconstruct the paths taken by the ice in recent years from the MOSAiC starting point. After about 180 days of drifting, it is safe to say that, up to now, the drift has actually occurred within the drift corridor based on the original data. However, a comparison of the data from previous years with the chronological course shows that the drift, since the turn of the year, has been fairly straightforward, first westwards and then southwards, without any major detours or loops.

23 April 2020 · Since Wednesday last week we experienced a real temperature roller coaster: starting at about −30°C (−22°F), temperature increased to above −2°C (28.4°F) before decreasing to about −20°C (−4°F) – and finally rising above the freezing point close to the snow surface. This warm air intrusion caused not only scientists sweating but also the snowpack on the floe. Due to the low temperatures between the warm phases, the snow went through so-called thaw-freeze cycles. These cause both a growth of the snow crystals and the formation of ice crystals within the snowpack. To observe these processes in more detail, we use computer tomography on board which allows us to get an amazing insight into the numerous snow crystals. Now, we are curious to track further changes of our growing crystals.

1 May 2020 · To observe the evolution of sea ice thickness and snow depth on different ice types on our floe, sea ice physicists walk regularly along fixed transect loops across the floe. Doing so, they pull a sledge with an electromagnetic measurement device to measure continuously the total sea ice thickness, meaning sea ice thickness plus snow depth. In addition, snow depth is measured every other step with a GPS-equipped snow probe. The results show nicely: despite snowfall and strong snowdrift over the last couple of weeks and months, snow depth increased on average by only 10 centimetres on second-year ice and 15 centimetres on first-year ice since October. However, the new snow mainly accumulated close to the various formed ridges and on leads on our floe.

To avoid air pollution from the ship, the Met City district with its two meteorological measuring towers and other instruments is situated far from the *Polarstern*. The two complex metal structures are fitted with various sensors. When a storm brought down the larger of the towers only a few weeks after it was raised, the news that its delicate sensors suffered no damage was a source of great relief. From a scientific point of view, this storm was one of the first highlights of the expedition, as for the first time the researchers were able to record the effects of such a storm on the Arctic climate system.

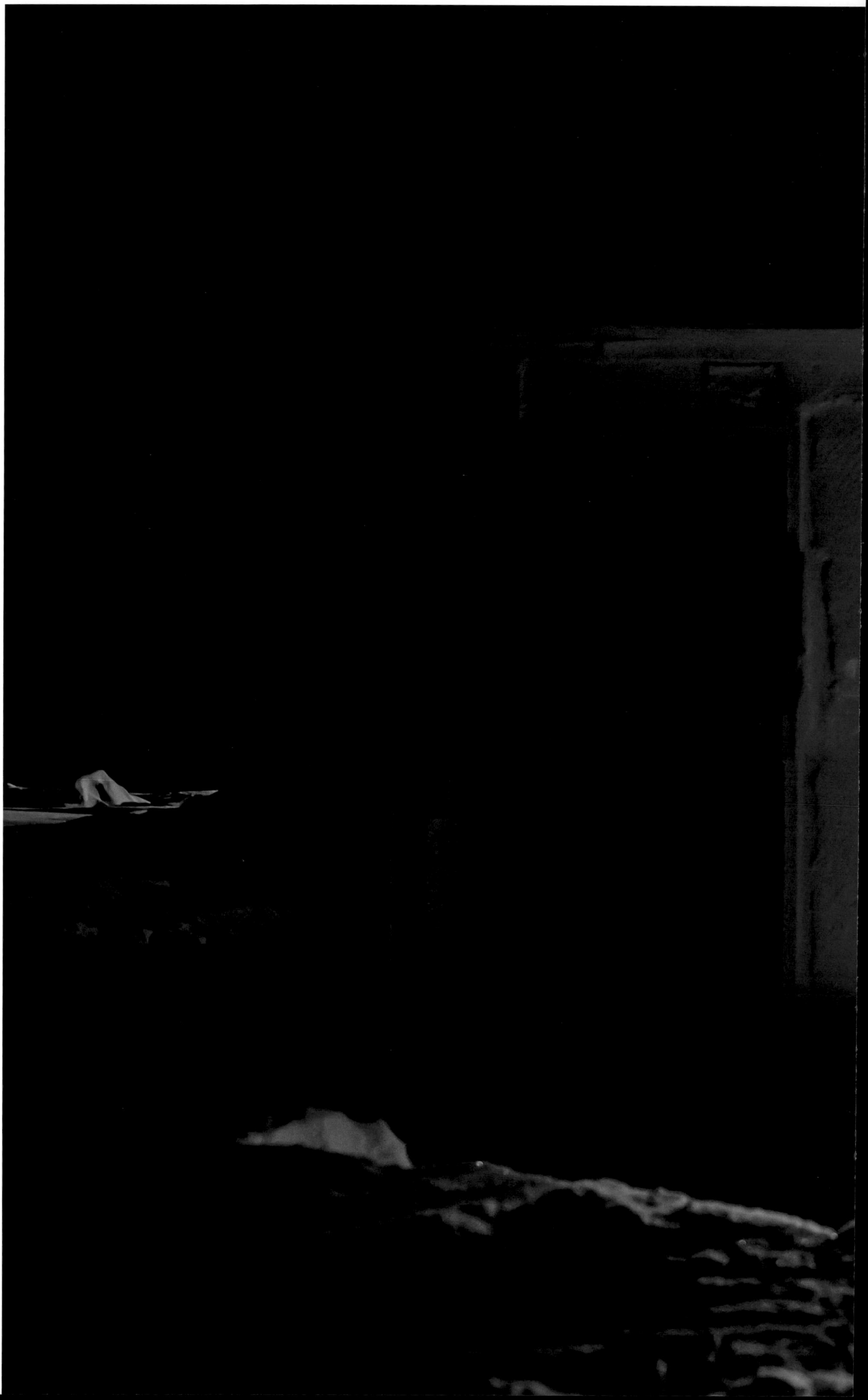

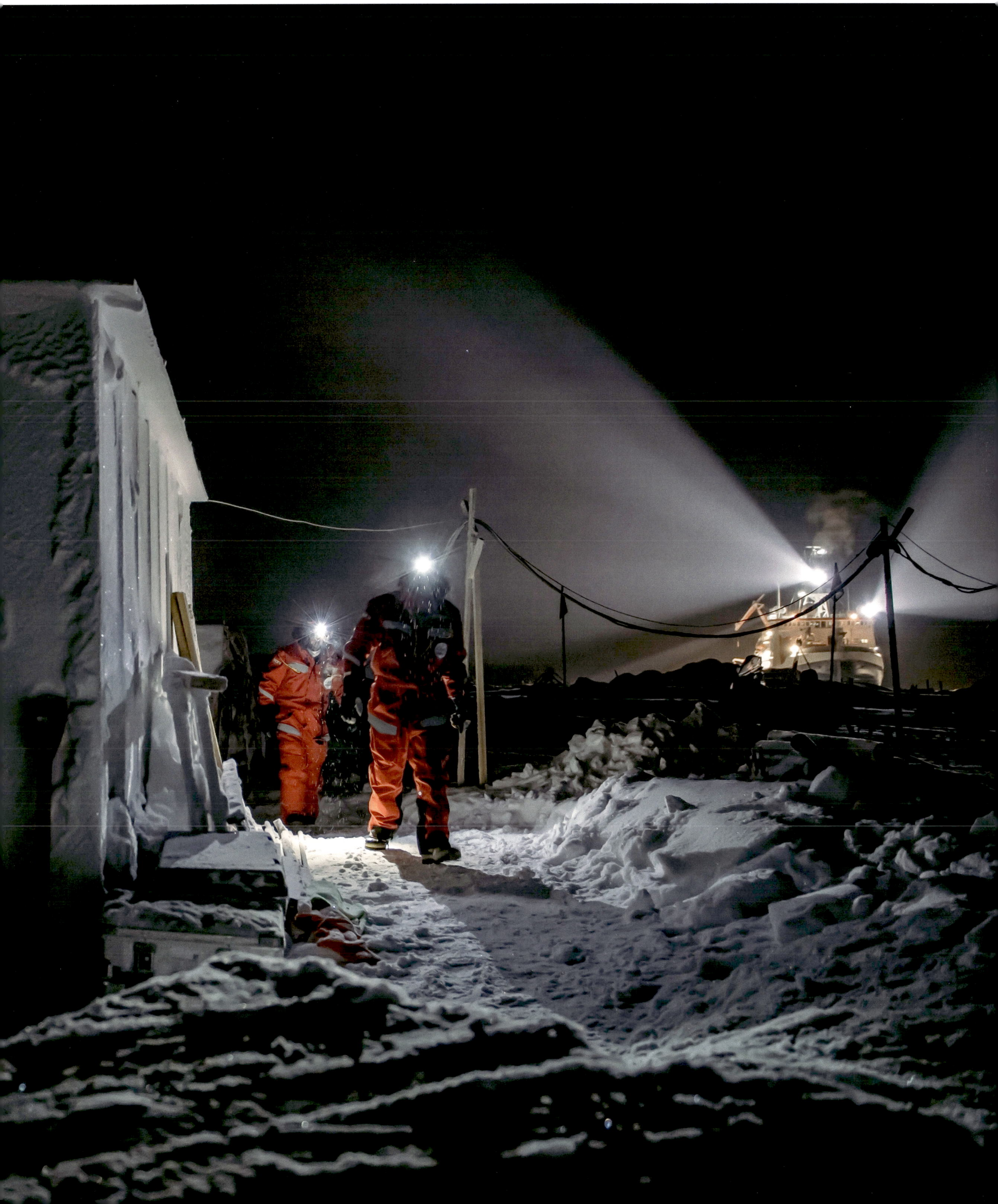

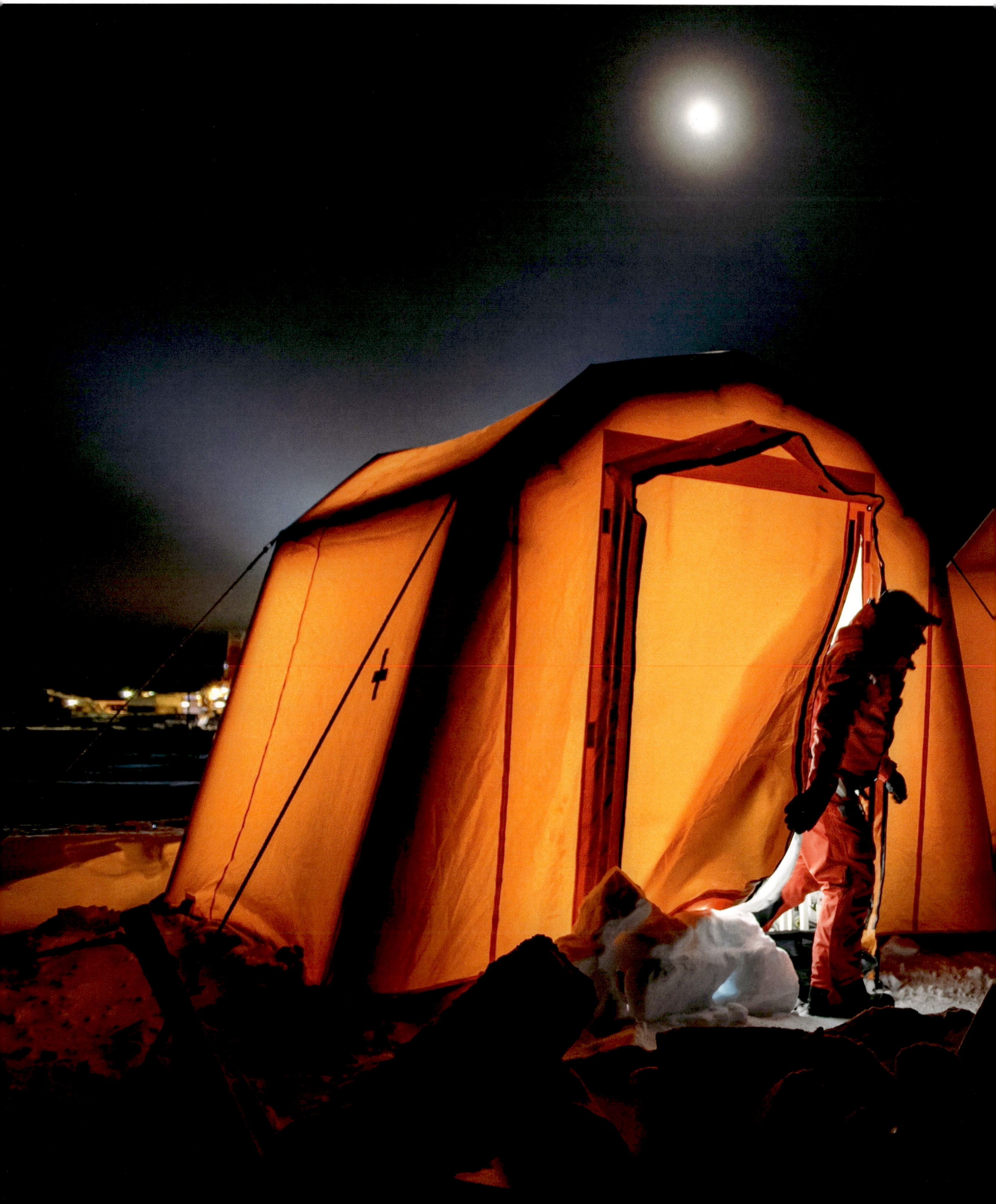

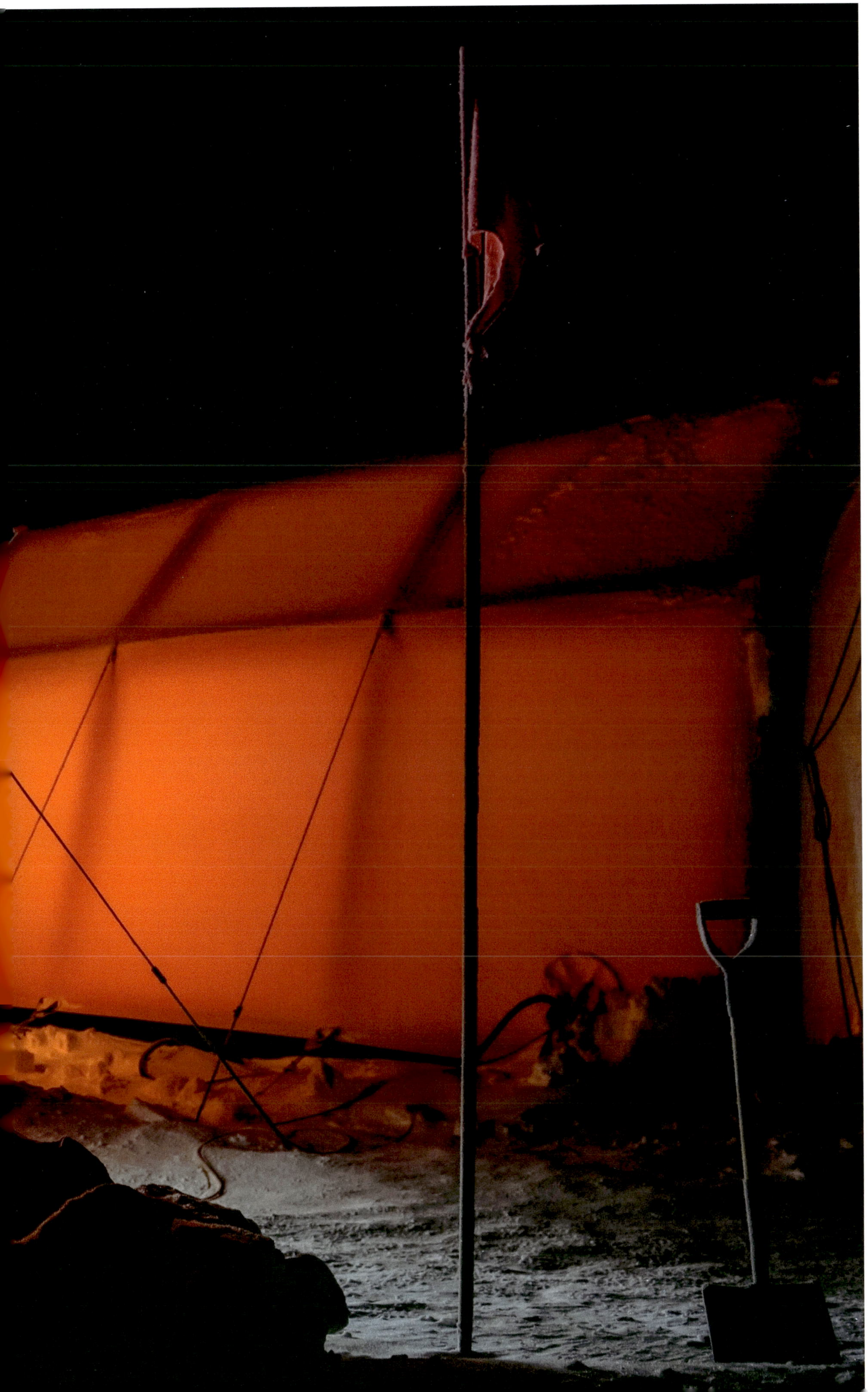

When members of the Atmosphere team arrived in Balloon Town one day, they were astonished: a bear's paw had ripped large tears in the bright orange sheets of their two tents. This had clearly satisfied the creature's curiosity, however, as the variously shaped tethered balloons awaiting deployment were undamaged. As soon as the weather allows, these helium-filled flying objects equipped with sensors and other devices will rise from Balloon Town to go hunting for data at various altitudes. For example, they measure wind speeds and cloud composition, or capture aerosol particles that float through the air.

Welcome to Ocean City, the meeting place for members of the Ocean, Biogeochemistry and Ecosystem teams. María Josefa Verdugo, a doctoral student from the Biogeochemistry team, is waiting in a tent that has an opening into the depths of the ocean: here a so-called CTD rosette regularly descends into the water through a large hole in the ice. This oceanographic measuring instrument, which looks like a bundle of gas cylinders, records, for example, conductivity (C) and temperature (T) in relation to the depth of the water (D). The CTD rosette also brings water samples to the surface for laboratory analyses. The tent is heated so that the samples do not immediately freeze in the icy air – which may be a further reason why Ocean City is such a popular rendezvous in the Ice Camp.

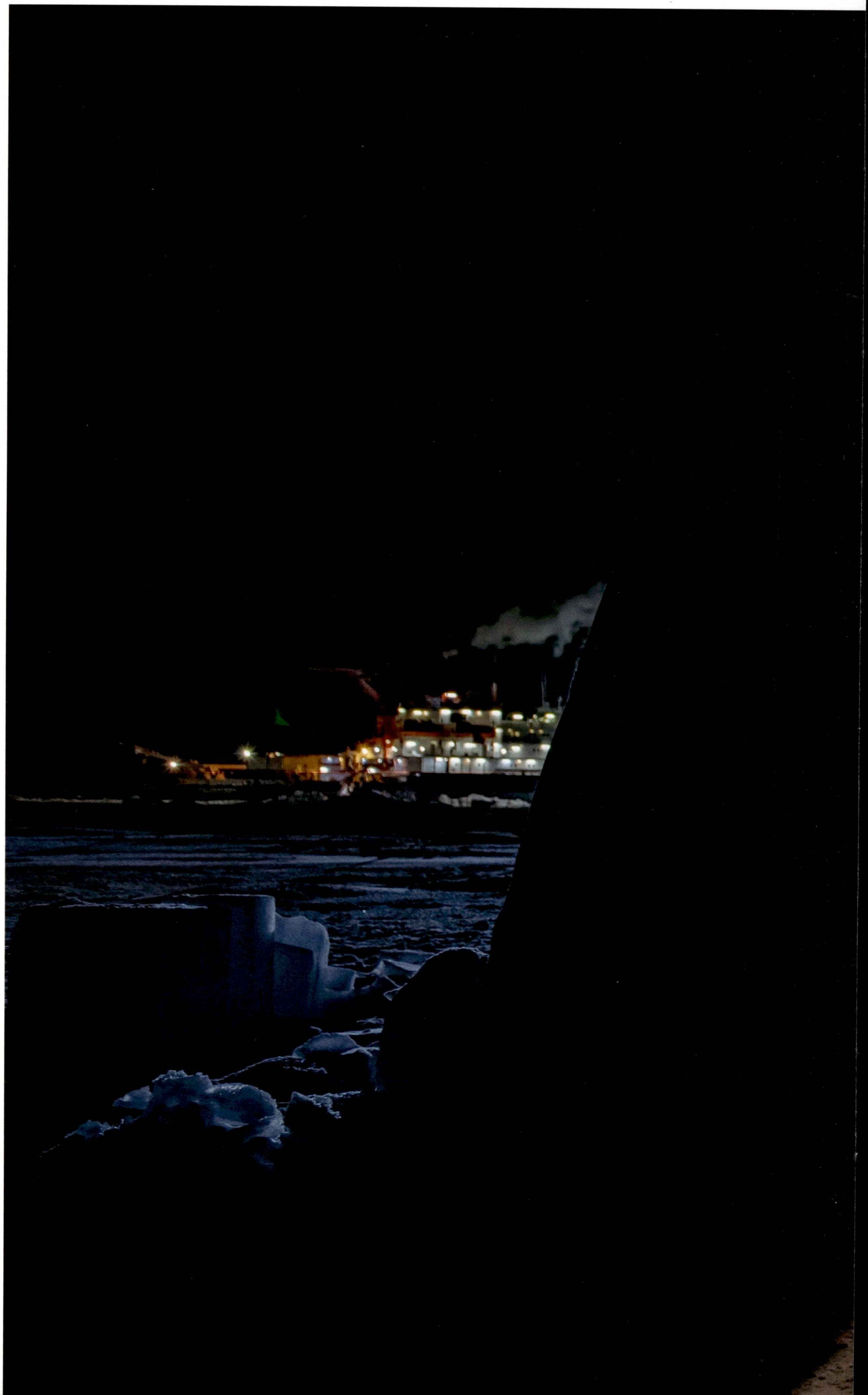

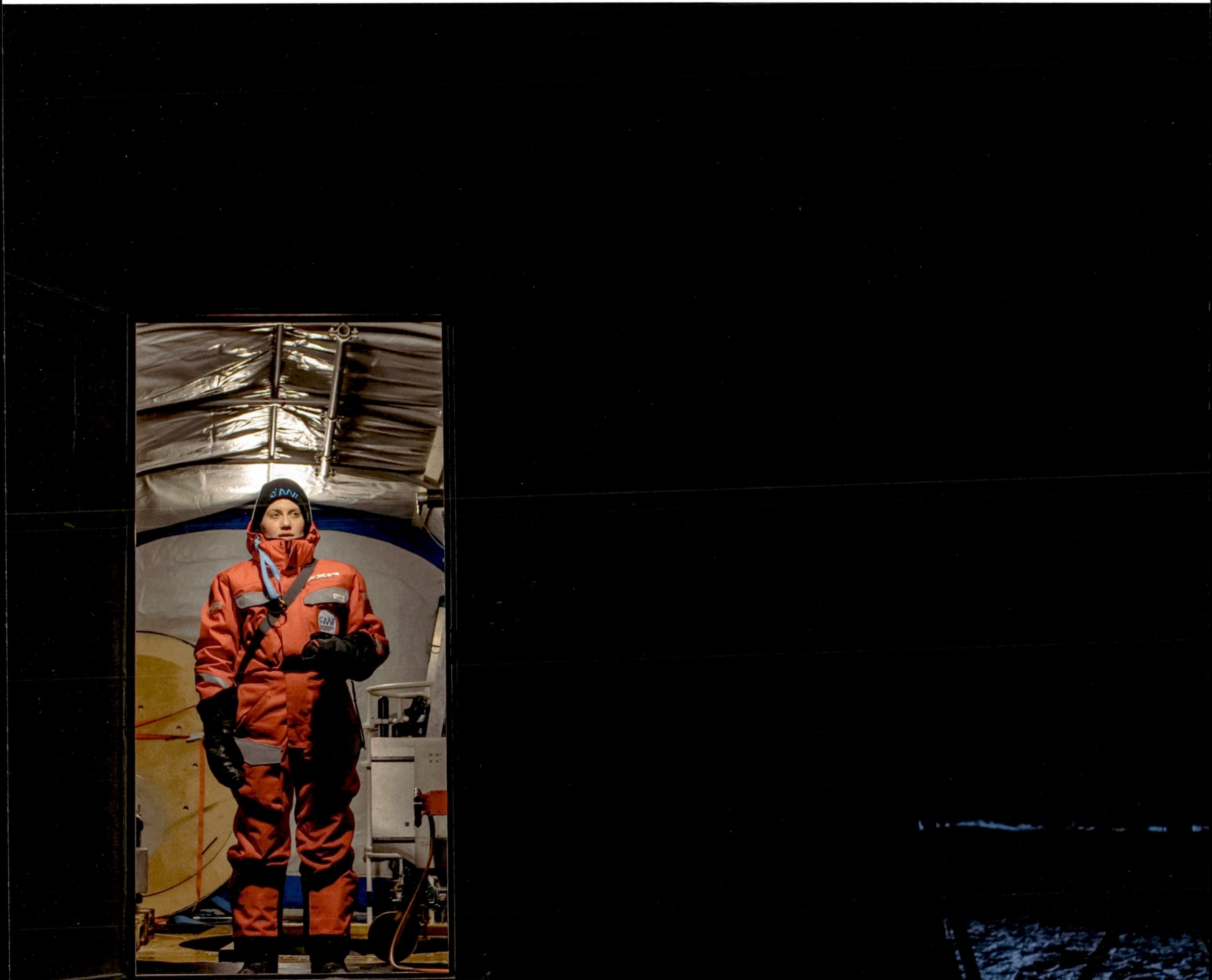

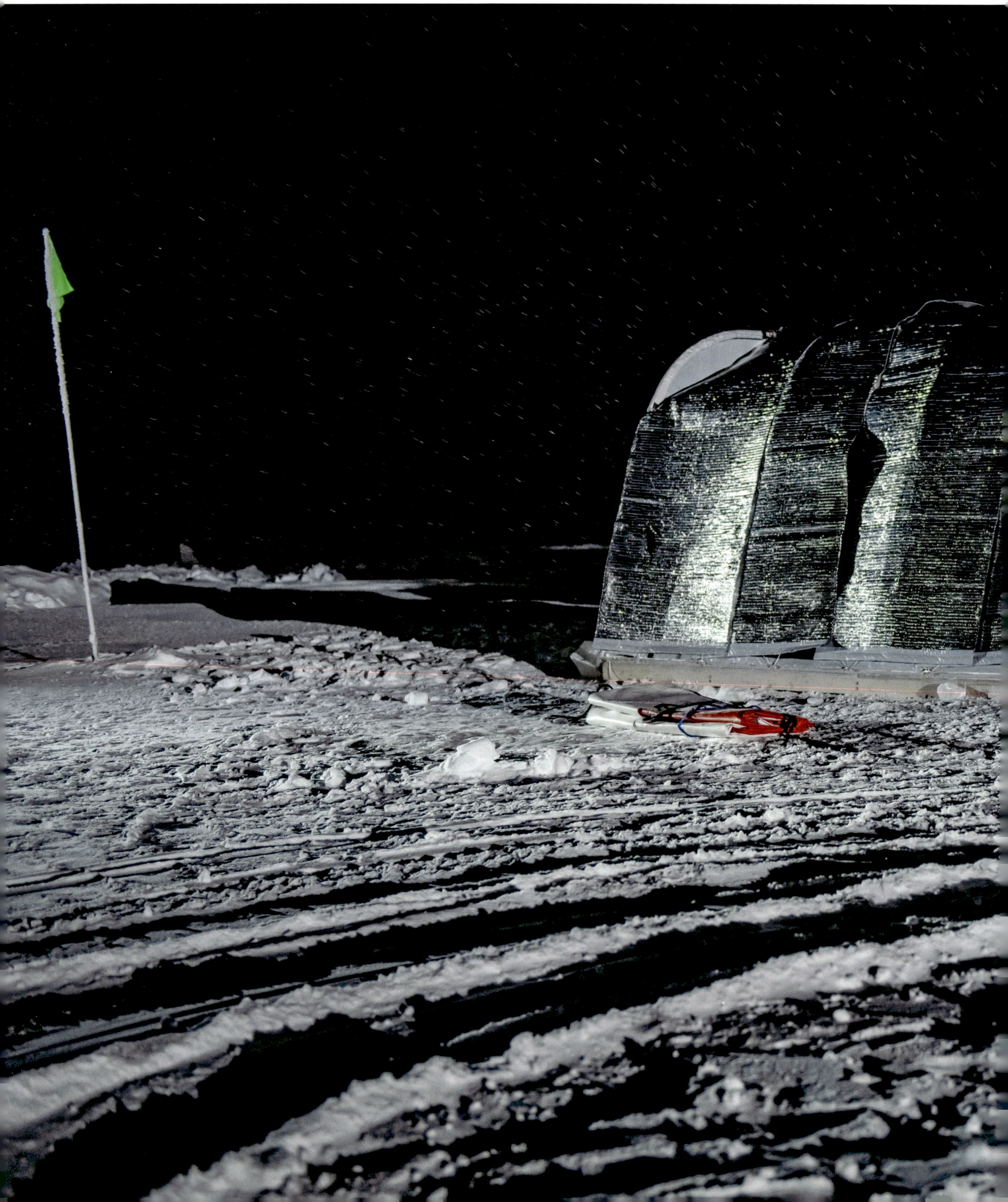

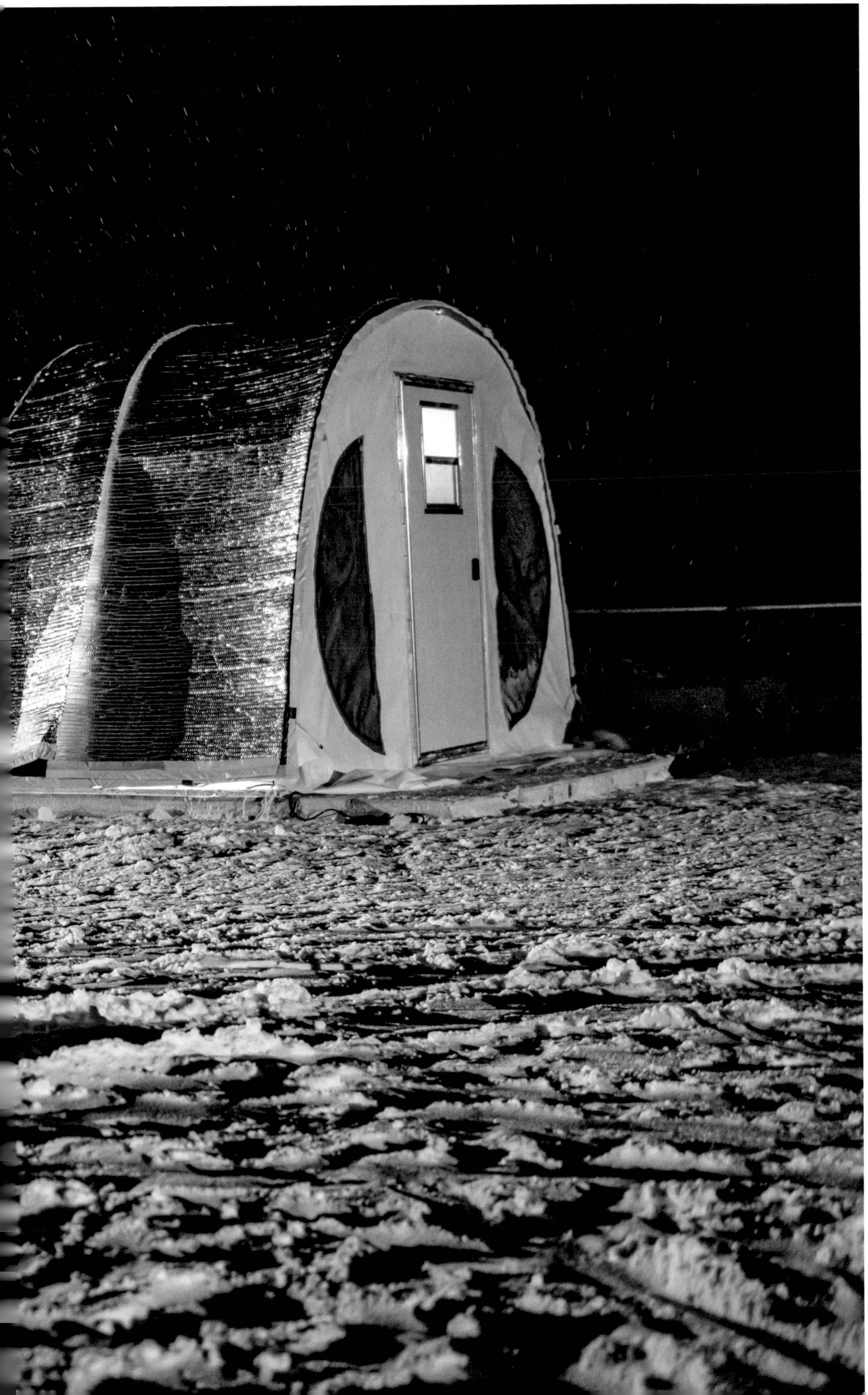

Shortly after it is erected, the tent of the so-called ROV Oasis is without its red-and-white outer cover. With its silver layer of insulation it would not look out of place on a moon base. However, the site chosen for the ROV Oasis quickly proves to be much more dynamic than any place on the moon: this part of the camp breaks off from the floe several times and drifts away, making salvage by helicopter necessary. The tent houses what is called the "Beast", a remotely operated vehicle (ROV) which undertakes long dives beneath the ice. Despite its impressive name, the yellow undersea robot has an inconspicuous box-like shape. The discoveries made by its camera during operation are all the more exciting: the underside of the sea ice is its own living environment for Arctic creatures. On its dives, the Beast even encounters a jellyfish, seals and fish in the middle of the polar night. The Beast also carries various sensors, such as for mapping the underside of the ice or investigating how much light can penetrate through snow and ice into the sea.

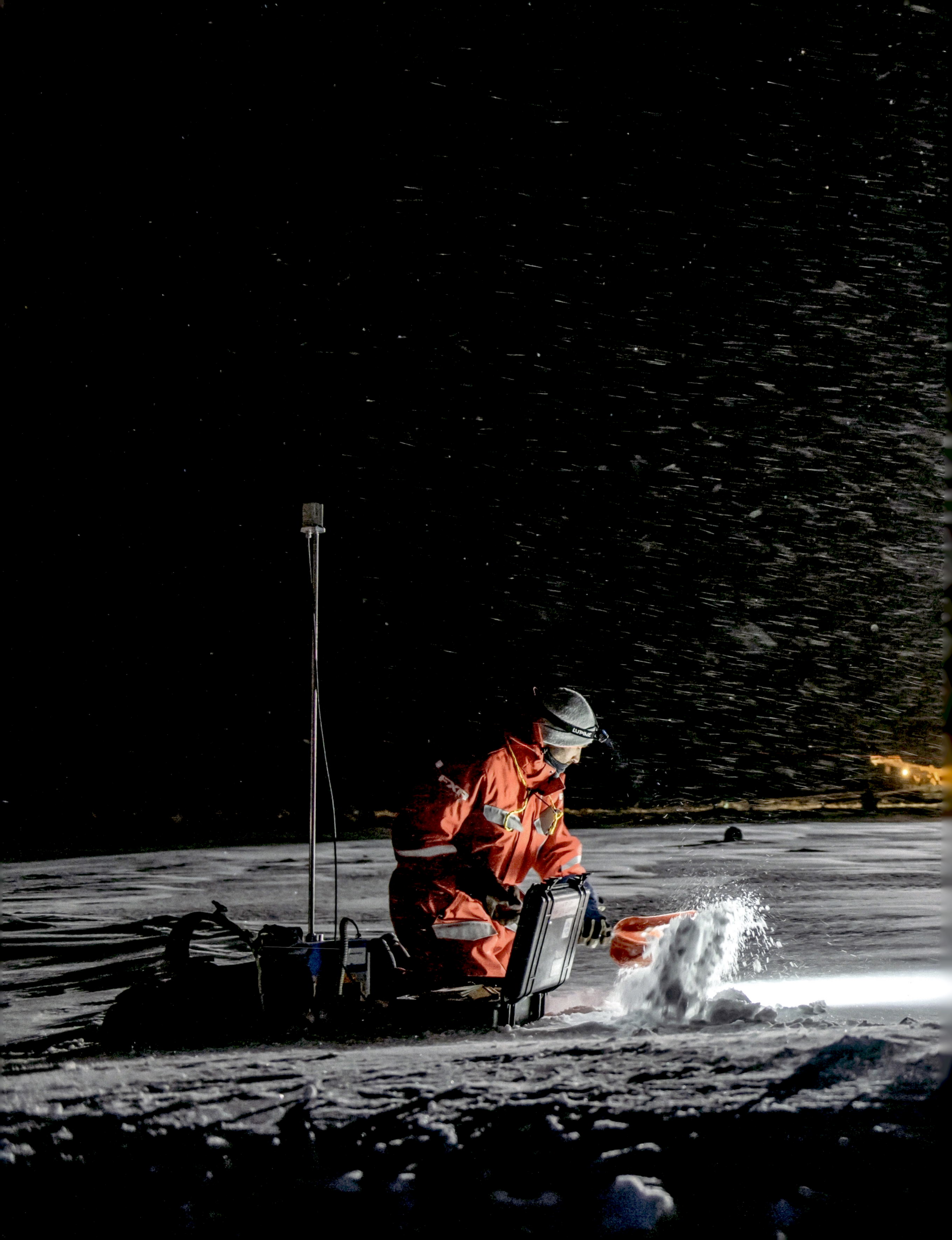
LUPINE

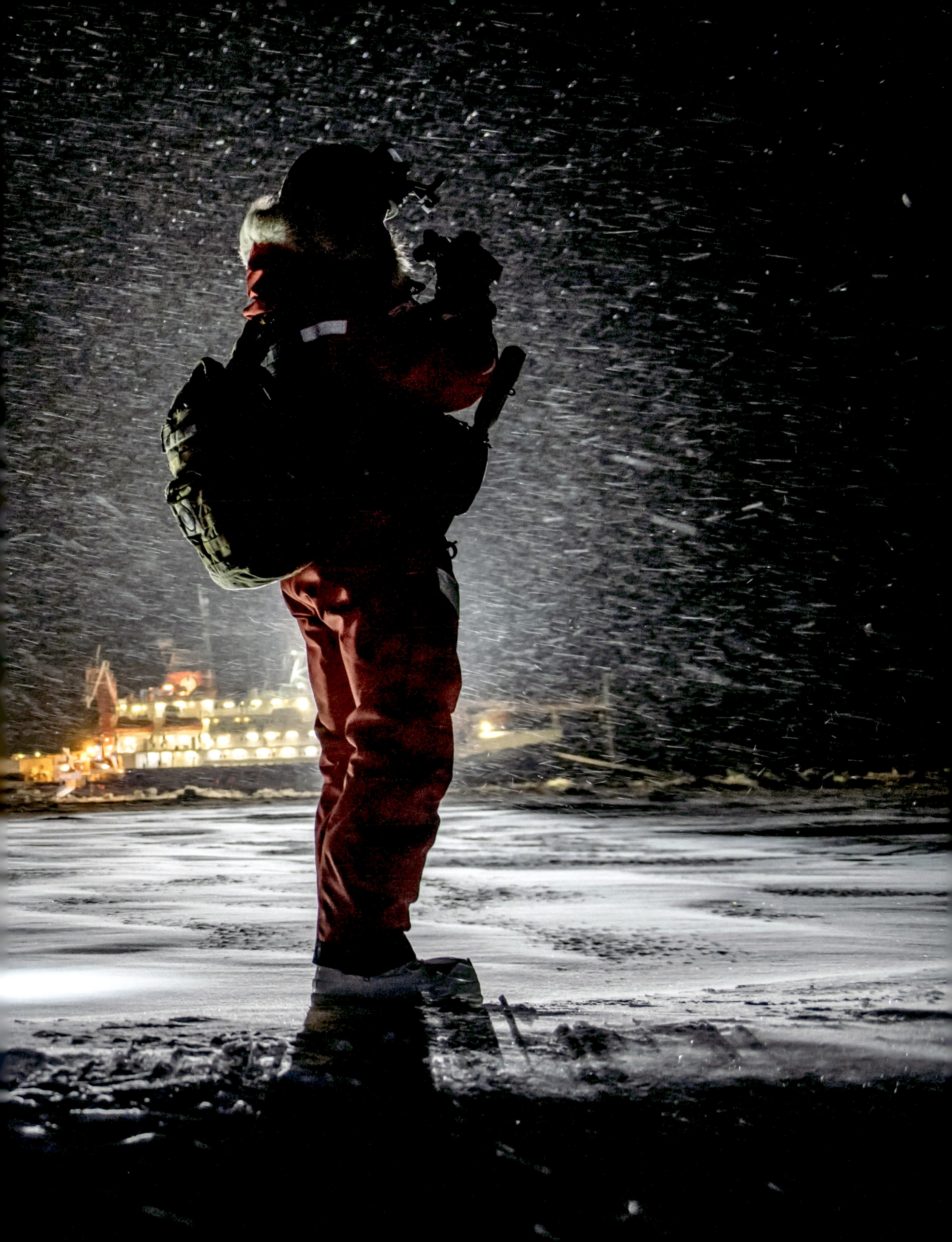

Below: The Norwegian bear guard Gaute Hermansen is at work with a chainsaw, as a hole in the ice is required to install an instrument.

Top right: As snow plays an important part in the formation of sea ice, the depth and temperature of the snow are monitored in various ways during the expedition. During one measurement, for example, a temperature difference of as much as 7 °C (12.6 °F) is recorded between the top and bottom of a layer of snow only 1 centimetre deep. In view of such differences, the effect of snow is hardly surprising: a deeper layer of snow insulates the ice against the cold Arctic air, which means that less seawater can freeze and become ice. Snow therefore makes the sea ice thinner.

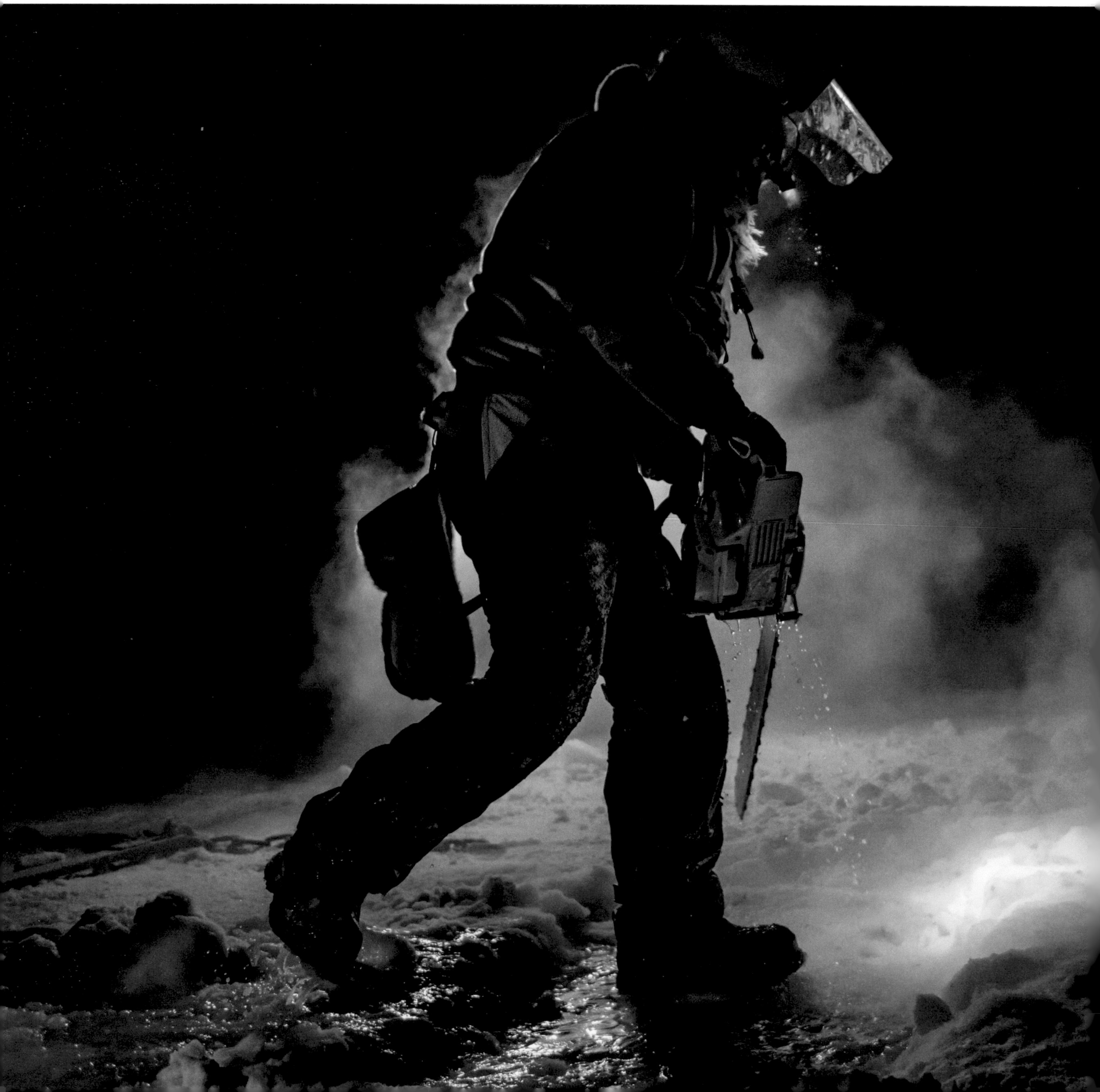

Bottom right: Ice thickness, another important factor in the Arctic climate system, plays a key role in many areas of research in the expedition. This instrument is known as a thermistor chain: the chain is suspended from the tripod apparatus through the air, snow and sea ice down into the ocean. It not only measures their thickness and "normal" temperature, but also has a further function: by heating up minimally and due to the heat conductivity of ice and snow, conclusions can be made about other properties. This includes the fact that snow has a stronger insulating effect than sea ice or water. The effect of 10 centimetres of snow is as great as that of 1 metre of sea ice.

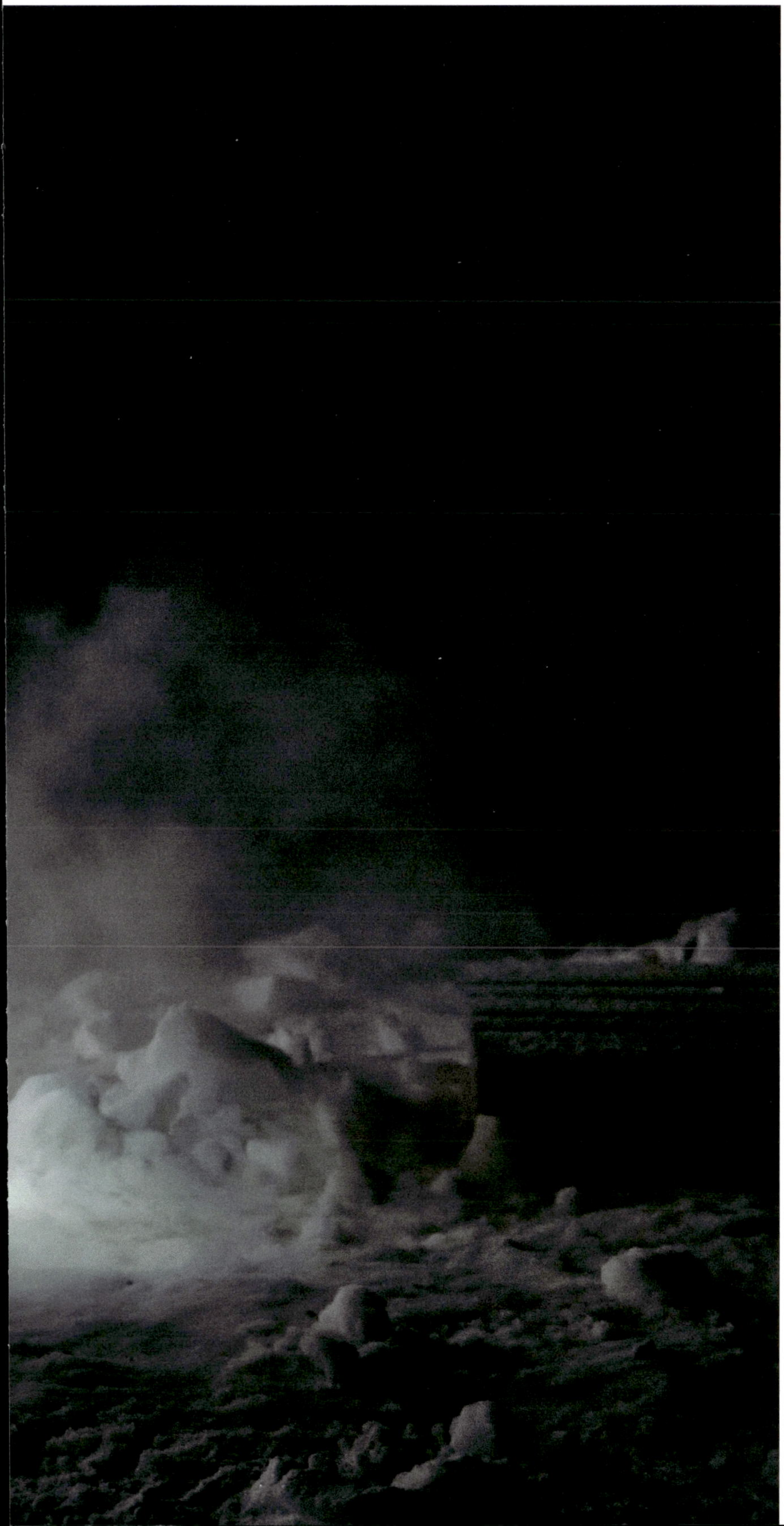

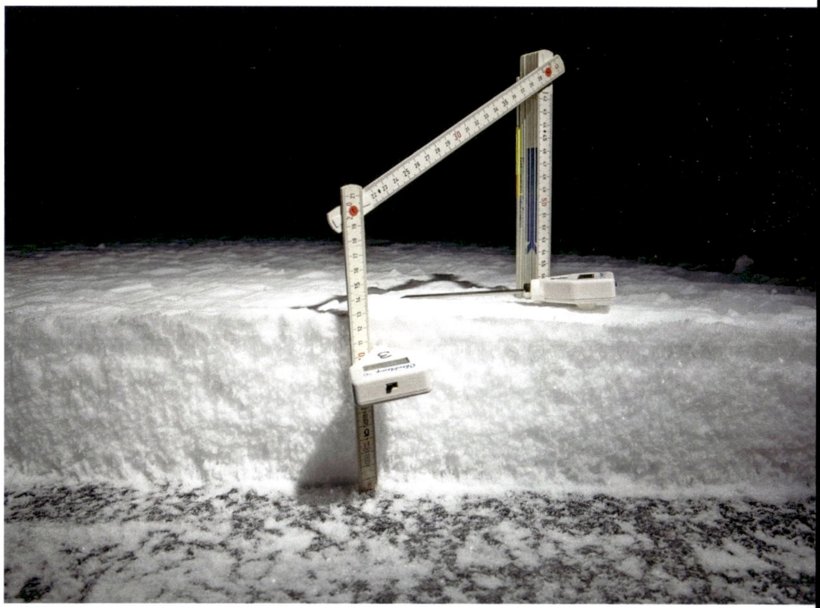

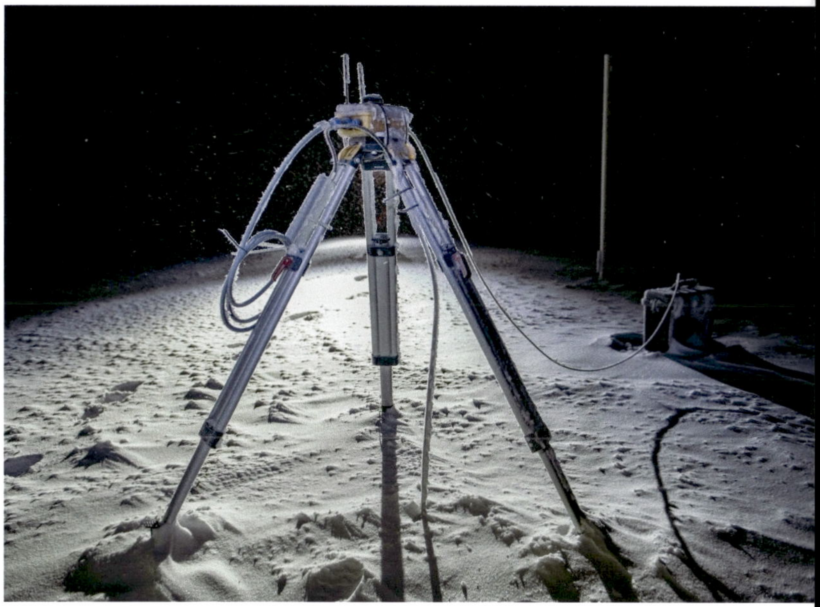

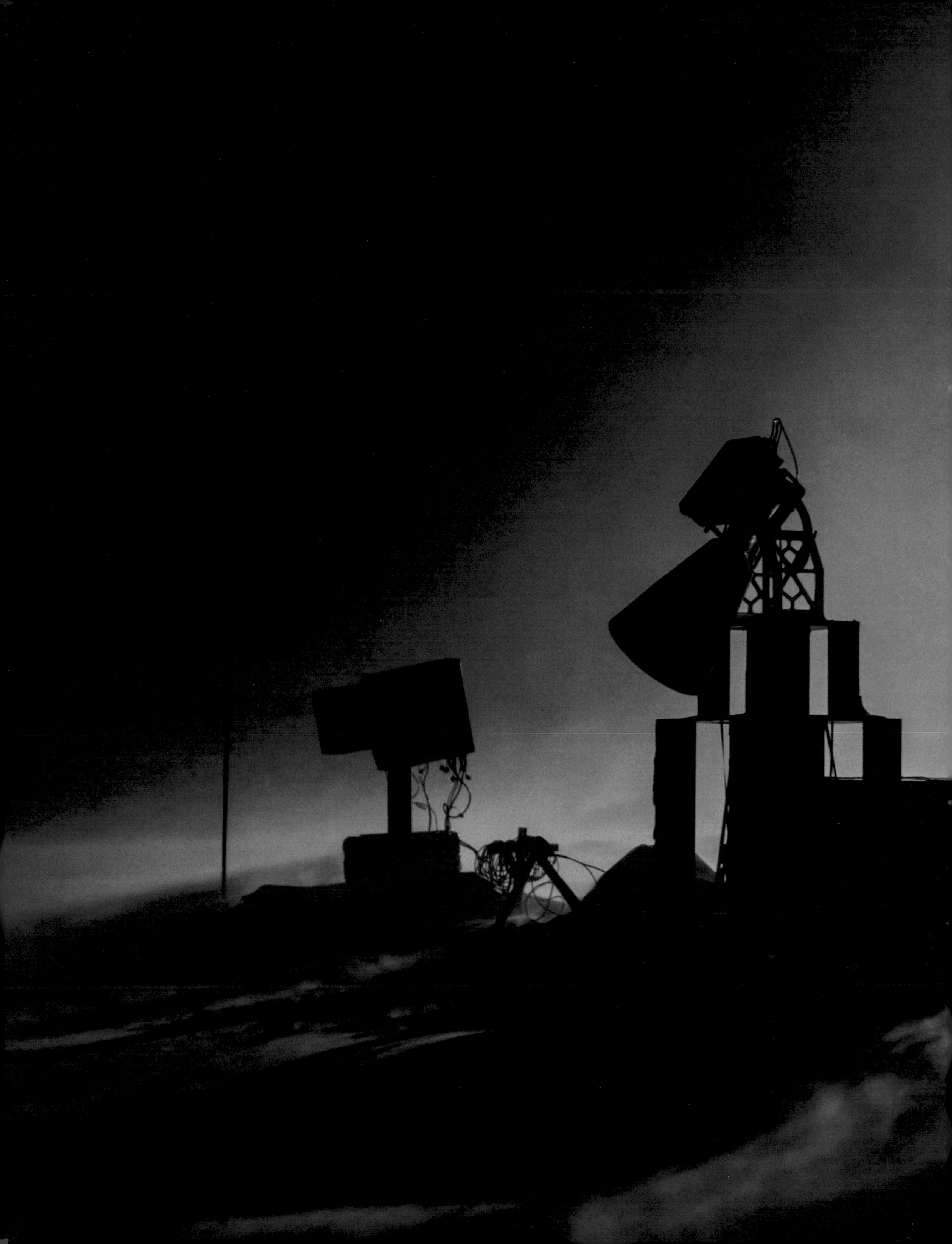

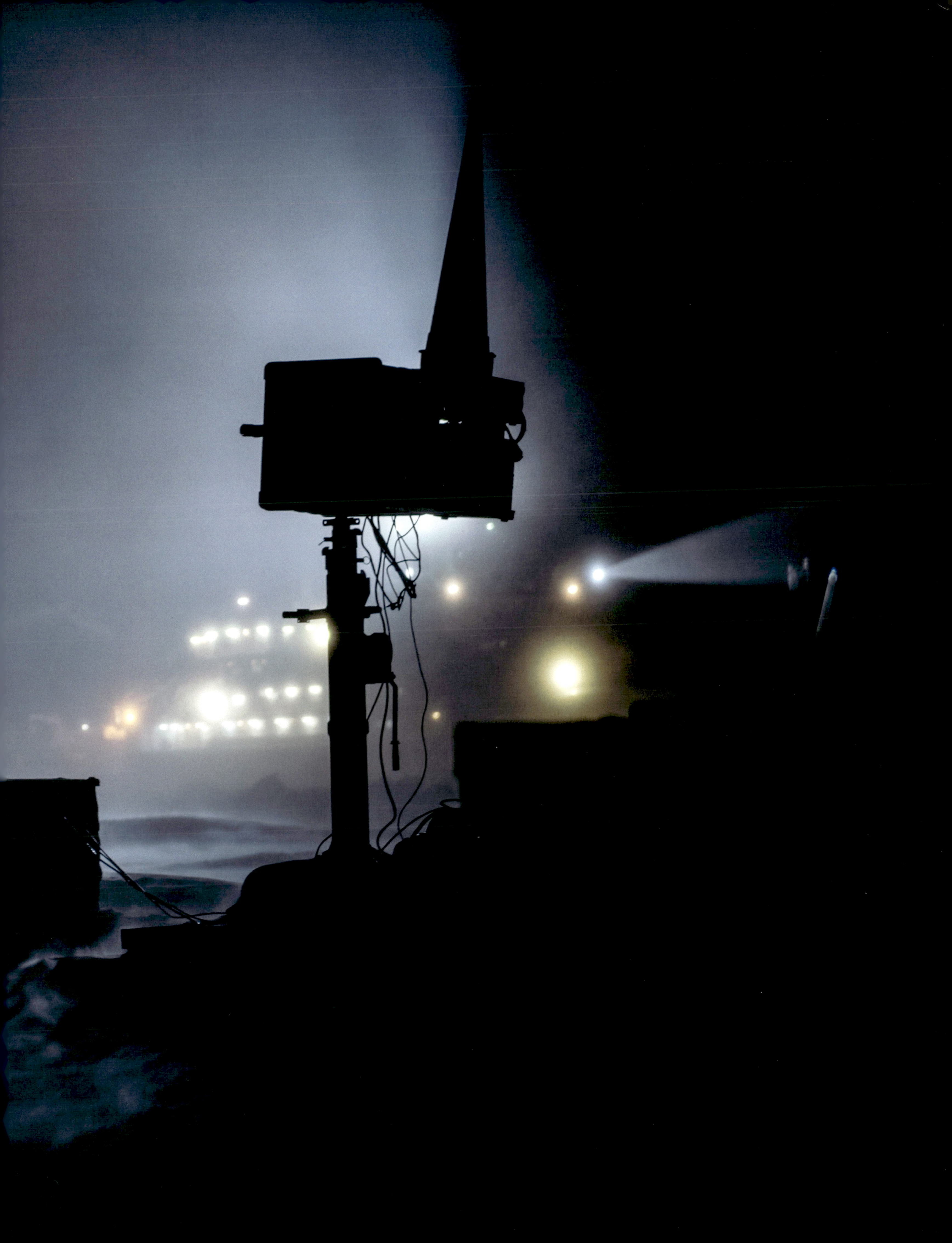

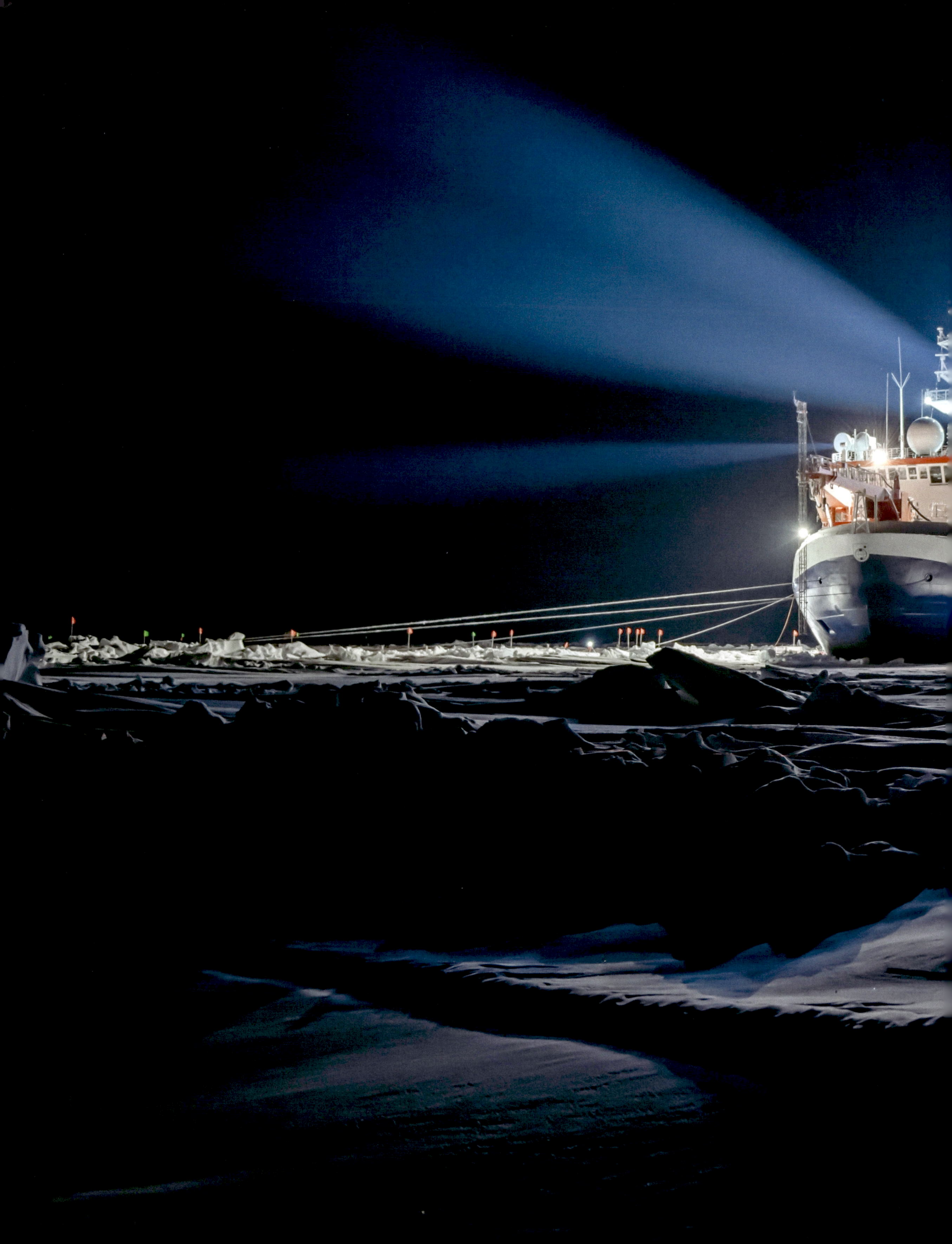

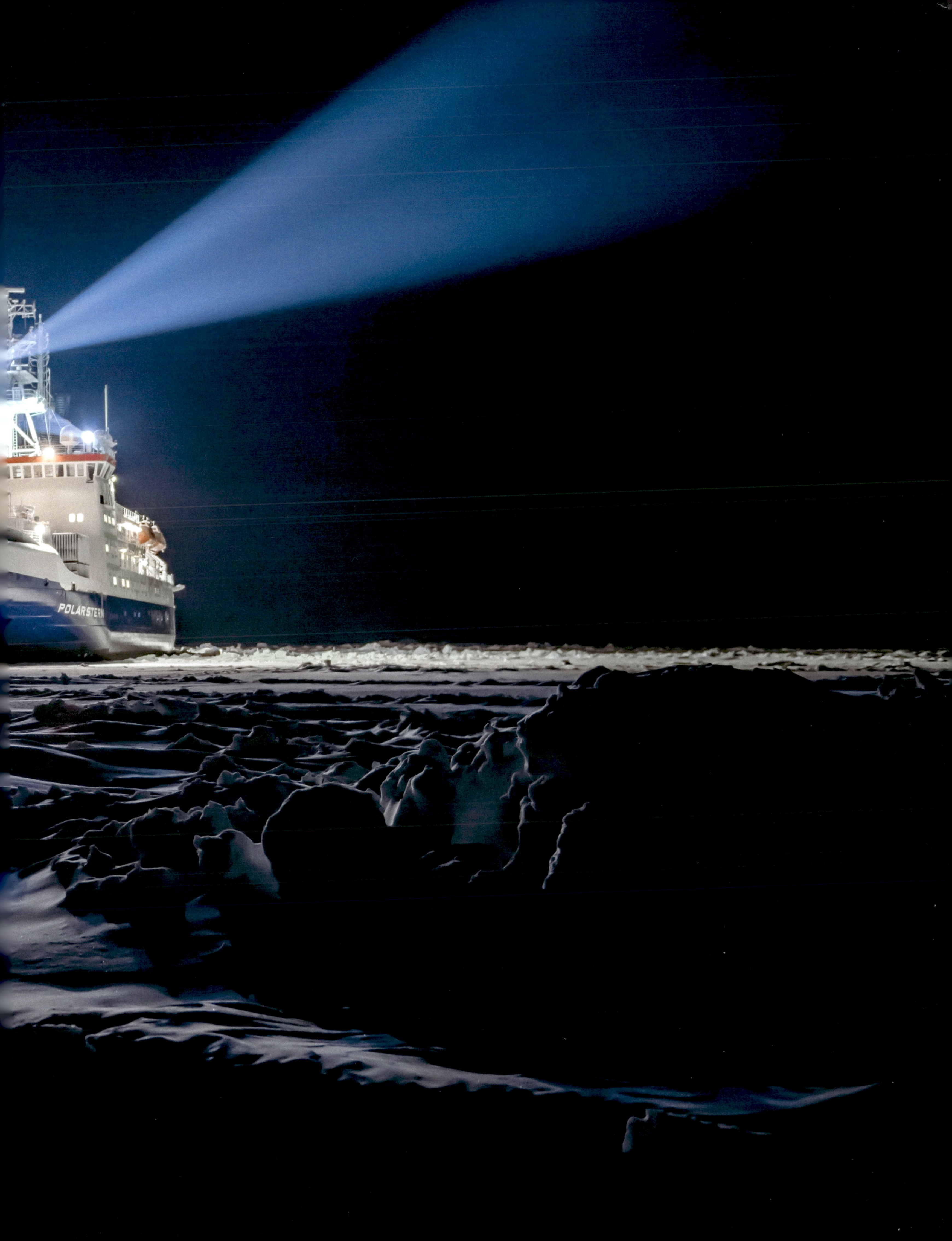
POLARSTERN

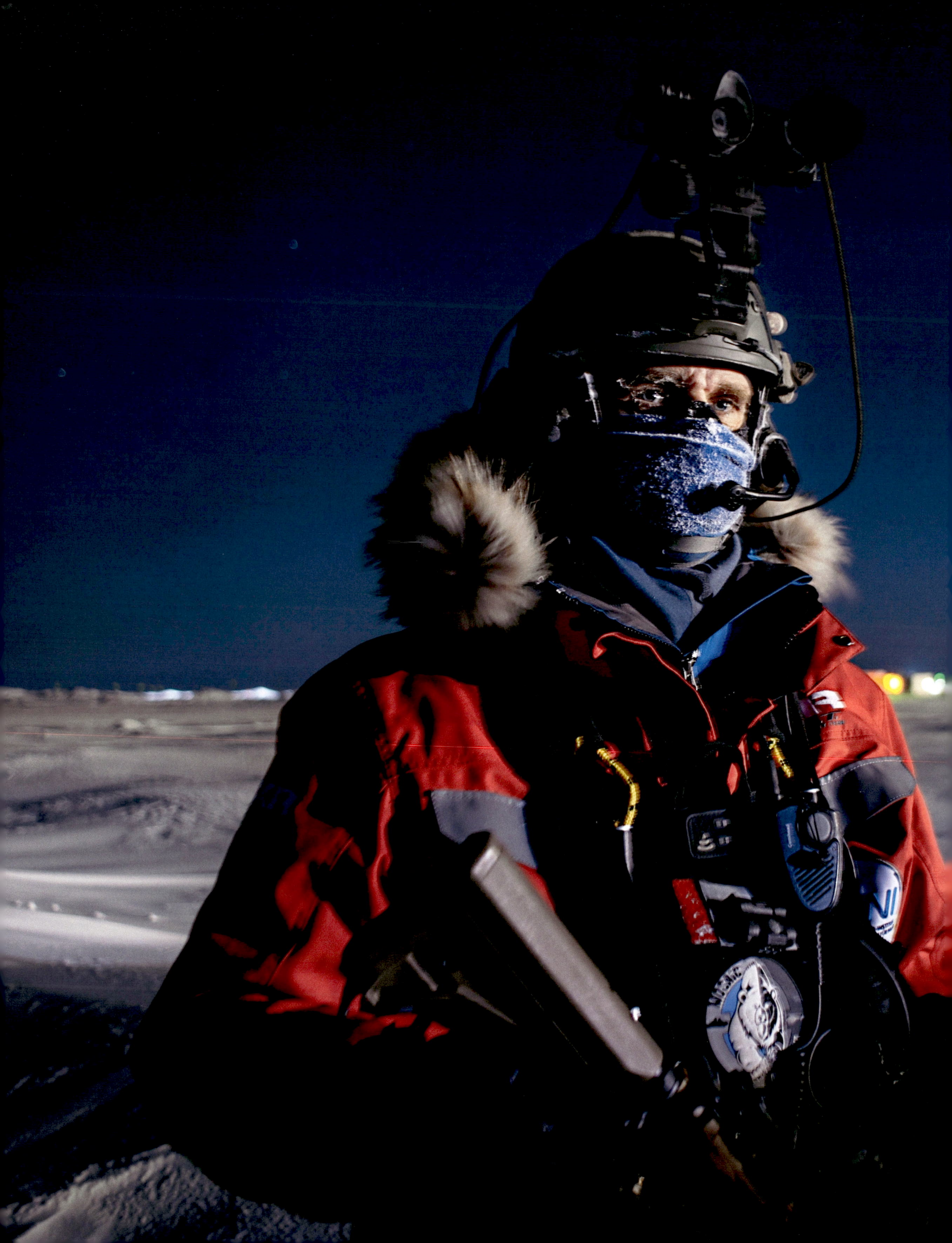

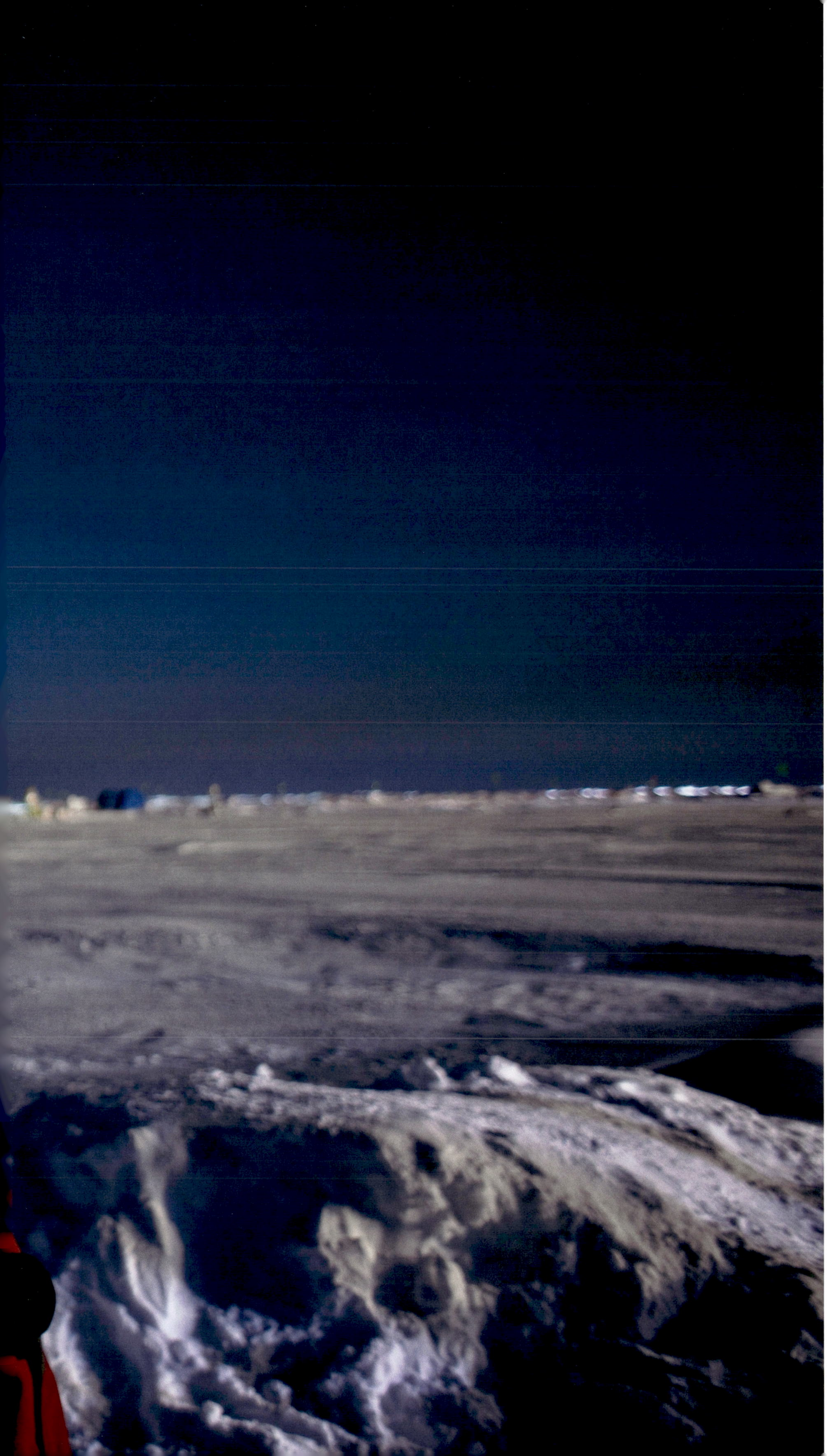

On clear nights, when the light of the moon and stars illuminates the surroundings, the vast expanse can easily be kept in view. However, on nights when clouds and fog restrict visibility, the bear guard Olaf Stenzel relies on the assistance of ultra-modern night vision devices. It is crucial to identify bears as early as possible, before they get dangerously close to the researchers on the ice.

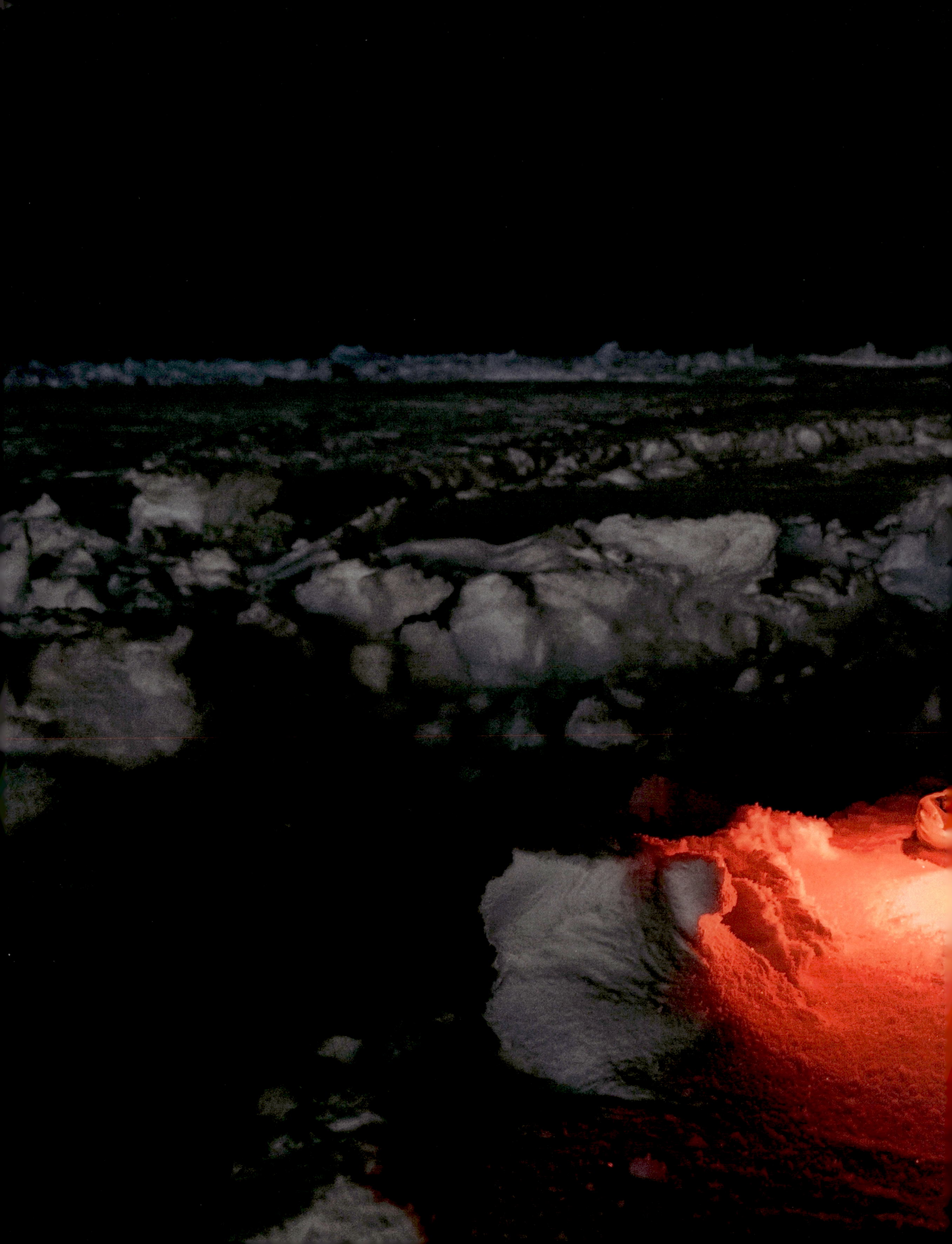

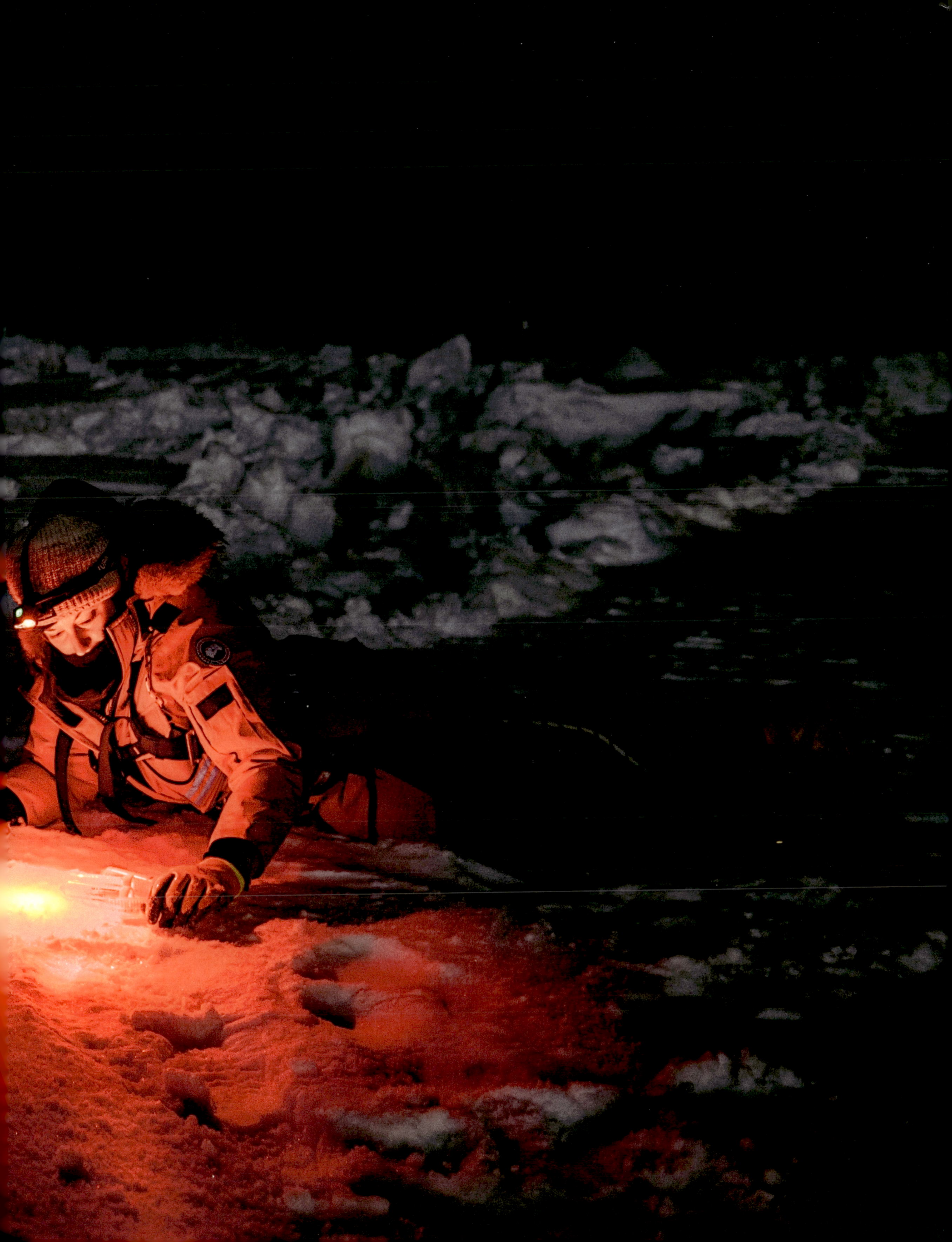

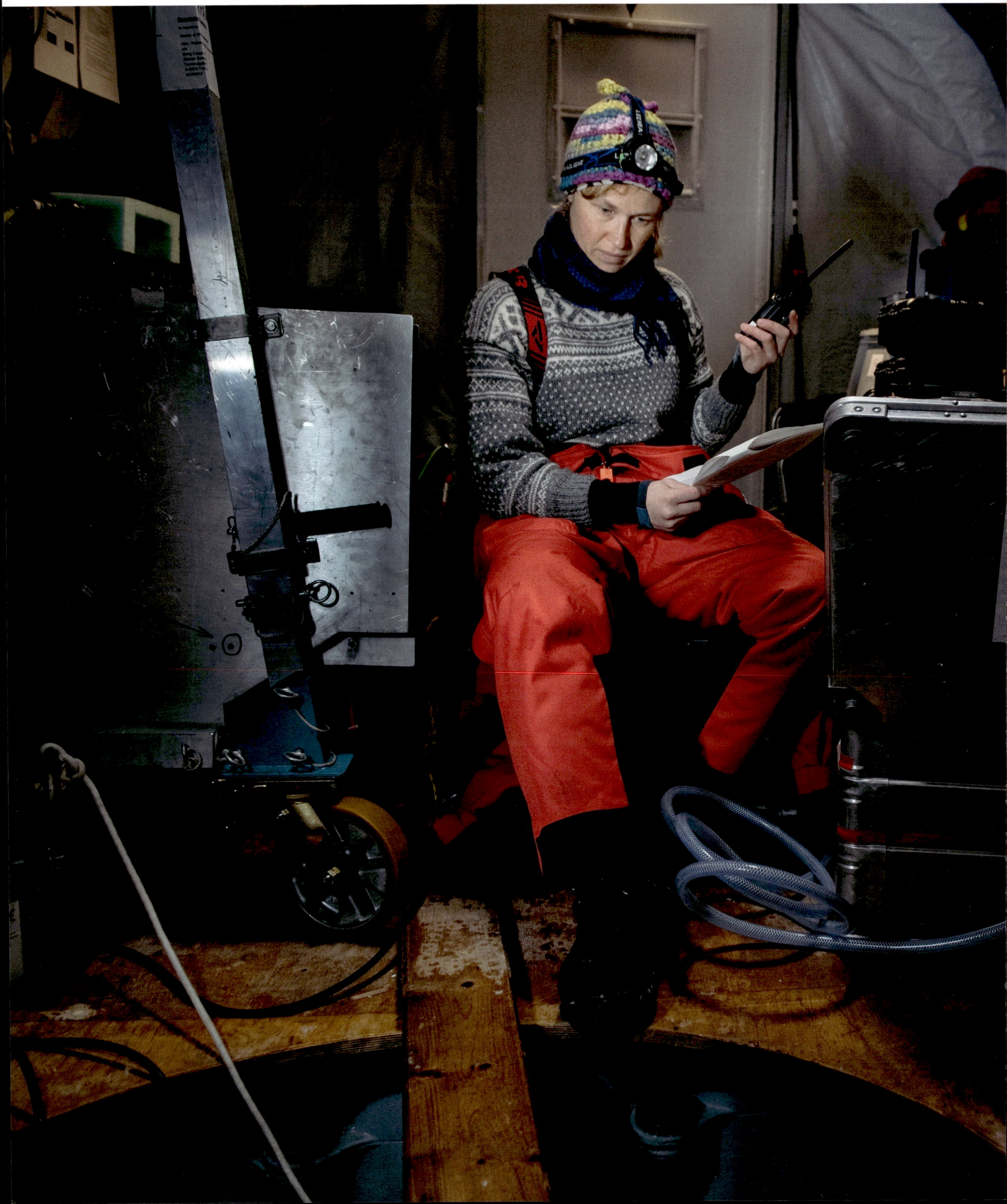

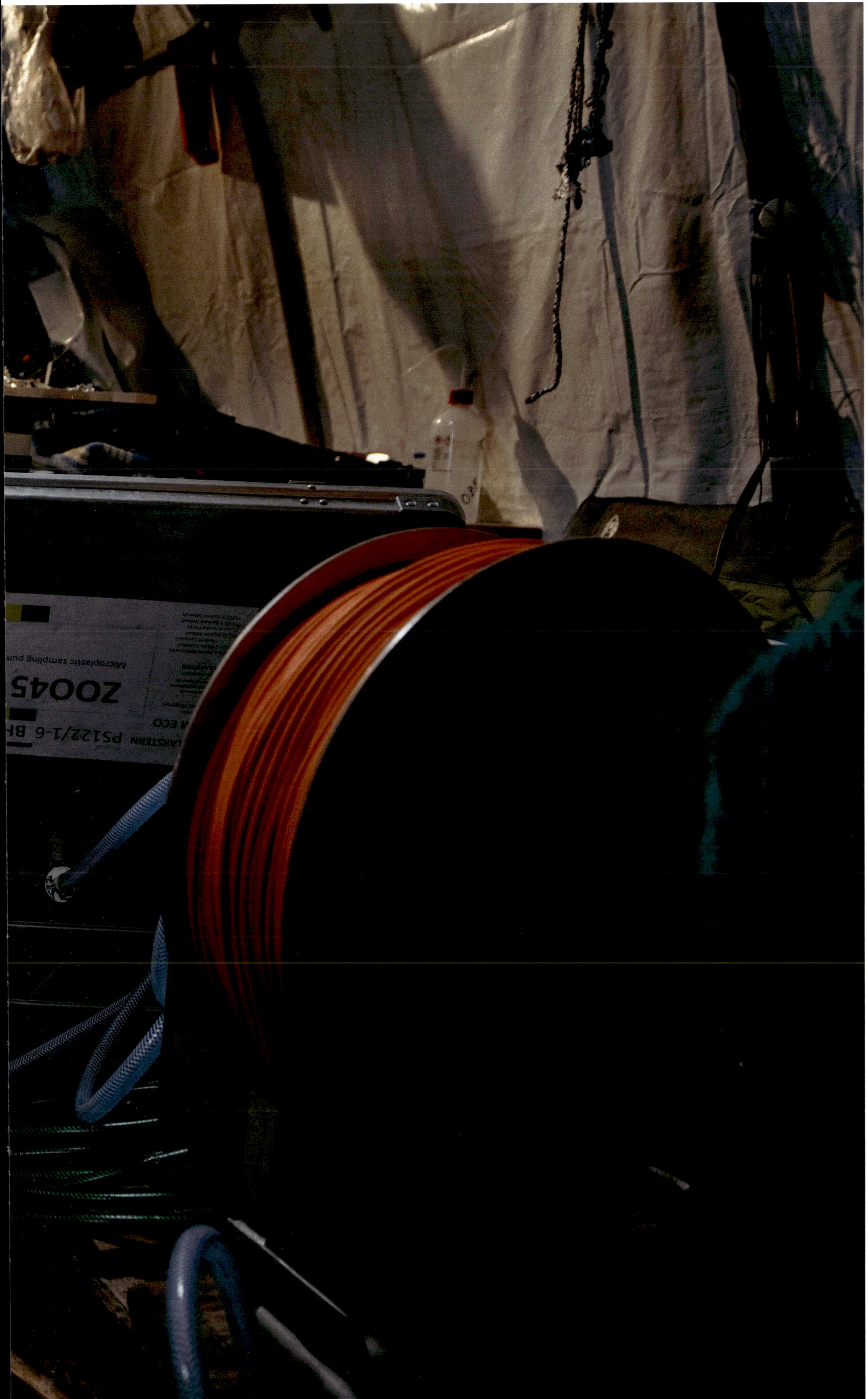

In addition to her role as safety expert, Bjela König has taken on scientific duties. Assisted by a colleague, she pumps 500 litres of seawater through a filter twice a week. Later analysis will be devoted to the question of whether the samples contain microplastics that have found their way to the central Arctic via wind and sea currents. Previous expeditions found microscopically small particles of plastic in the Arctic: an environmental problem caused by humanity that has now reached the most remote regions of planet Earth.

The scientist Ian Raphael uses a traditional – that is, manual – method of measuring the growth of snow. With the help of a rod firmly thrust into the ice, the growth of the snow can be well monitored. The coexistence of various methods of measurement during the MOSAiC expedition, as well as comparability with earlier studies, makes it possible to draw reliable and comprehensive conclusions about the conditions and changes that characterize the Arctic climate system in the early twenty-first century.

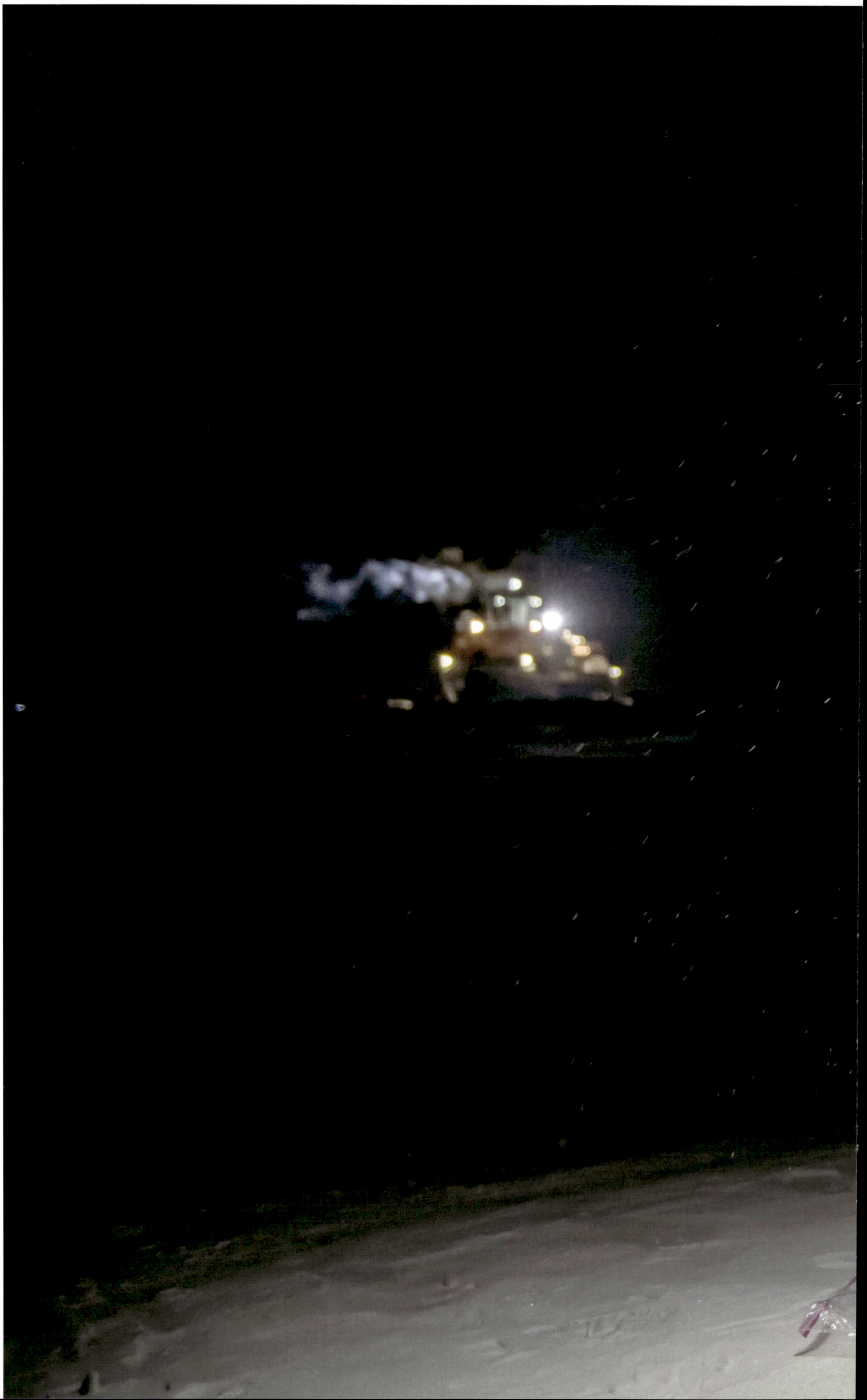

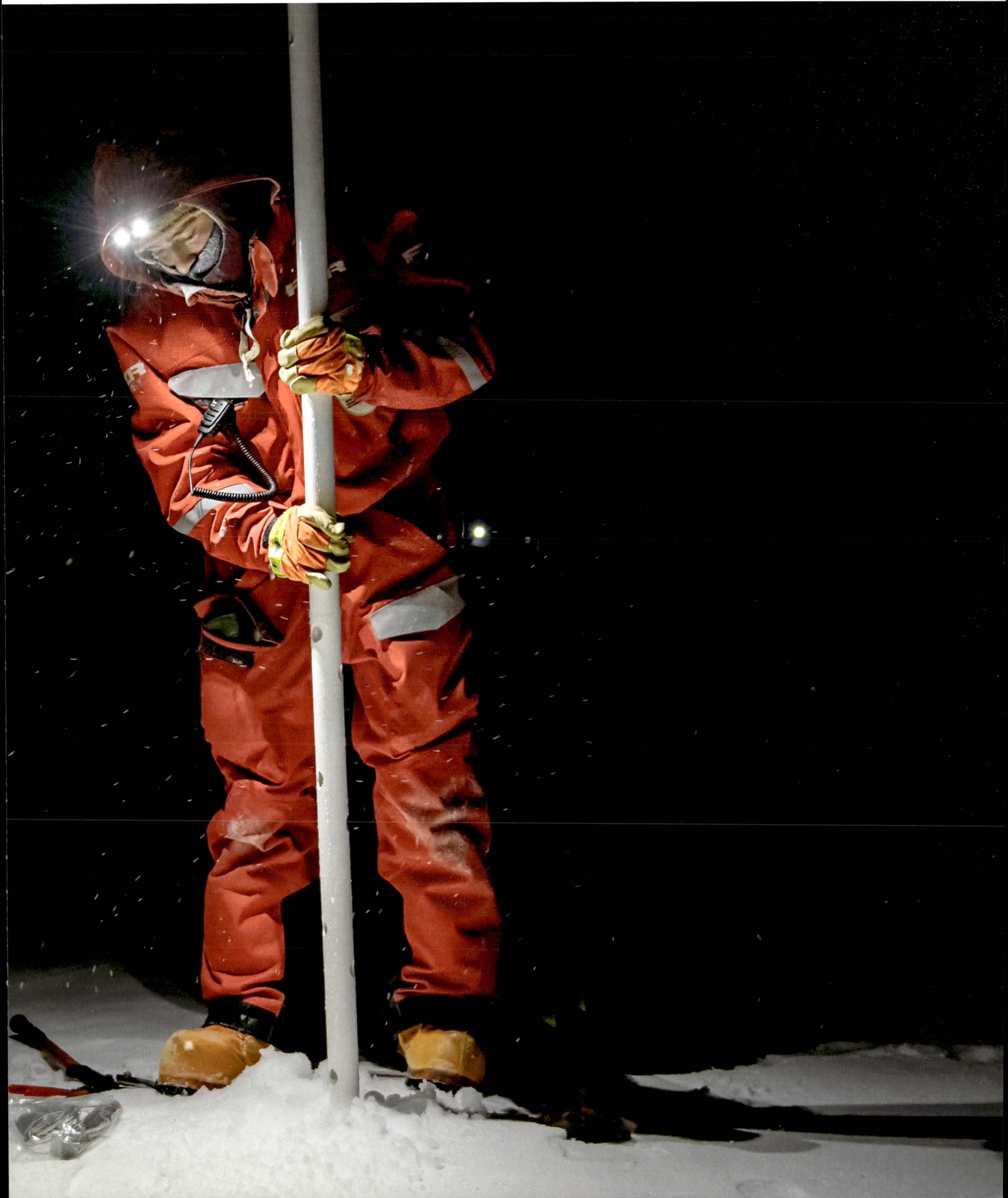

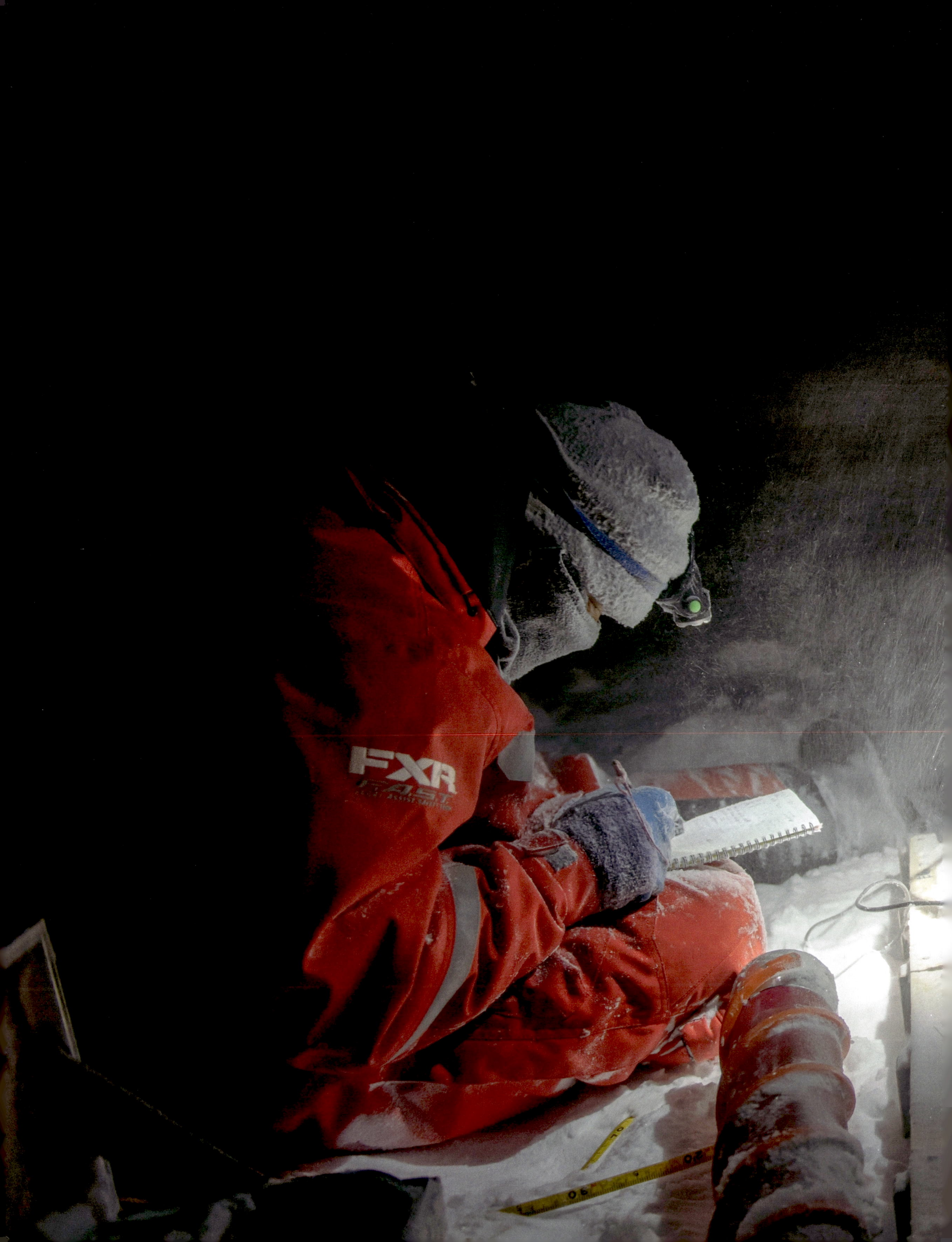
FXR

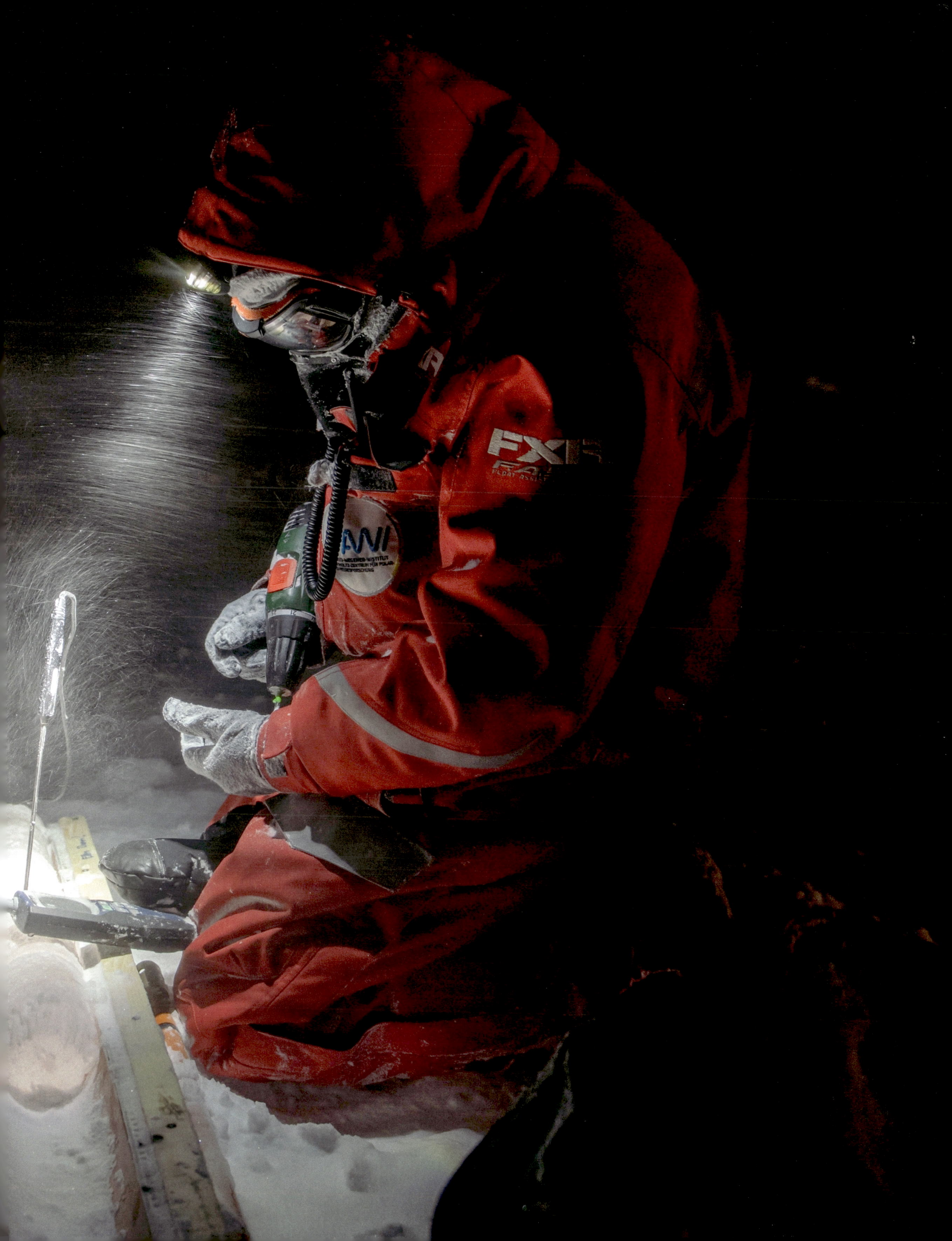
FXR
AWI

The Arctic Sea Ice

Cracks in the cold store

An endless desert of ice covers the Arctic Ocean. Here and there, thick, metres-high pressure ridges rise to the sky. Only through tremendous effort can icebreakers penetrate this region. From our perspective it is thus hard to imagine how fragile the sea ice really is. Yet appearances are deceptive. If we could view the ice and snow in a cross-section through the earth, we would see a paper-thin layer, in areas measuring less than a metre, separating the ocean with its depth of 4,000 metres from the atmosphere with its high-altitude winds at over 15,000 metres. More and more cracks are appearing in this layer. The gigantic ice layer is changing rapidly – with climatic consequences far beyond the Arctic. During the MOSAiC expedition the scientists therefore want to investigate, for the first time, the changes taking place in the sea ice on the same ice floe over a whole year. They are also looking to closely examine the snow that covers the ice floes, another layer that divides the ocean from the atmosphere and also influences the Arctic climate system.

Sea ice is formed when the surface of the ocean freezes. Due to its salinity, this happens at temperatures of −1.8°C (28.6°F). Ice crystals on the water surface only a few millimetres in size combine into a continuous film that is, at first, practically invisible. On satellite images, for example, this white layer often looks like a large, seamless blanket. In fact, the sea ice is more like a gigantic mosaic made of thousands of individual floes and ice formations, and it is in constant motion due to ocean currents and winds. Because the mass of water in the ocean does not increase when sea ice melts, its disappearance does not cause a rise in sea level. Yet the ice has enormous influence on the climate of our planet.

Sea ice constitutes a significant proportion of the so-called cryosphere, all of the forms and deposits of ice and snow on our planet – its cold stores, as it were. In terms of its mass, the cryosphere is the second-largest component of the earth's climate system after the ocean. The sea ice constantly interacts with both the atmosphere on one hand and the ocean on the other. Light-coloured surfaces reflect a high proportion of sunlight back into space. For this reason, the surfaces covered by sea ice are warmed much slower than the dark water of the open seas. If global warming leads to increased melting of sea ice, less solar radiation is reflected. The surface as a whole would then have a smaller reflective capacity, also known as albedo. As a consequence, the Arctic Ocean absorbs significantly more heat. The ice therefore melts both as a result of the higher air temperature and also through the increased warmth of the water in the ocean – a constantly increasing effect. Since this process is particularly pronounced in the polar regions, it is referred to as polar amplification, which means climate change is especially noticeable in the Arctic.

Like a heart beating in slow motion, the Arctic sea ice expands in winter and contracts again in the following summer. It reaches its annual minimum in September, when the ice-covered surface is only half as large as in winter. Alongside this seasonal natural phenomenon, there is another trend, however, which is linked to global warming: since 1979 scientists have observed from satellite images how the expanse of sea ice has become increasingly smaller, especially in summer. These observations are disturbing: since 2007 the September minimum has been smaller every year than in the first decades of satellite observation. In 2012, the lowest level ever recorded was reached, with a surface of 3.4 million square kilometres, and in September 2019, for the second time on record, the sea ice shrank to an area lower than 4 million square kilometres. By way of comparison, in the early 1980s the sea ice in September still covered over 7 million square kilometres of ocean, an area as large as Australia.

This observation alone does not allow a final judgement on the extent of the loss of sea ice. In order to calculate the total volume of the ice, it is necessary to consider its thickness, which is often highly variable and can reach several metres down into the ocean. In addition, the snow that lies on the sea ice makes it even more complicated to measure the ice's thickness. In recent decades both the extent and the thickness of the ice have been decreasing. In the

Among all the instruments, technologies and diving robots of the Ice Camp, no station resembles a science-fiction film as much as the Remote Sensing Site. Matching this impression, the site, in cooperation with research institutions from all over the world, collects the data that are required to improve the satellite measurements of ESA and other space agencies and to enable even more precise observation of the important climate variables of snow depth, ice thickness and ice surface from the wider perspective of space. The scene at the Remote Sensing Site with its extraterrestrial appearance mainly serves the very real purpose of finding out how much the Arctic sea ice is retreating and what is the context for this change.

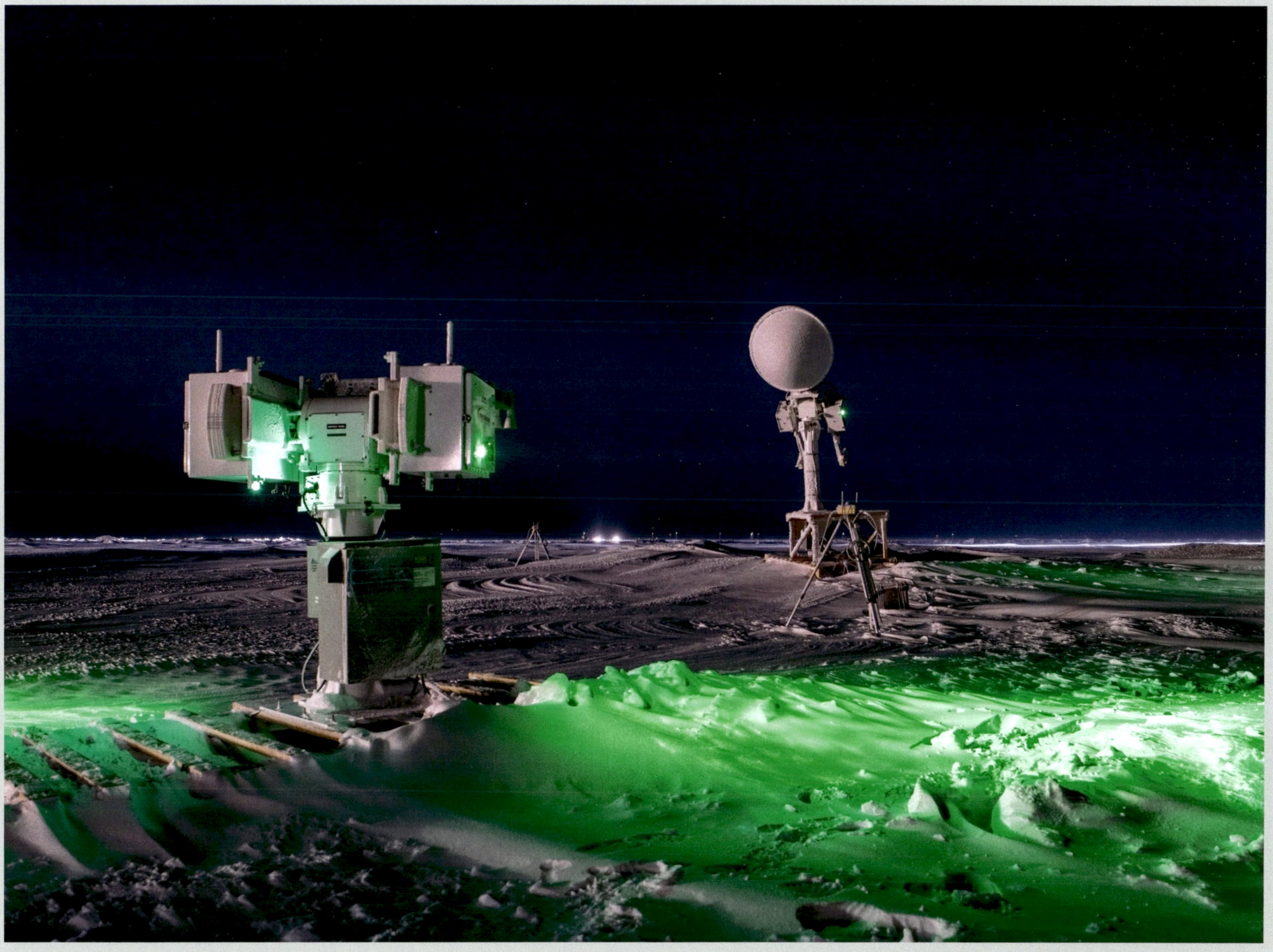

1960s, the sea ice was still about 3 metres thick at most points that were measured; since then, most measurements currently reveal a thickness of less than a metre.

One of the primary reasons for this phenomenon is that today sea ice is not as old as it used to be. Scientists distinguish between seasonal sea ice and perennial or multi-year sea ice. The latter survives for at least one summer, becomes thicker the following winter and usually attains an age of two to five years. The proportion of thick multi-year ice has been declining steeply for years, and now accounts for less than half of the total. The thinner seasonal ice is characteristic of the Arctic today. The frozen ocean is now significantly different from that which we know from Fridtjof Nansen's notes. The movement of the ice surface from wind and ocean currents is much greater and the ice is thus considerably less stable.

Daily Routine

Life at the end of the world

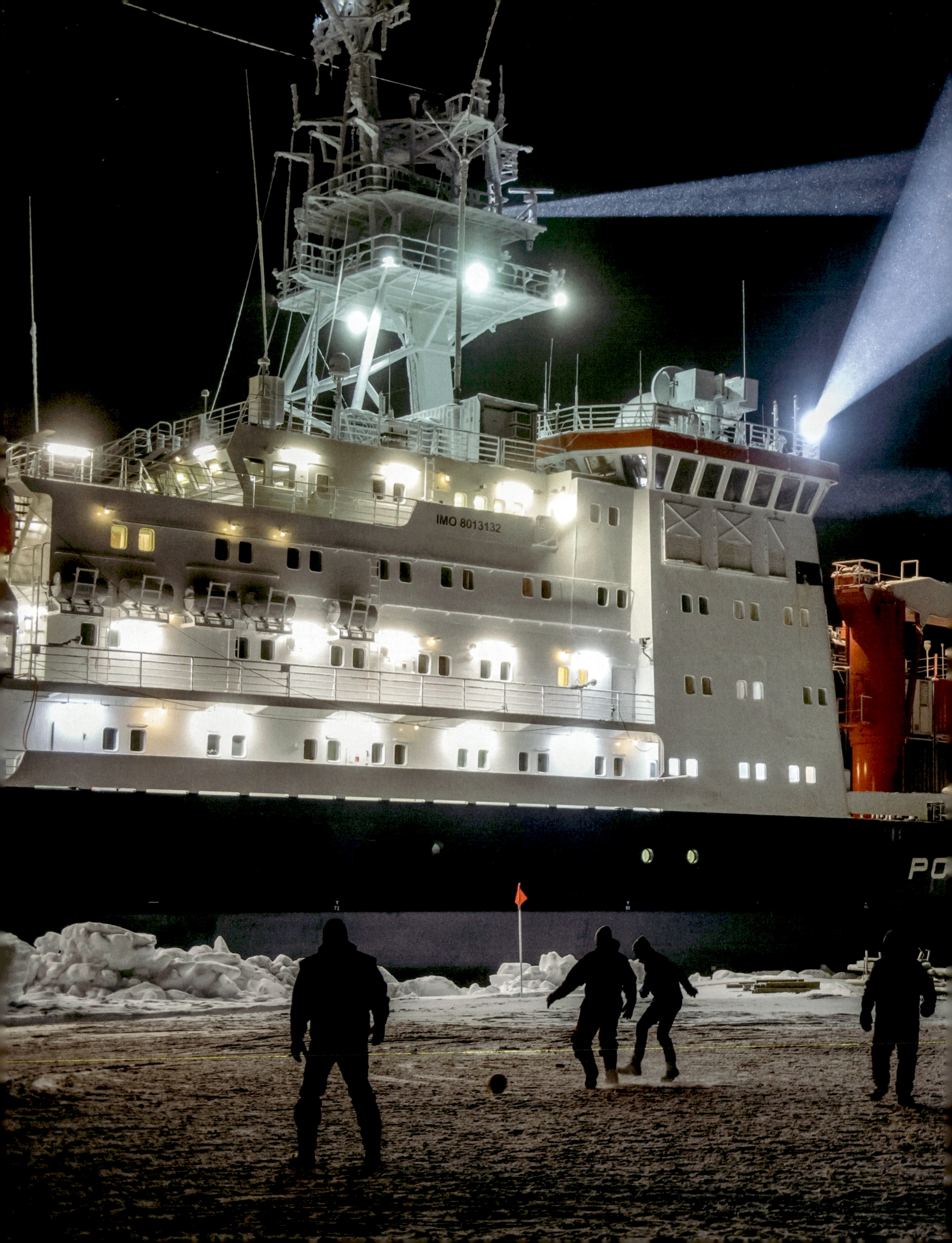
IMO 8013132

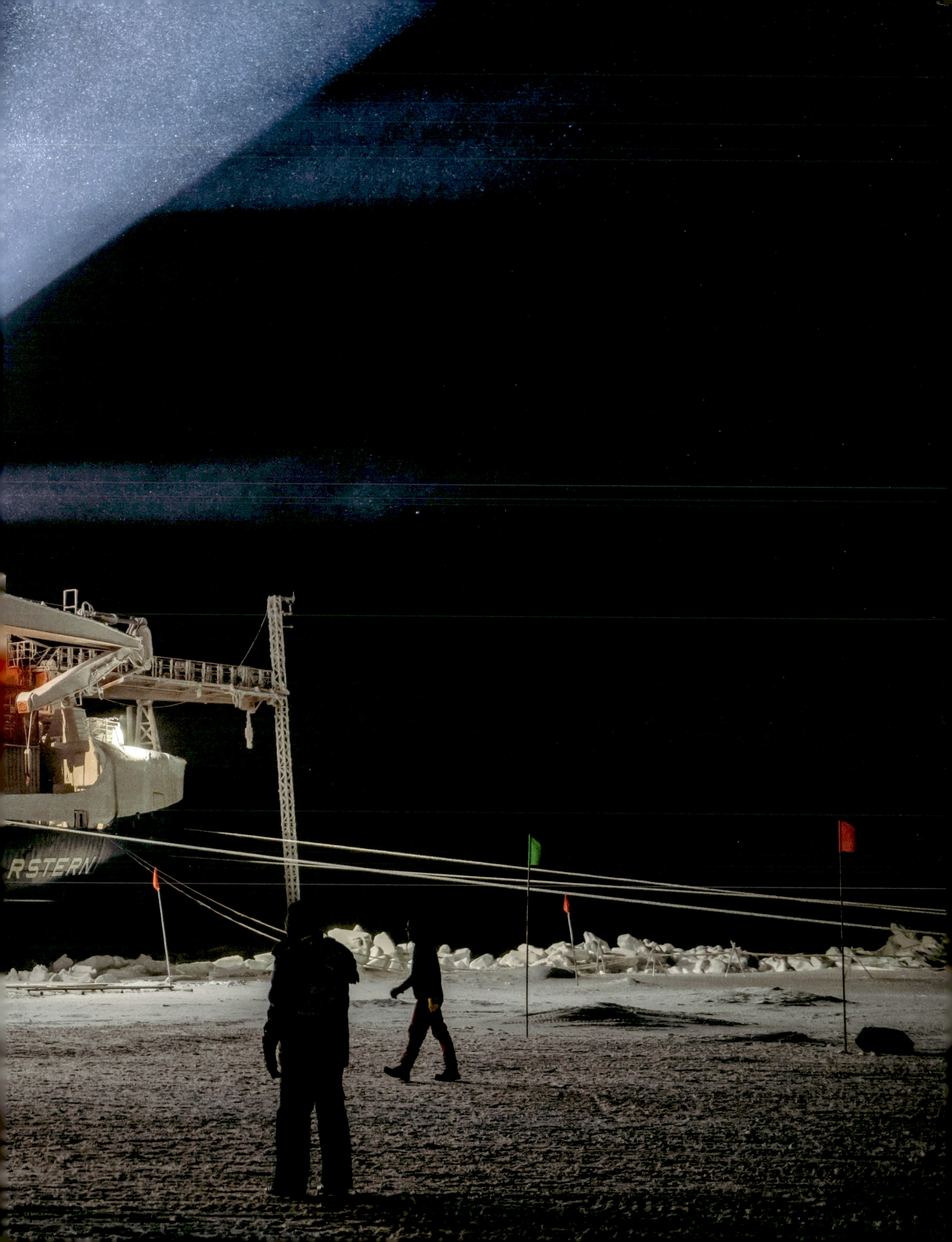
RSTERN

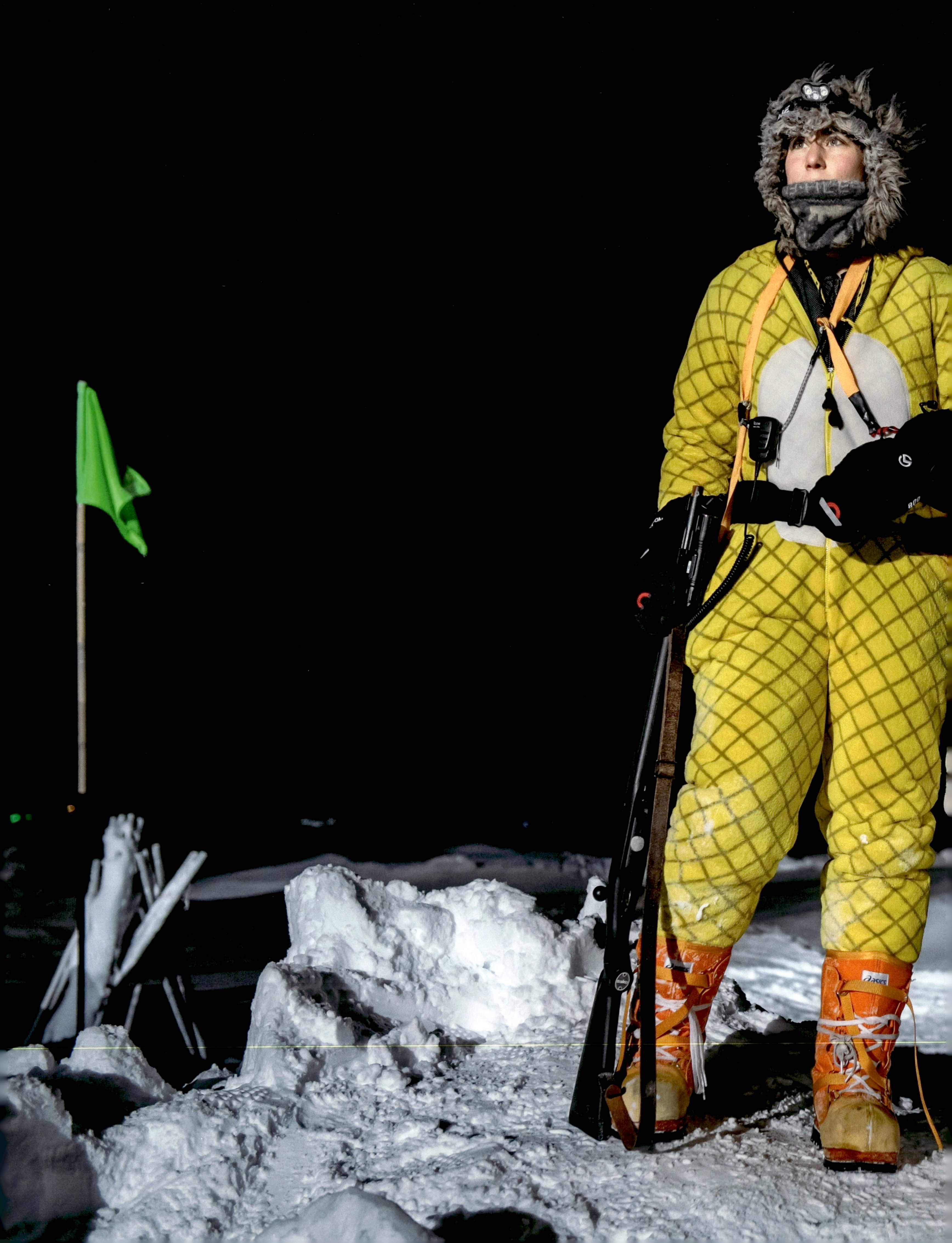
asics

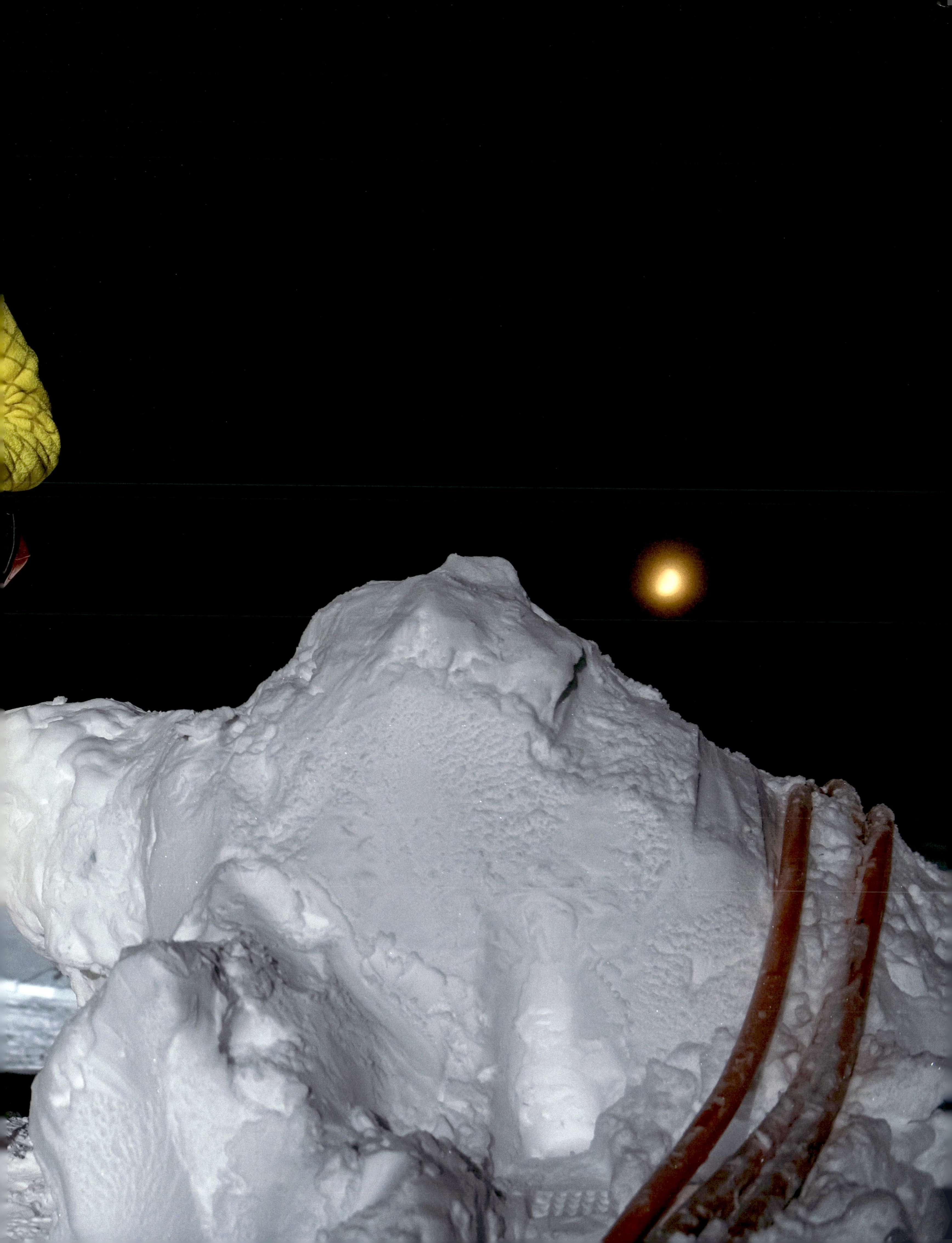

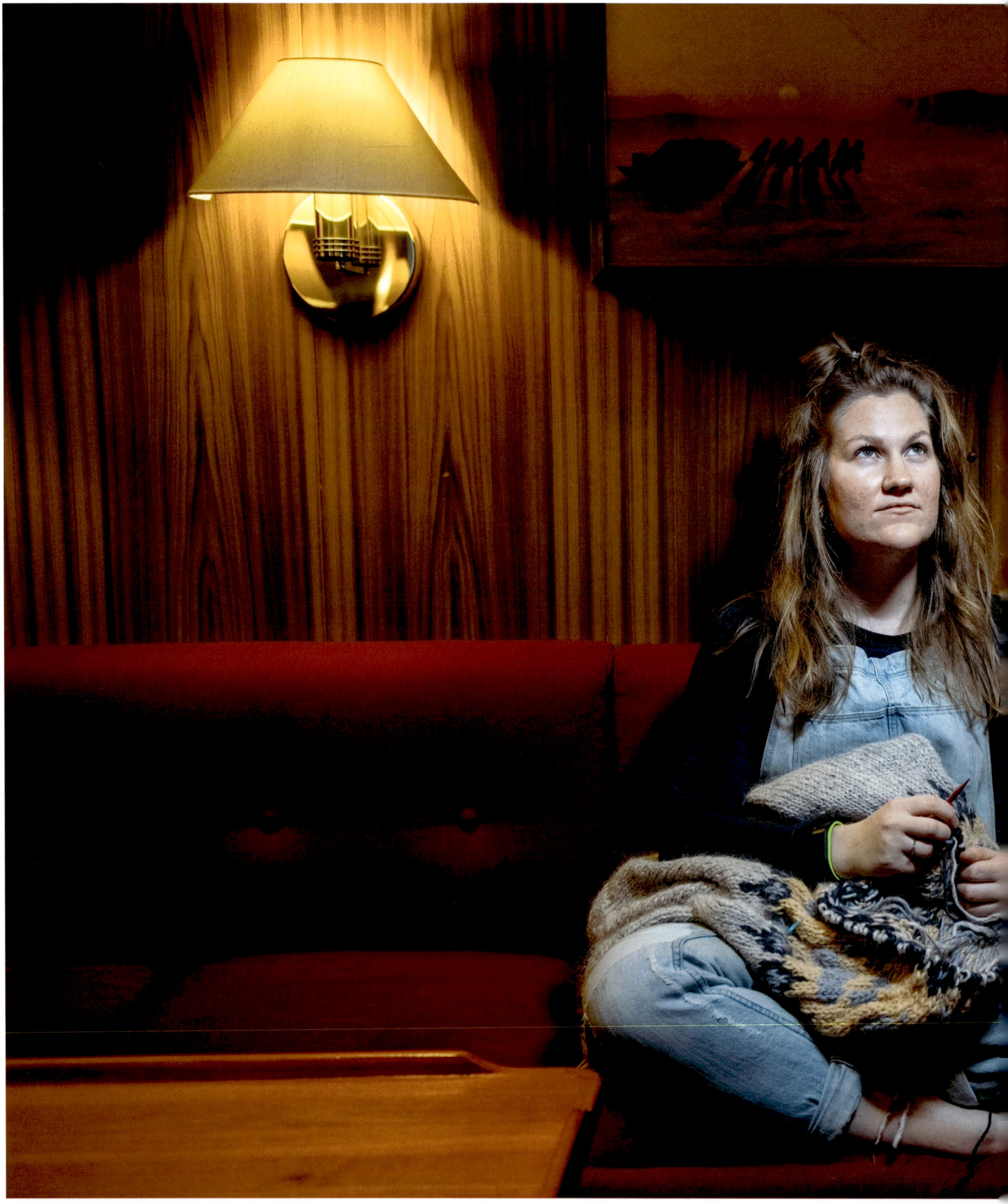

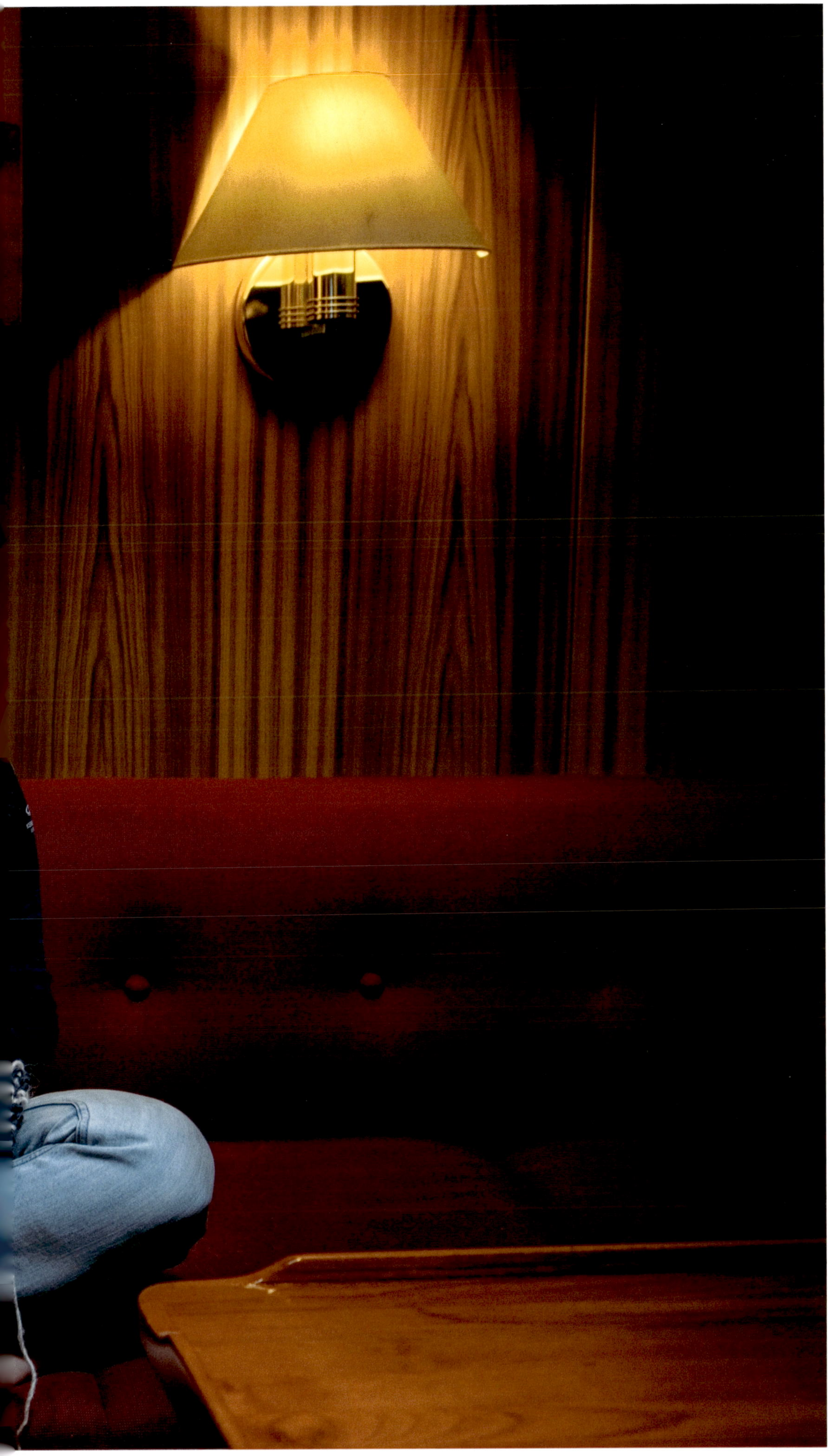

The research camp has been set up, working routines have been established, and the group is growing closer together. After the intense first weeks of the expedition, the participants are becoming increasingly creative in how they spend their free time. Directly in front of the ship a small recreation area has been created. Here it is even possible to play a football match in the beam of the searchlight – so long as some polar bear guards volunteer to keep the pitch safe. One of the expedition's full-time bear guards is Trude Hohle. In the evening she can often be found knitting pullovers and hats in the Red Saloon.

O

On the ice, the longing for warmth can grow without limit. Doing research in extreme frost, or doing clearing work in a biting wind that blows the cold deep into your bones, or hours of searching the dark horizon for polar bears: work on the floe is exhausting. But however laborious the hours on the ice, at the end of the day warm rooms inside the *Polarstern* always await the team.

In the evening, when the people on board pull up the gangway, which they call the drawbridge, their ship is like a castle in the ice, where they can hand in their safety equipment and rifles, and leave the dangers of the Arctic behind them on the floe. After liberating themselves layer by layer from all of their polar clothing, only the marks of wind and cold on their faces reveal the region they are currently inhabiting.

Like a protective shell, the *Polarstern* then separates the members of the expedition from the unrelenting Arctic outside. During the three warm meals that are taken daily in the two mess rooms, they talk about past or upcoming events of the day. Where has a new crack split the ice? Which measuring instrument has produced astonishing results? What has happened on the ship in the meantime? What are tomorrow's plans? Several times a day, the messes become hubs for exchanging important information. The food on board is home-style. Stews, fried potatoes or cold platters – these variations on German dishes were probably served forty years ago when the *Polarstern* set off on her first expeditions to polar regions. Especially on Thursdays, the so-called seaman's Sunday in the German tradition, the meal is hearty. The shipping company has planned long in advance for this expedition so that there is enough food on board. To give one example: on the first day of the expedition, about 10,000 eggs fill the cold stores so that they can be turned into omelettes or cakes in the coming weeks. Most of the expedition participants greatly appreciate the food that the galley serves up. At the same time, doing justice to the dietary habits of a hundred people from different countries invariably involves compromise. For this reason some have brought with them favourite foods from home that they definitely do not want to go without.

Leisure time in the central Arctic means, above all, offline time. The opportunities to communicate with friends and family are extremely restricted. The satellites orbiting the Earth are not primarily oriented to the needs of a small group of people at the North Pole. It is almost a luxury to be able to send emails or texts via Messenger with a microscopic data volume. In other ways, too, the end of the worldwide web has been reached here – no social media, no news, no streaming. Instead of surfing for hours, long-forgotten leisure activities, reminders of life before the internet, come into their own. After lunch, for example, the expedition members hold table-tennis tournaments. A few of them have brought their musical instruments along. In the common rooms you see researchers knitting hats or sitting on yoga mats. For two hours in the evening a small area of ice next to the ship is available for leisure activities, so football matches can be played. These hours in the evening are important for switching off, as there are no weekends on a research ship, and measurements have to be taken even on holidays. For those who have spent the day standing on the ice, these few free moments are an opportunity to recharge their batteries and form friendships. But talk about science goes on even after work – in the corridors, over a meal, in the Red Saloon or at a lecture in the cinema.

From time to time the routines are interrupted, at least a little. On the first Sunday in Advent, the crew lovingly adorn the messes with Christmas decorations. Biscuits, nuts and dates are eaten in an unusually festive atmosphere. For Christmas Eve the members of the expedition have brought lots of presents on board. After a classic north German Christmas dinner – potato salad and bockwurst sausages – everyone appears in fine clothes, some of which were packed

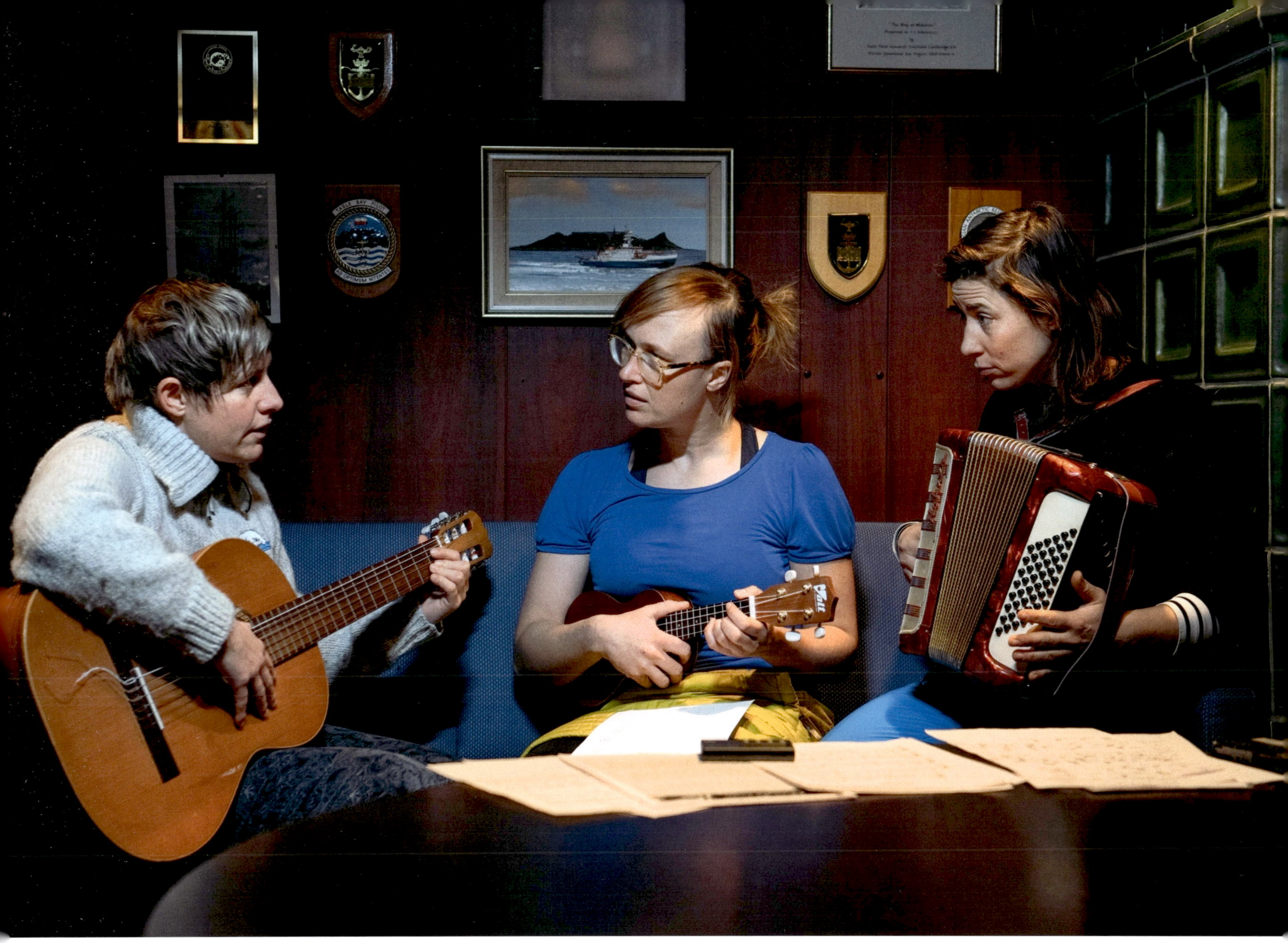

In addition to the equipment mandatory for the exhibition, participants take various things that they do not want to be without for four months in the ice. Antonia Immerz, Bjela König and Verena Mohaupt (from left to right) are playing music in the Blue Saloon, practising pieces they want to perform at the Christmas party. Occasions exist for small concerts, and not only on holidays, but sometimes spontaneously after work in one of the laboratories.

specifically for this evening. The captain makes a likewise festive speech in the Blue Saloon. Christmas near the North Pole is a special situation, he says, and the sense of community on the *Polarstern* is special, too, as the people here are the best alternative to family at home. And at the end of this holiday, he quotes Fridtjof Nansen with an excerpt from the Norwegian's diary describing how he spent Christmas in the pack ice with his comrades 125 years before. With small gifts for the crew, the scientists also express their thanks for the superb support that they have received.

Even on a day as special as this, free time does not mean leisure for everybody. Work is always going on somewhere on the ship. In the middle of the night, the galley in the bowels of the ship becomes a bakery. A few levels further down, the crew of the engine room keep the electricity, heating and water supply in operation. Just like the crew on deck and on the bridge, they work round the clock in shifts. In the metal workshop someone is welding heavy anchors from train rails. On C Deck the drums of half a dozen washing machines are rotating, and in the hospital the ship's only doctor is always available for an emergency consultation. They all ensure that the *Polarstern* works like a small town and that the scientists, far from the inhabited world, can concentrate entirely on their research.

Putting on and taking off polar clothing can take forever. It starts with thermal underwear and ends with safety overalls – on the principle of layering like an onion, because even at double-digit temperatures below zero, hours of strenuous work on the ice sometimes make the expedition members sweat. After a few weeks, most of them have found their optimal combination of expedition clothing.

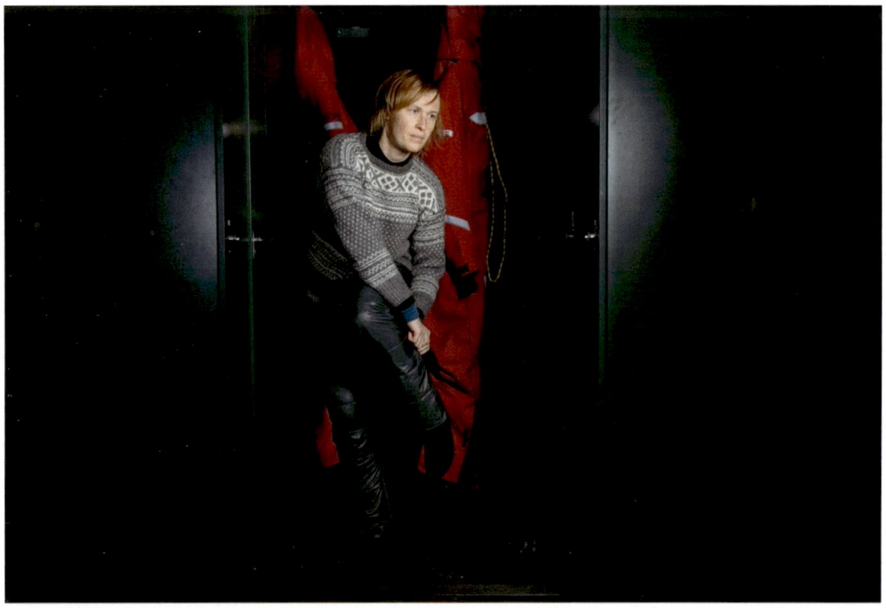

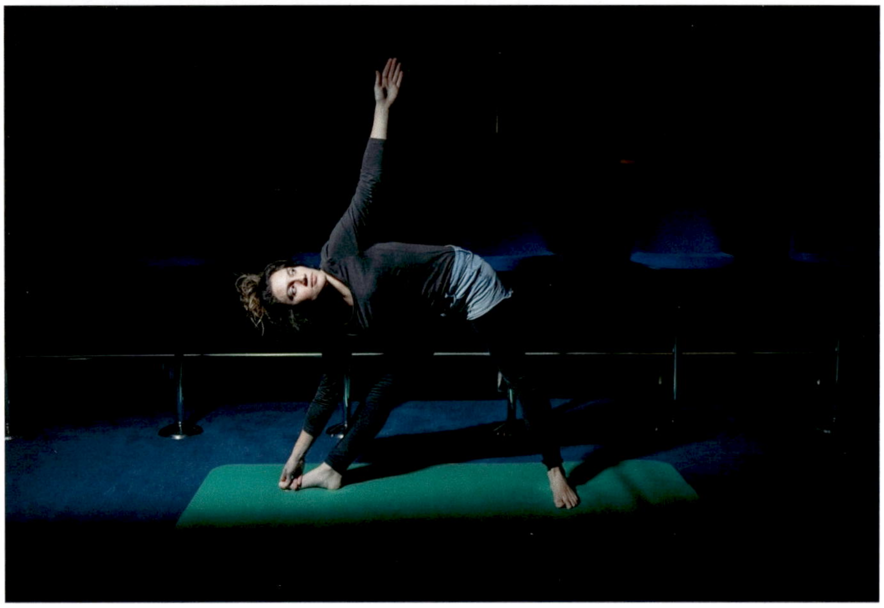

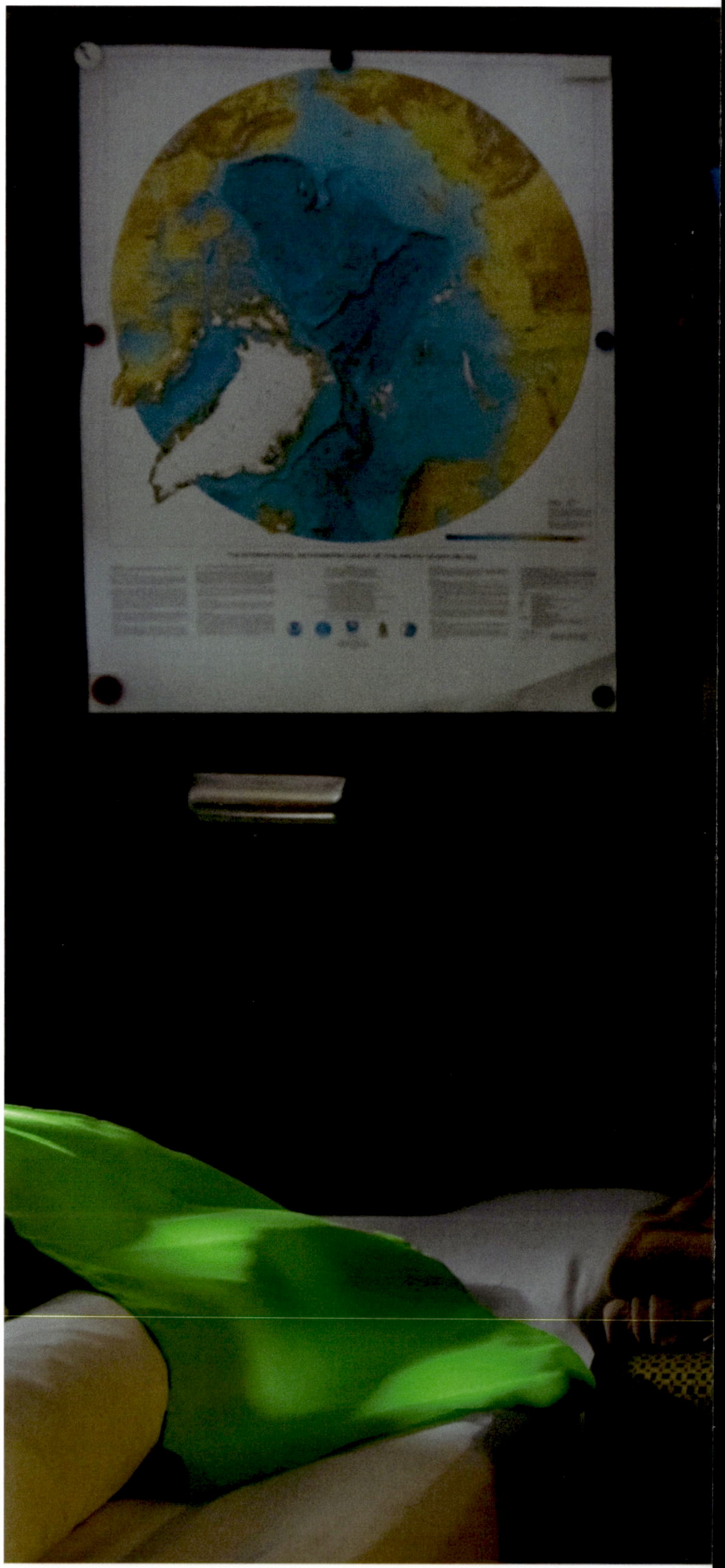

For Carsten Zillgen, on some evenings work continues in his cabin. Unlike all of the handmade caps and socks knitted during the expedition, his work has nothing to do with polar fashion. Instead he sews flags to mark landing places for the two on-board helicopters.

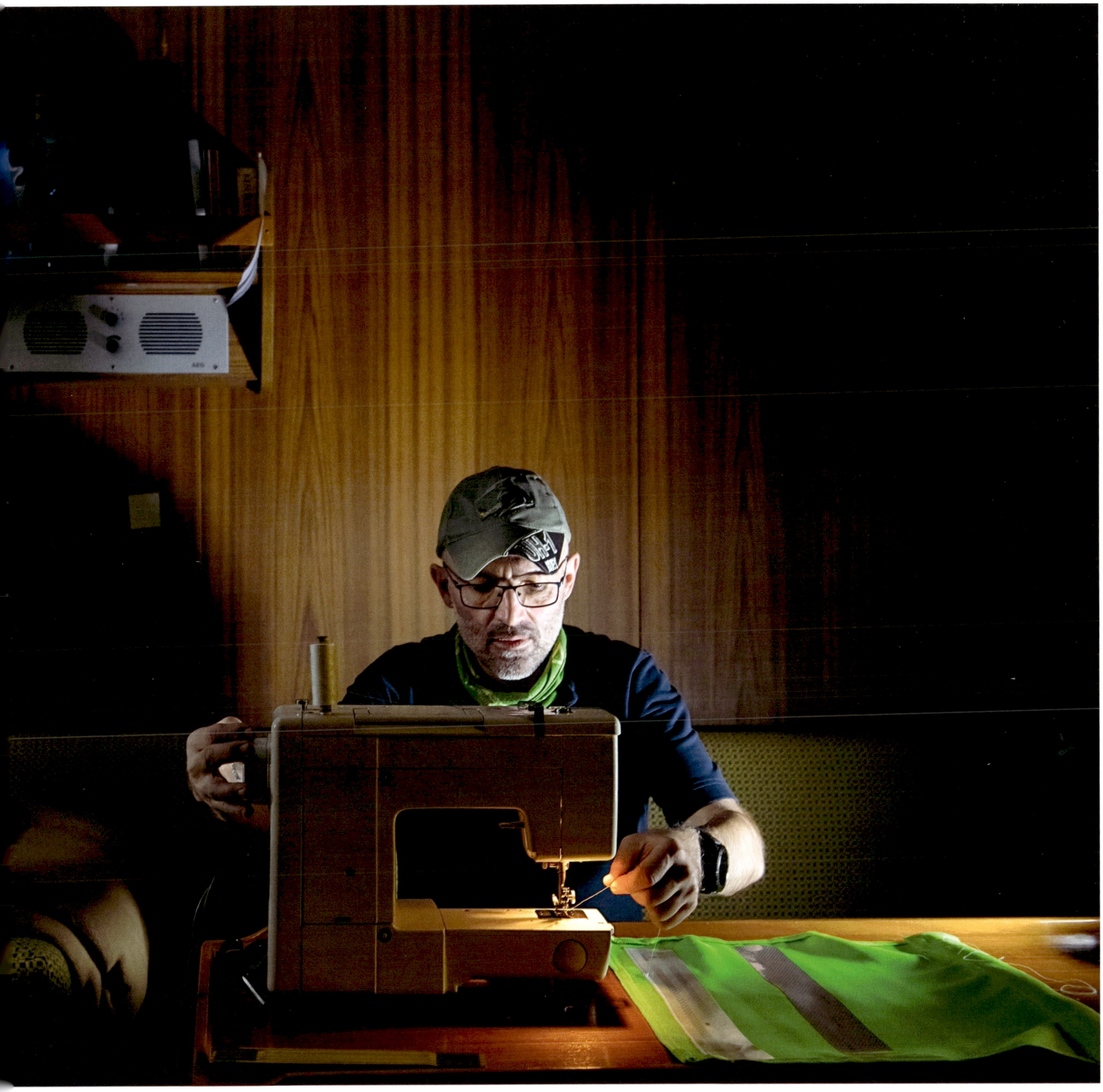

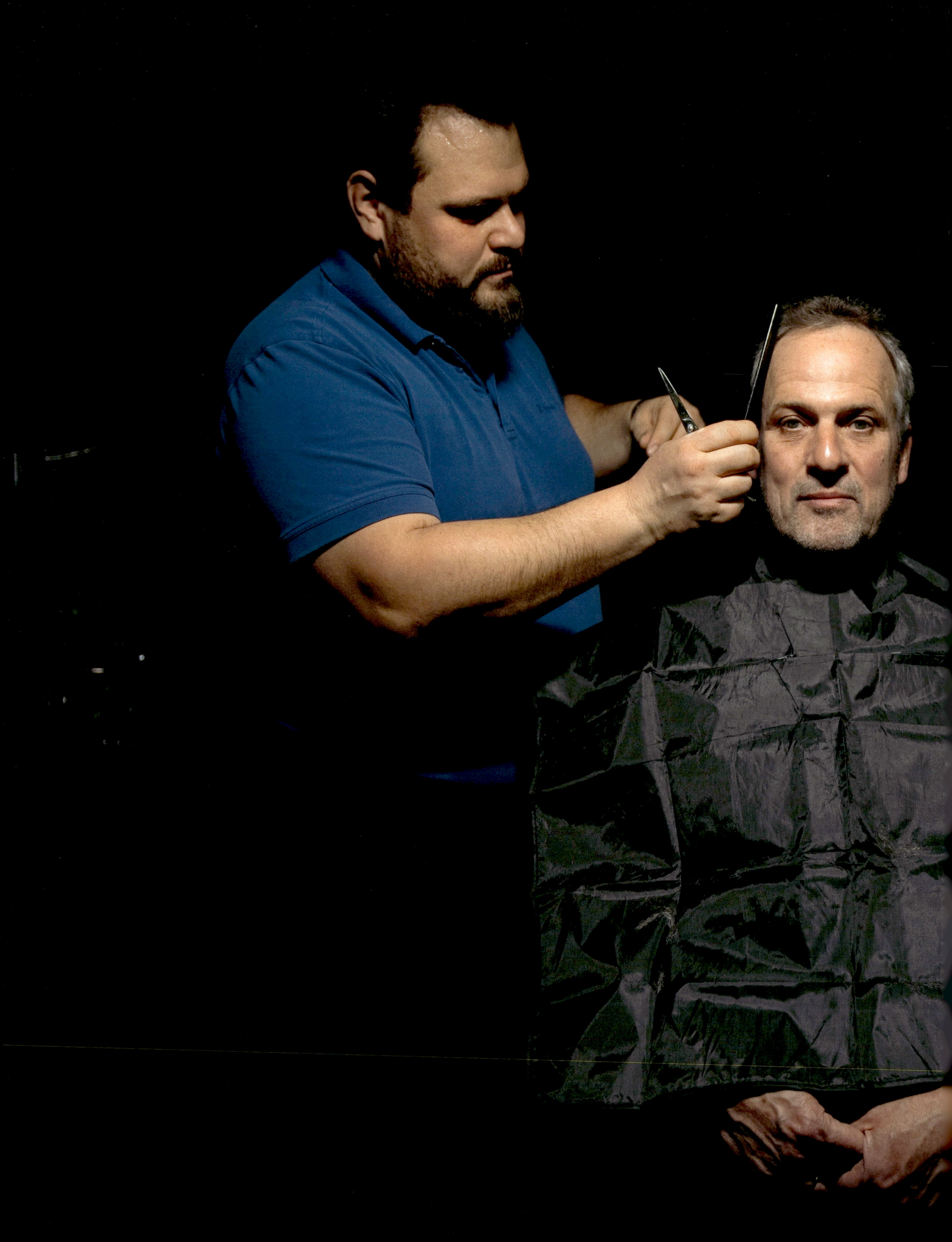

Boatswain Sebastian Brück has a hobby that some expedition members greatly appreciate: when the day's work on deck is over, in the evening he opens his improvised hairdressing salon in one of the workshops. After a month or so he can be sure that customers will start showing up. After scheduling an appointment the haircut is free.

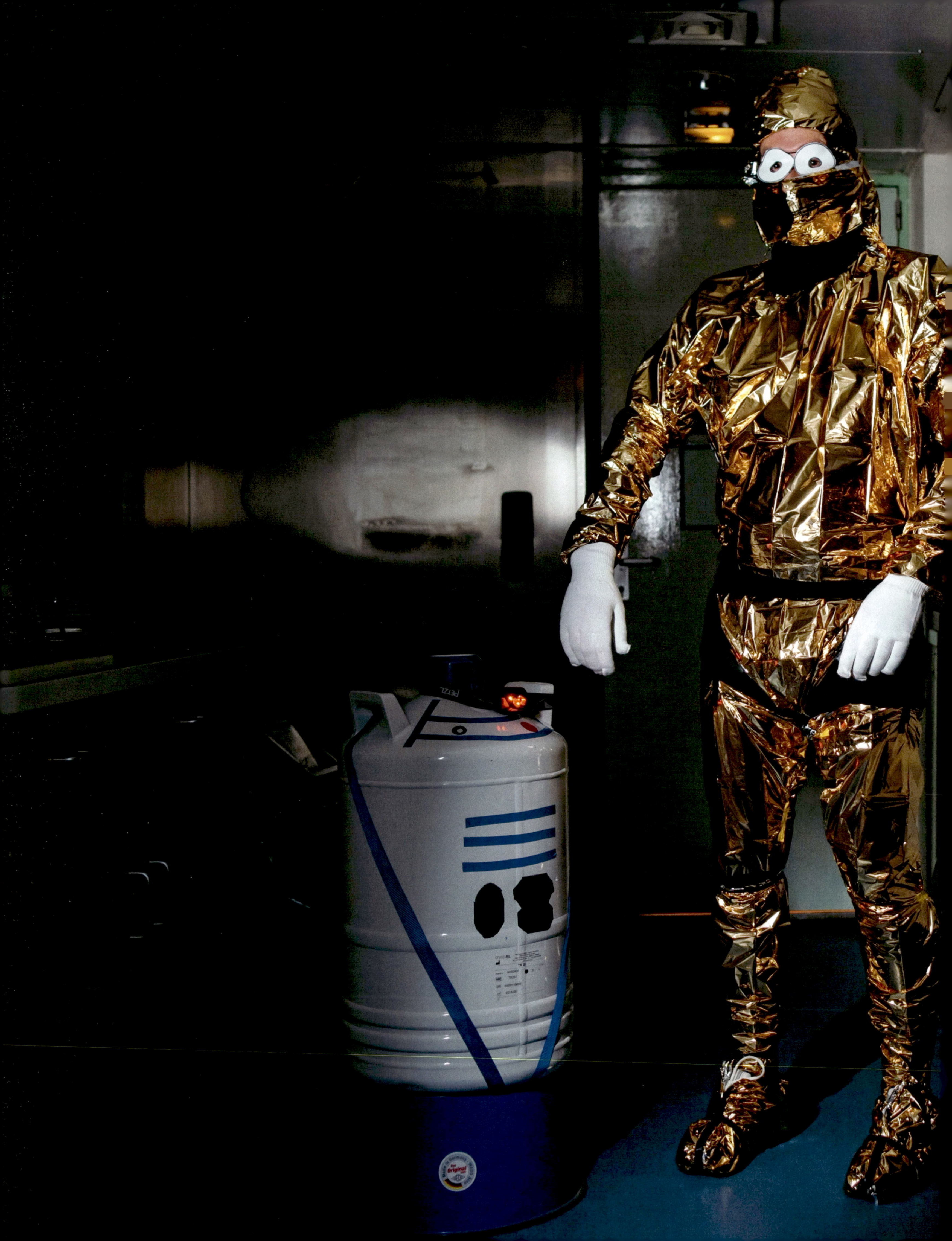

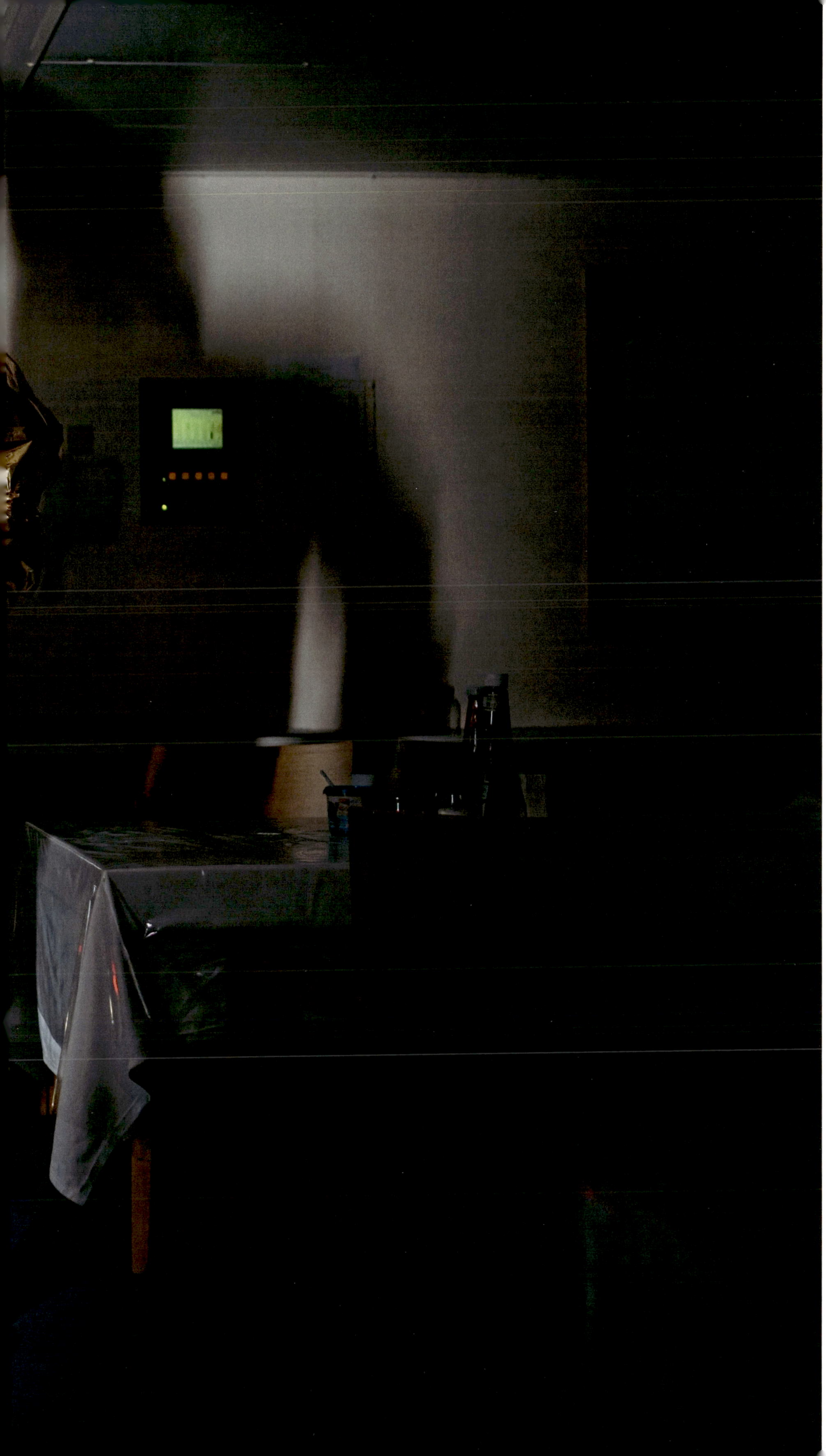

How do you go to a Halloween party when you can't order a costume quickly online? Ronny Engelmann is able to improvise a costume as C-3PO, the humanoid robot from *Star Wars*. But his research also seems as futuristic as his party costumes: with a green laser beam, called a lidar, he measures aerosol particles and clouds at heights of over 20 kilometres.

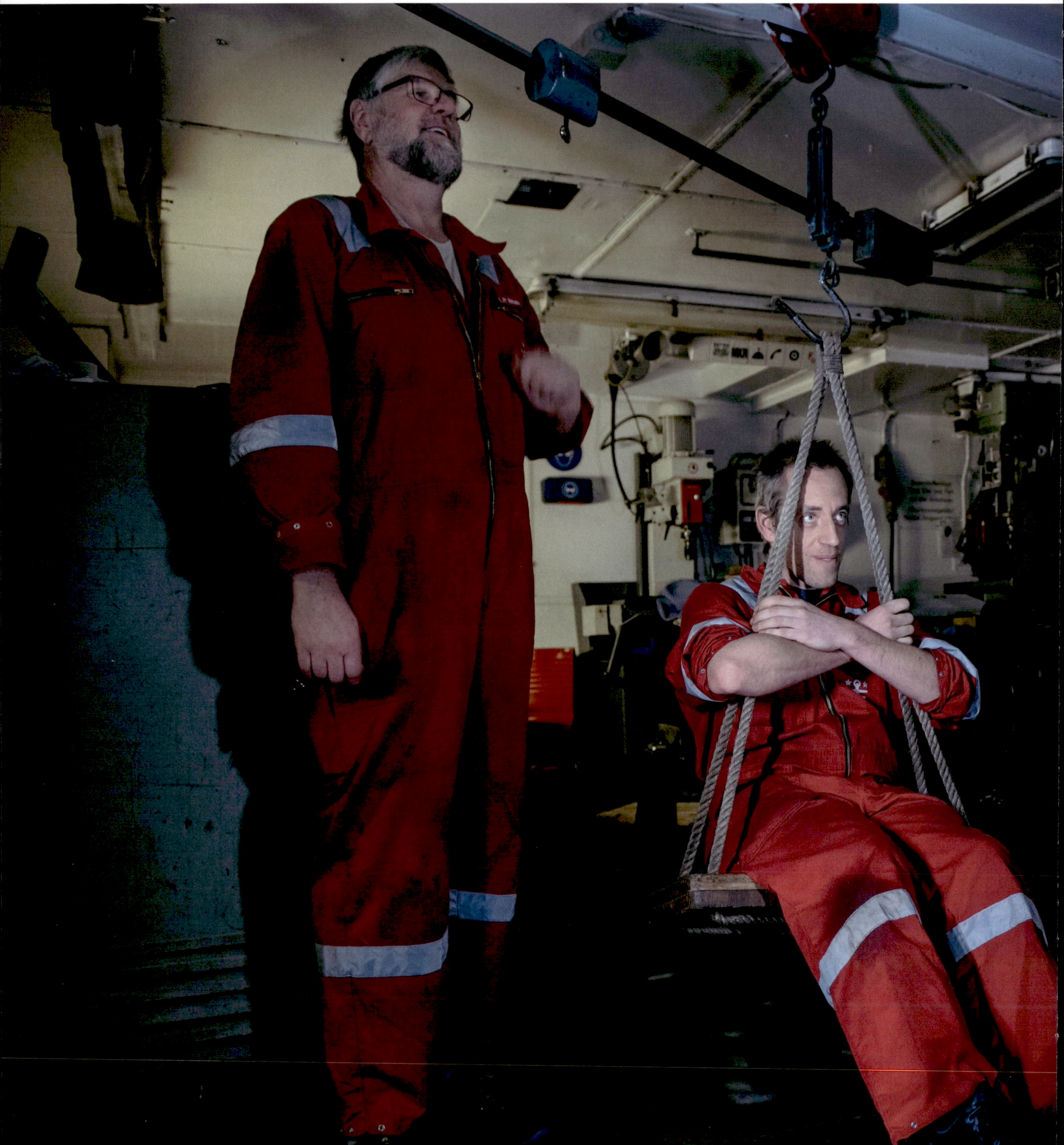

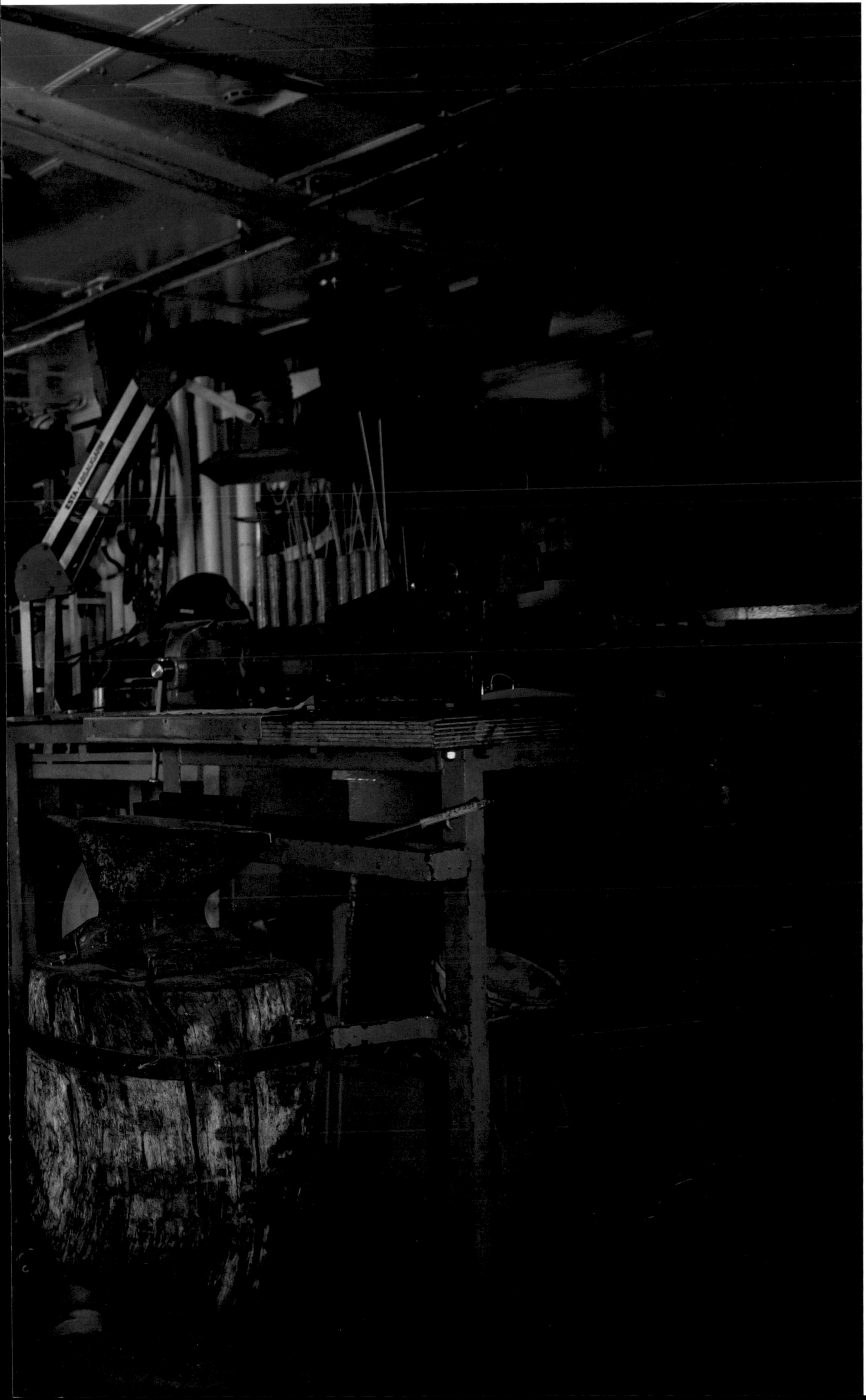

On Sundays the expedition team and crew meet in the workshop for an old *Polarstern* tradition: the weight watcher's club. Before each person hops on the scales, they guess whether they have lost or gained weight or stayed the same over the prior week. Those who guess incorrectly pay fifty cents into the kitty. After the expedition, the money is donated to a good cause.

Every morning Maren Zahn gets up at 3:45 am to bake rolls in the galley and prepare breakfast with the rest of the kitchen team. Sometimes she uses the quiet morning hours to conjure up a cake, which is often on the menu on Thursdays: at sea, Thursday is known as "seaman's Sunday", as the German tradition is to eat a particularly hearty meal on this day.

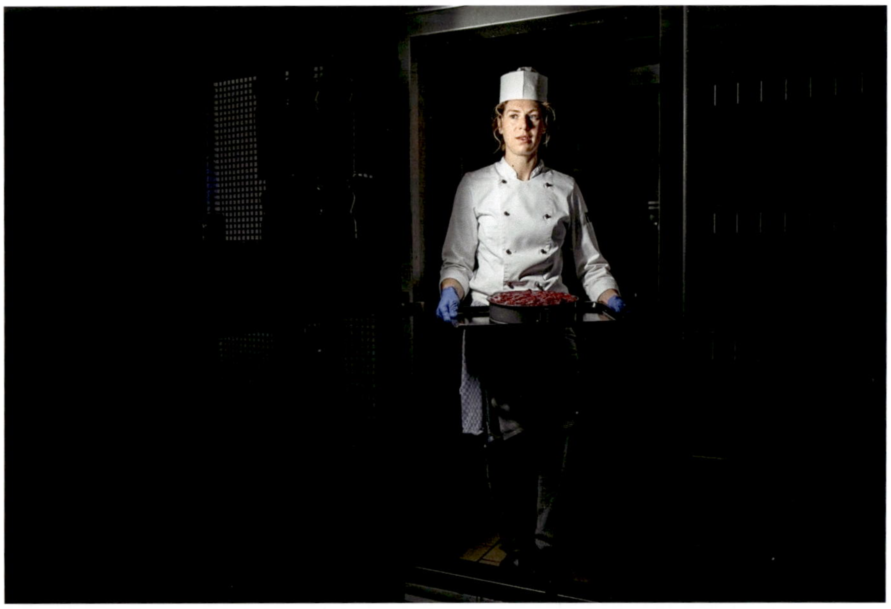

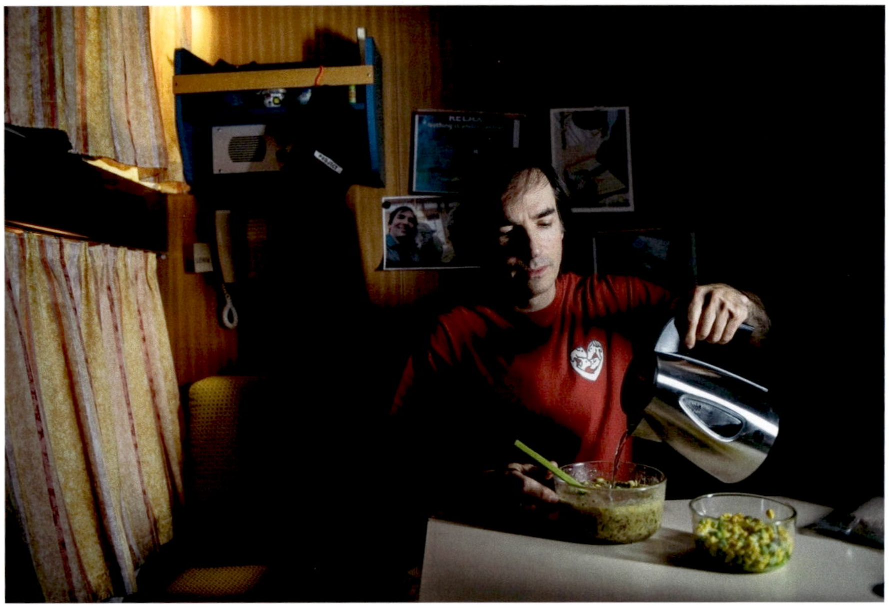

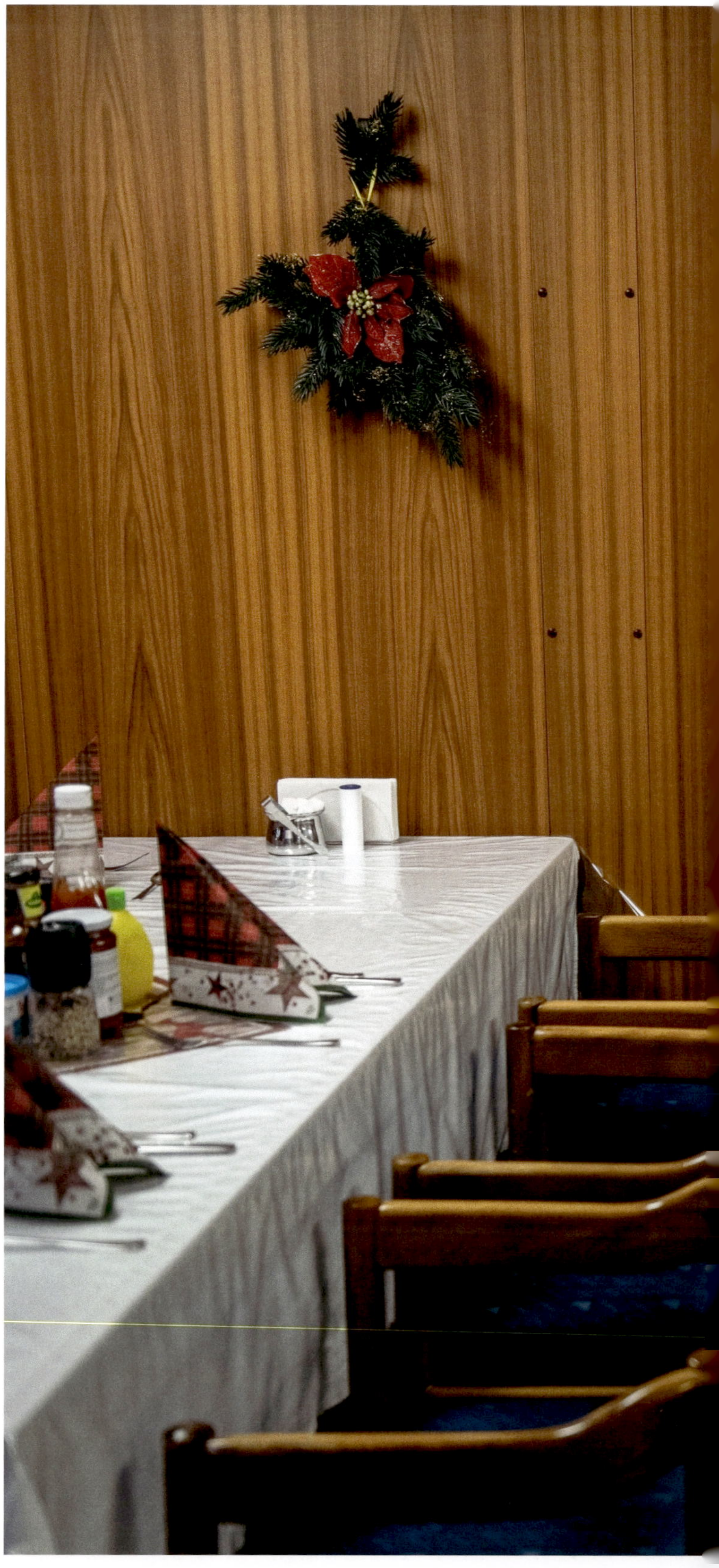

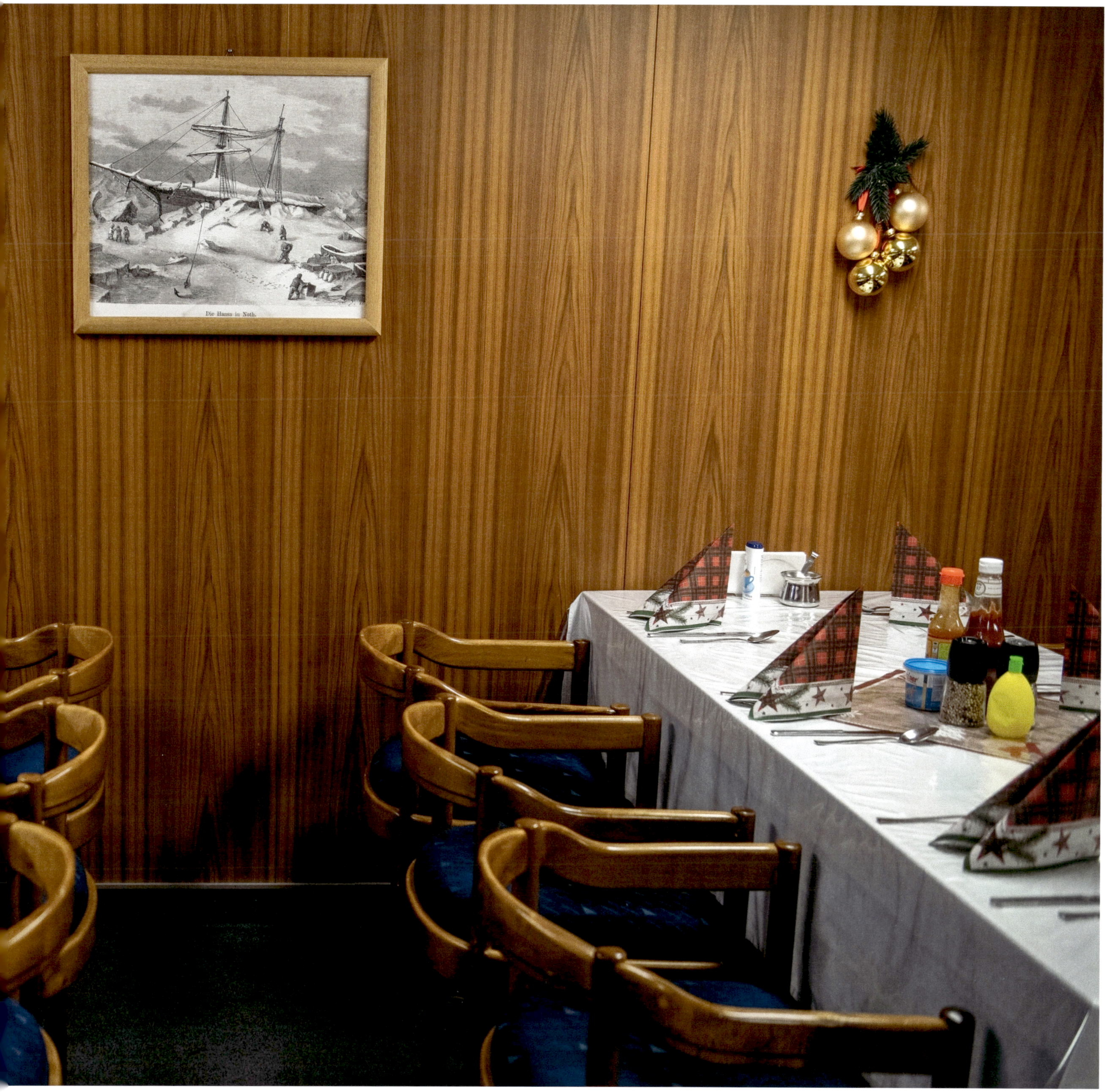
Die Hansa in Noth.

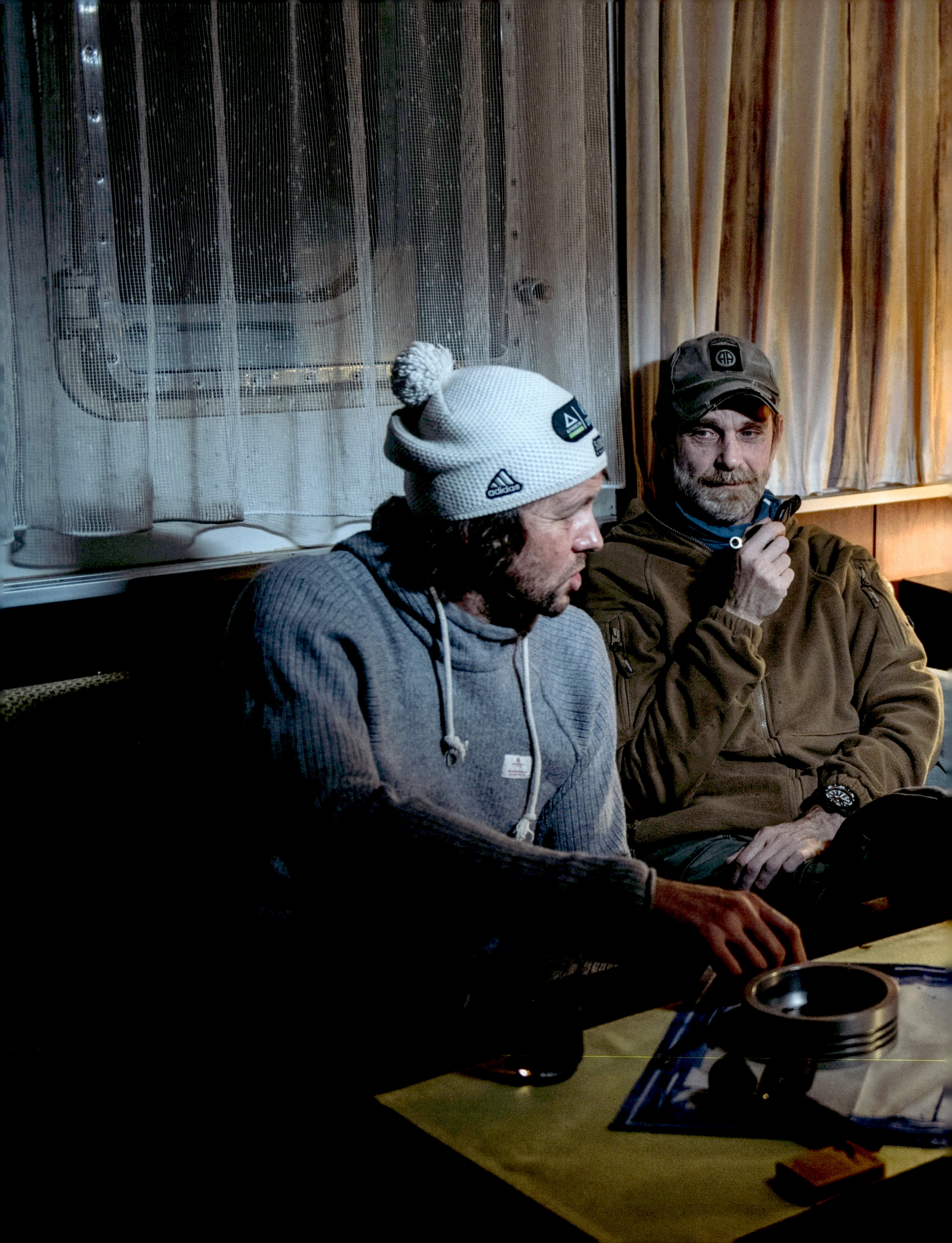
adidas

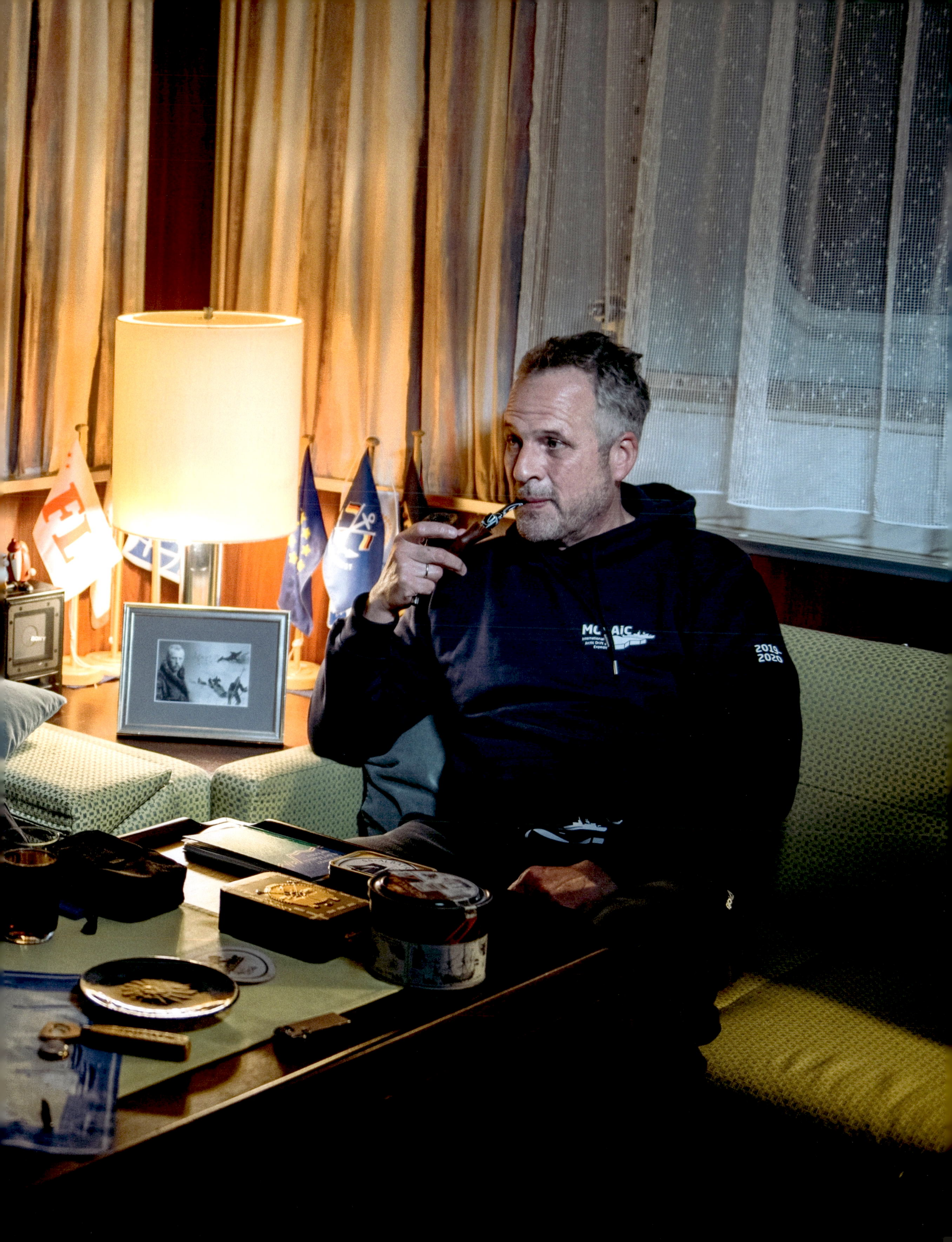
MOSAIC
2019-2020

The Ecosystem of the Arctic

Life below freezing

The uncrowned king of the Arctic and a frequent visitor to the research camp: the polar bear is the undisputed ruler of one of the world's most extreme regions. Perfectly adapted to its surroundings, beneath its white fur it has black skin that absorbs sunlight and keeps it warm in the cold. Although its scientific name, *Ursus maritimus*, marks it as a bear of the sea more than a bear of the ice, this Latin term fails to reveal how much the animal, although an excellent swimmer, is dependent on the ice. Thus the white bear has become a symbol for the life-threatening extent of climate change. Global warming is literally melting away the foundation of this hunter's livelihood from beneath its paws – but it is not only the bears. While the polar bear may be the best-known inhabitant of the Arctic, it is by no means alone out there. Life in the Arctic takes place around the sea ice that determines its rhythm with cycles of melting and freezing – on, beneath and right inside the ice. Because even the extremely saline brine channels that run through the sea ice at very small scales are home to tiny organisms that have adapted to this extreme environment. Using ice and water samples along with underwater cameras and acoustic devices, the research team sets to work during the expedition to explore the diversity and activity of Arctic life. There is a long list of questions: How do organisms survive the long, pitch-black winter when not a single ray of sun appears for months? Do some of them even use the light of the moon to carry out minor metabolic processes? What exactly happens to the ecosystem when light returns in spring, and how does life in the Arctic interact with the annual cycle of thawing and freezing of the sea ice? And what are the effects of dramatic environmental change as the sea ice disappears and temperatures rise in the Arctic Ocean?

From large to tiny, the range of polar life forms a delicate food web, a dense network of dependencies between the inhabitants of the Arctic Ocean. Seals, themselves the preferred food of polar bears, go hunting for fish even during the polar night. Arctic foxes and seabirds cross the icy desert. Narwhals hunt for polar cod, which, like many fish in polar regions, have a kind of natural antifreeze in their bodies that prevents their blood from congealing in the icy ocean. Through this adaptation, polar cod are the best example of a life form that survives not in spite of but in coexistence with the ice. And they are smart: these extreme fish not only use the small holes and cavities on the underside of the ice as hiding places in their young years, but probably also enter the sea ice off the Siberian coast to use it a vehicle, letting the transpolar drift carry them across the Arctic Ocean as far as the Fram Strait.

The polar cod shares its favourite source of food with marine mammals such as the bowhead whale, which filters the water for zooplankton: a world of the most minuscule animals in bizarre shapes, including crustaceans such as copepods, or pteropods such as the beautifully winged sea slug *Clione limacina*, better known as sea angels. The zooplankton in turn feeds on the foundation of the Arctic food web: microalgae. The sun inaugurates this buffet. When the first light shines through the snow and ice cover after the long polar night, a spark of life is ignited in the sea ice and ocean, and algae begin to grow amidst this frosty world. Even the most delicate glow wakens ice algae – such as microscopic diatoms – in the brine channels within and on the underside of the ice. They are one of the main sources of nutrition for life in the Arctic. When the sea ice finally melts and the sun rises higher, an increasing amount of light penetrates more deeply into the water column of the ocean, where likewise awaken phytoplankton, single-cell algae that float in the water and are the very foundation of the food web in the seas. As sunlight enters, the microalgae absorbs matter for photosynthesis, above all carbon dioxide, processes it, and grows. And it consumes more than a little: thanks to the busy activity of the phytoplankton, the oceans are able to absorb huge quantities of carbon dioxide while releasing oxygen for us to breathe. These tiny algae play the major role in circulating our planet's carbon and sinking it into the depths of the sea. In spring and summer the magnificence of the Arctic algae becomes colourfully visible, when melt ponds on the ice turn bright turquoise. As soon as copepods and other minute animals in the marine plankton graze on the blooming algae, Arctic fish fall on these tiny creatures – and, across all species, a huge feast takes place in the light of polar day.

Zooplankton gathered from the diving robot, or ROV, the Beast, waiting for evaluation in the ROV Oasis.

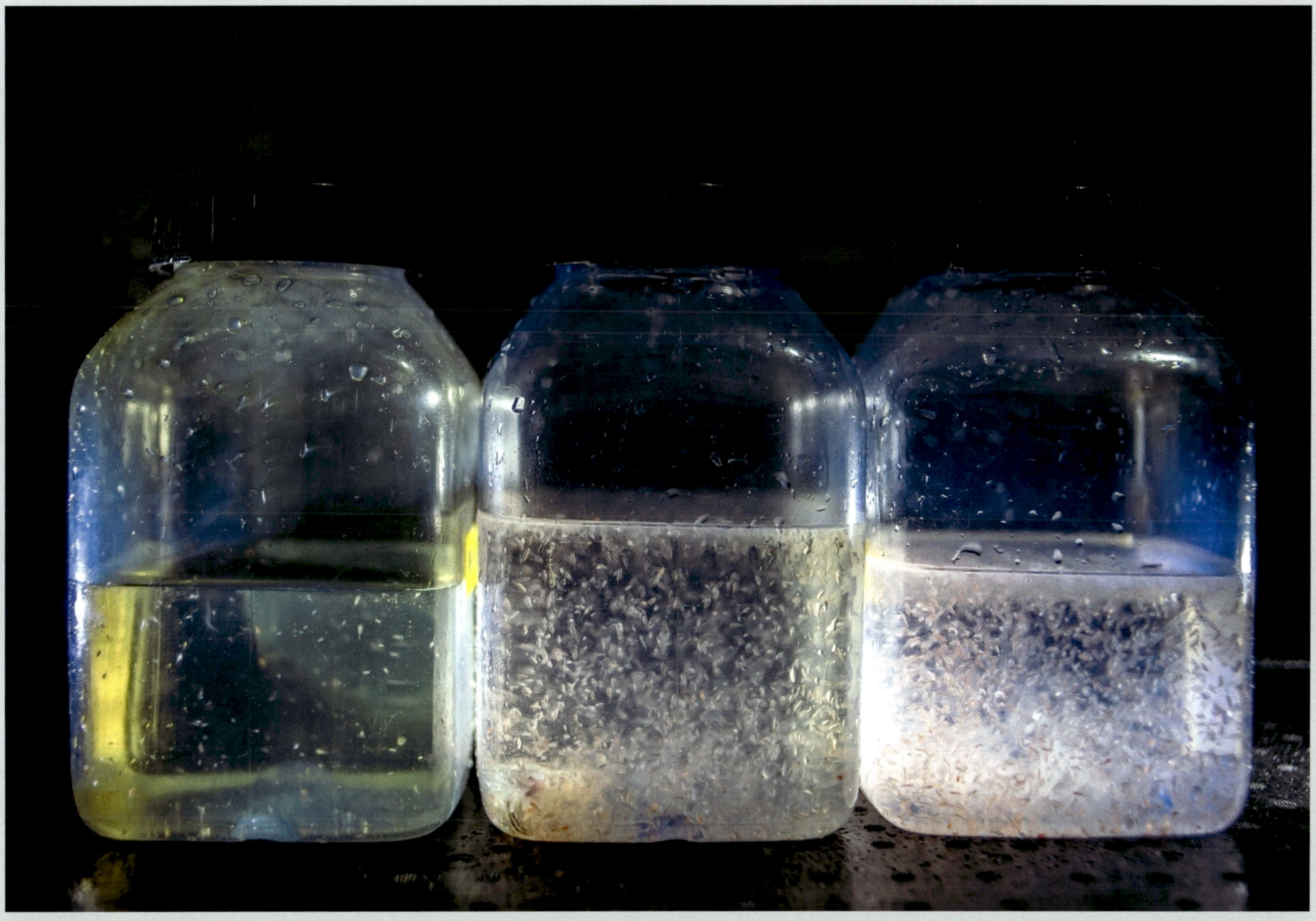

In view of the disappearing sea ice, the long-term outlook for ice algae and all of the creatures that depend on it is bleak. Phytoplankton may be a different case. Extreme algae bloom under the ice may have become more frequent in recent years due to the thinning of sea ice and its earlier perforation by melt ponds. And this is happening while its concentration in most other seas around the world has declined by an average of 40 per cent since 1950 – an alarming development if one considers that more than half of the oxygen in the atmosphere derives not from plants on land but from marine phytoplankton. One of every two breaths we take is supplied by oxygen from the sea. But what does this literally blooming algae life mean for the fragile ecosystem and climate system of the Arctic? Does the zooplankton benefit from this richly laid buffet in the water column, or does this consequence of climate change disturb the interaction of the ice algal bloom and animal life? The drift of the *Polarstern* enables the Ecosystem research team during the MOSAiC expedition to explore precisely how global warming affects the delicate food web and the inhabitants of the Arctic over a complete annual cycle.

The Research of Team Biogeochemistry

Hunting for traces

Another research team is also interested in the algae and microorganisms of the Arctic – or rather in what they produce. By emitting gases, some of these microorganisms have a profound influence on the climate system. They can even be co-responsible for the formation of clouds. The Biogeochemistry team investigates such interactions and is thus an important link between the other research fields of the expedition. There is a constant exchange between the ocean and the atmosphere: gases relevant to climate such as carbon dioxide and other so-called trace gases, like environmentally harmful halogens, move from the air to the water and vice versa. By causing the sea ice to diminish, global warming also brings about increasingly frequent and extensive contact between the sea and the air. It stands to reason that this in turn alters the exchange of gases between the ocean and the atmosphere. These processes must be observed and understood in order to incorporate them into climate models.

The phytoplankton in the oceans enriches the atmosphere with oxygen. But algae affect marine gas production in other ways as well. As a metabolite they release the chemical compound dimethylsulfoniopropionate (DMSP). This molecule is then further broken down by marine microbes into dimethyl sulphide (DMS), a sulphur compound responsible for the typical smell of the sea. What happens next is still a subject of scientific discussion. It is possible that this gas remains trapped in the Arctic Ocean when sea ice covers the water. If the ice cover is absent, however, it rises into the atmosphere. Here, dimethyl sulphide can have an effect like an aerosol, that is, like a particle floating in the air: water vapour condenses on its surface and a cloud is born. If more DMS-producing algae grow as a result of climate change, they could create more clouds, which could in turn influence the Arctic climate system. This leads to many questions, including: Do algae shape the climate?

The researchers of the Biogeochemistry team are looking for answers to questions such as this. They take water and ice samples and analyse their gas constituents in the laboratories on the *Polarstern*. By doing so, the research team is also tracking another, particularly influential greenhouse gas: methane. This gas is about twenty-five to thirty times more harmful to the climate than carbon dioxide. It could lead to severe changes in the climate system and is thus regarded as an compound that could tip the global climate balance – but its effects have not yet been adequately considered in previous climate models.

With regard to methane, researchers in the Arctic face a complex situation. Its biggest sources are the wetlands and permafrost in high northern latitudes, for example in Siberia and Canada. As a consequence of global warming, the thawing in these areas is reaching great depths – and thus animal and plant remains in the ground are also thawing and being decomposed by microorganisms. The result of this decomposition is the release of huge amounts of greenhouse gases, including methane. Some of the methane goes directly into the atmosphere. Other quantities of this greenhouse gas, on the other hand, are transported into the Arctic Ocean by rivers, such as the great rivers of Siberia. Dissolved in seawater, the methane is then carried further – or enclosed in the ice when the seawater freezes. The researchers of the MOSAiC expedition are following the traces of this methane by also being carried by sea-ice drift from the Siberian coast to the central Arctic. Thanks to the data that was recorded on their months-long journey, they can better understand how much of this climate-damaging gas remains in and is consumed by organisms living in the ocean or sea ice, and how much eventually rises into the atmosphere – data that Team Atmosphere collects. While the *Polarstern*, Arctic creatures and climate gases like methane drift together with the sea ice through the polar year, the different research teams on board thus have a unique opportunity to jointly examine and discuss these diverse and interconnected processes, in order to gain a better understanding of the complexity of the Arctic and the global climate system.

Top: Using ice drills the research teams remove ice cores needed for a great number of measurements. For this purpose, a group often sets off in snowmobiles for the so-called Dark Sector. Here, instead of glaring white artificial light, only red light may be used so that Arctic organisms are not affected during the darkness of the polar night. Back on the *Polarstern*, some of the drilled ice cores are kept in a frozen state for physical observation. Others, which serve to investigate specific organisms such as algae or methane and other gases, are cut up and thawed.

Below: In the laboratory the ice cores are then checked for temperature (left) or divided up after thawing for various analyses (right), for example to investigate the nutrients that they contain or the various kinds of organisms in the ice. These investigations in the lab are also carried out in red light to avoid artificial stimulation of the organisms.

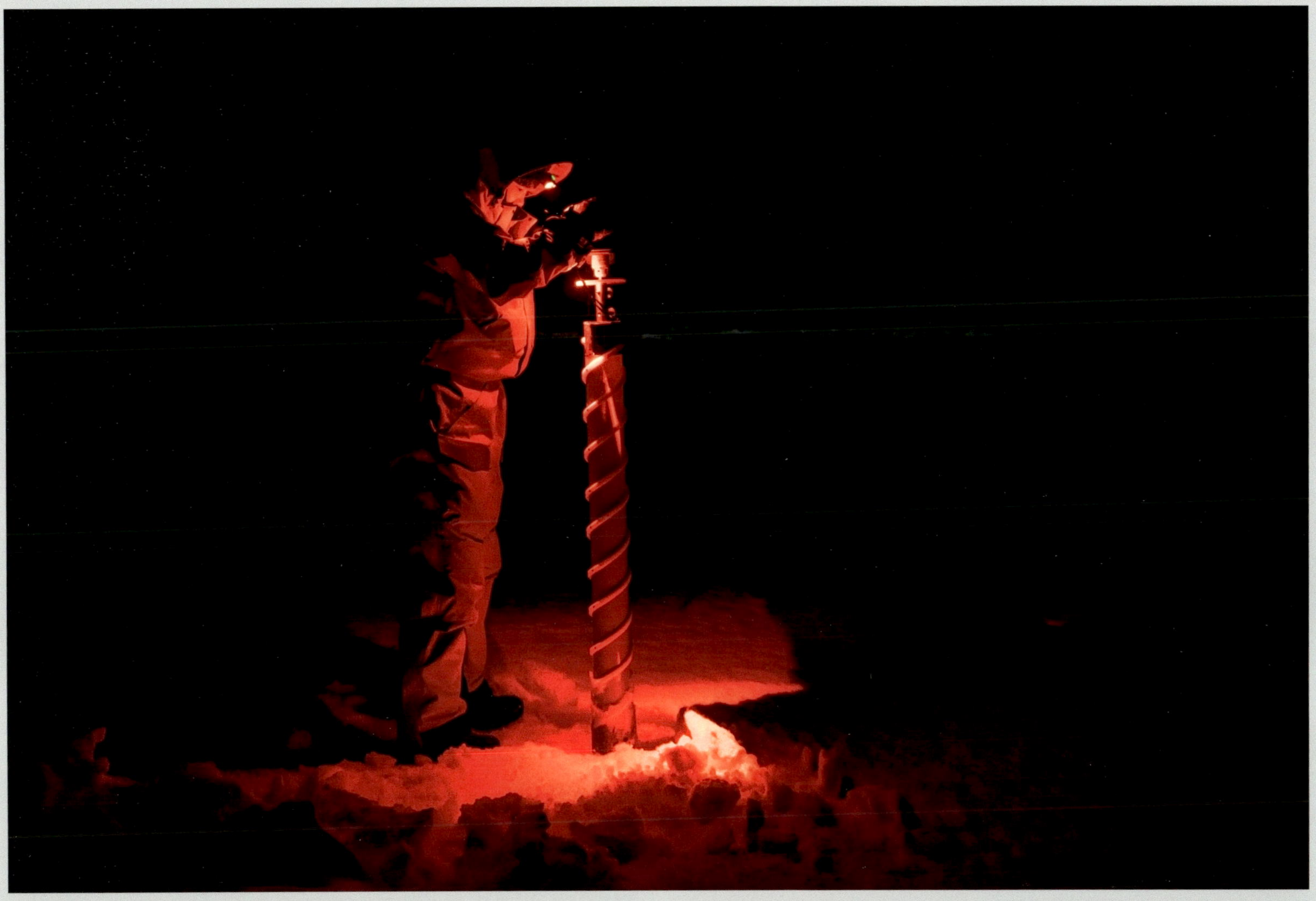

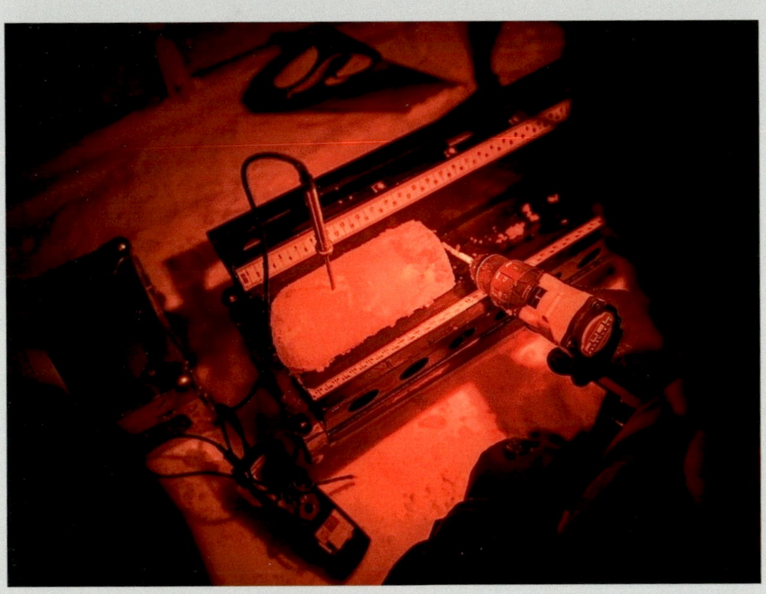

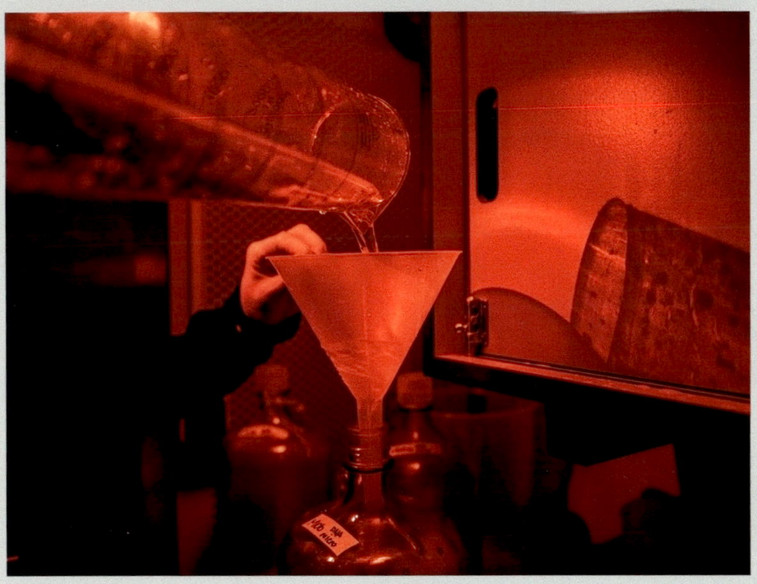

Night Shift

Research in twenty-four hours of darkness

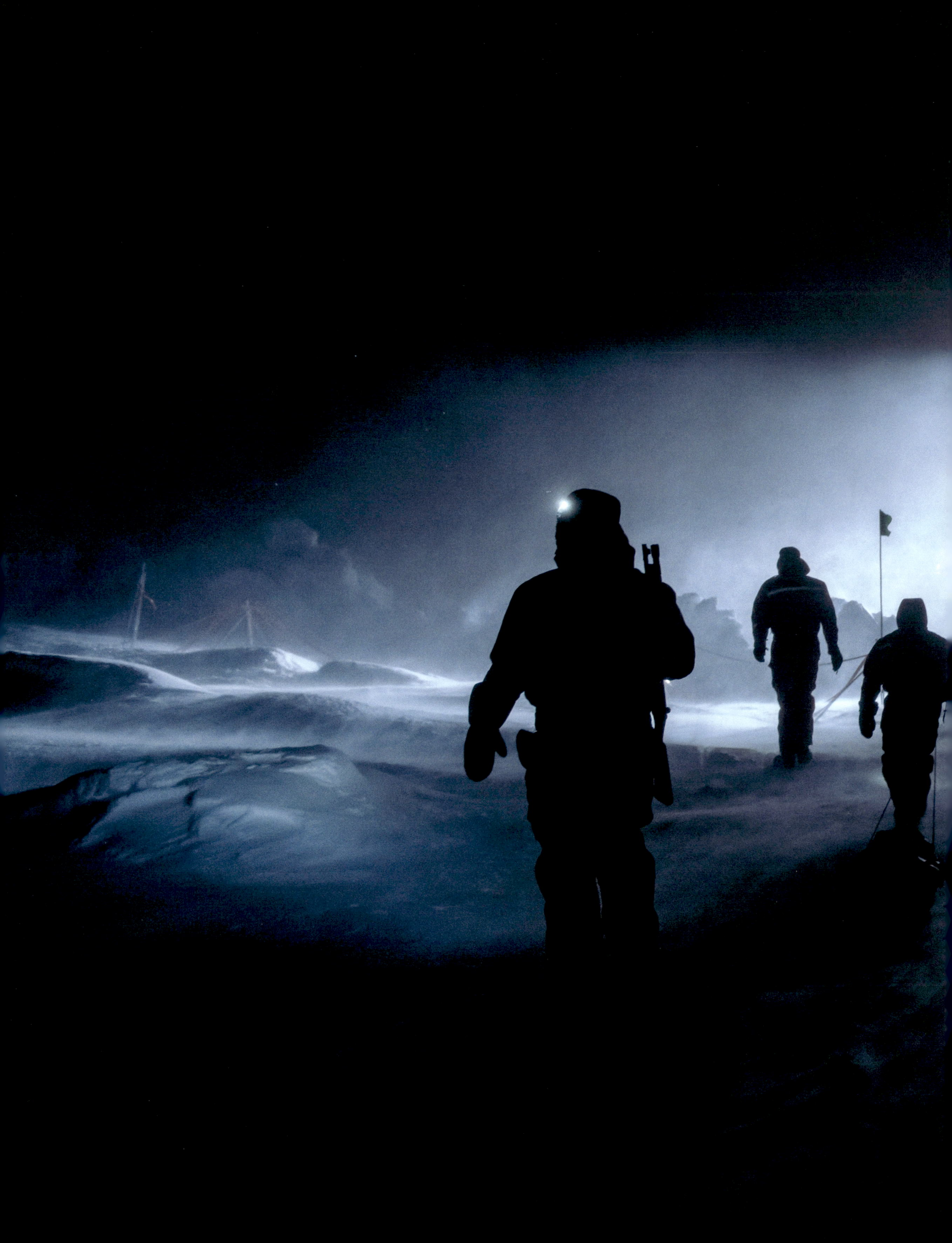

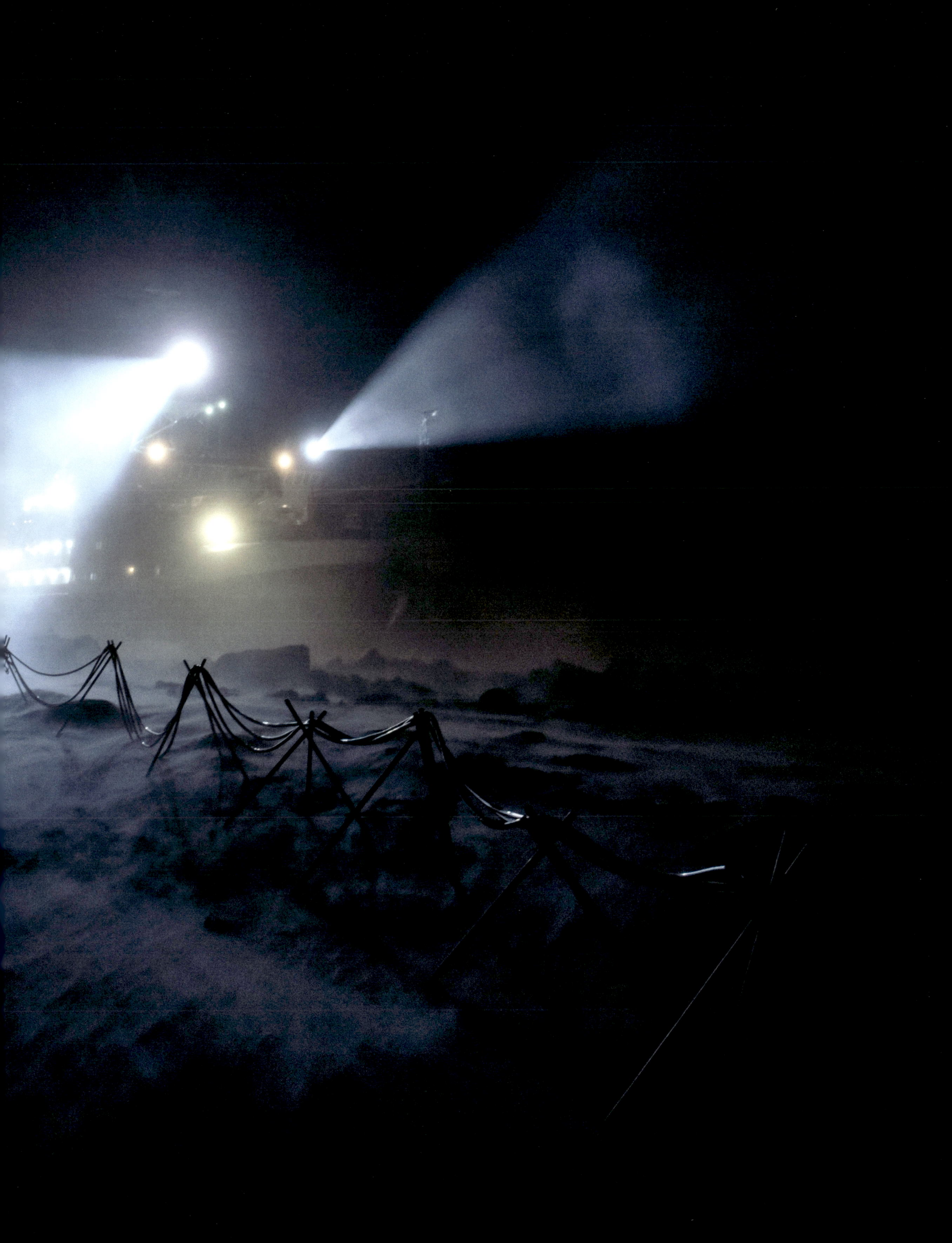

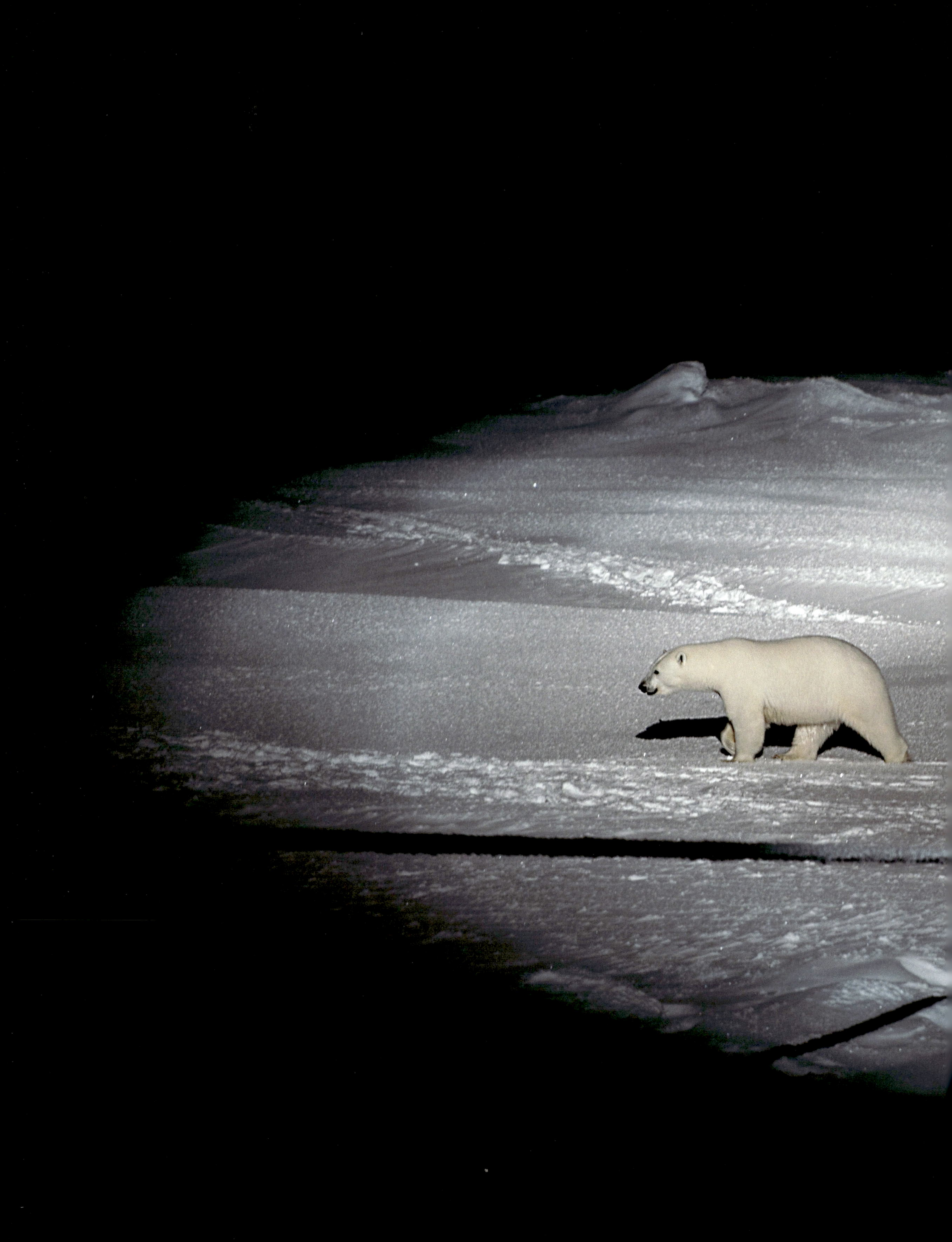

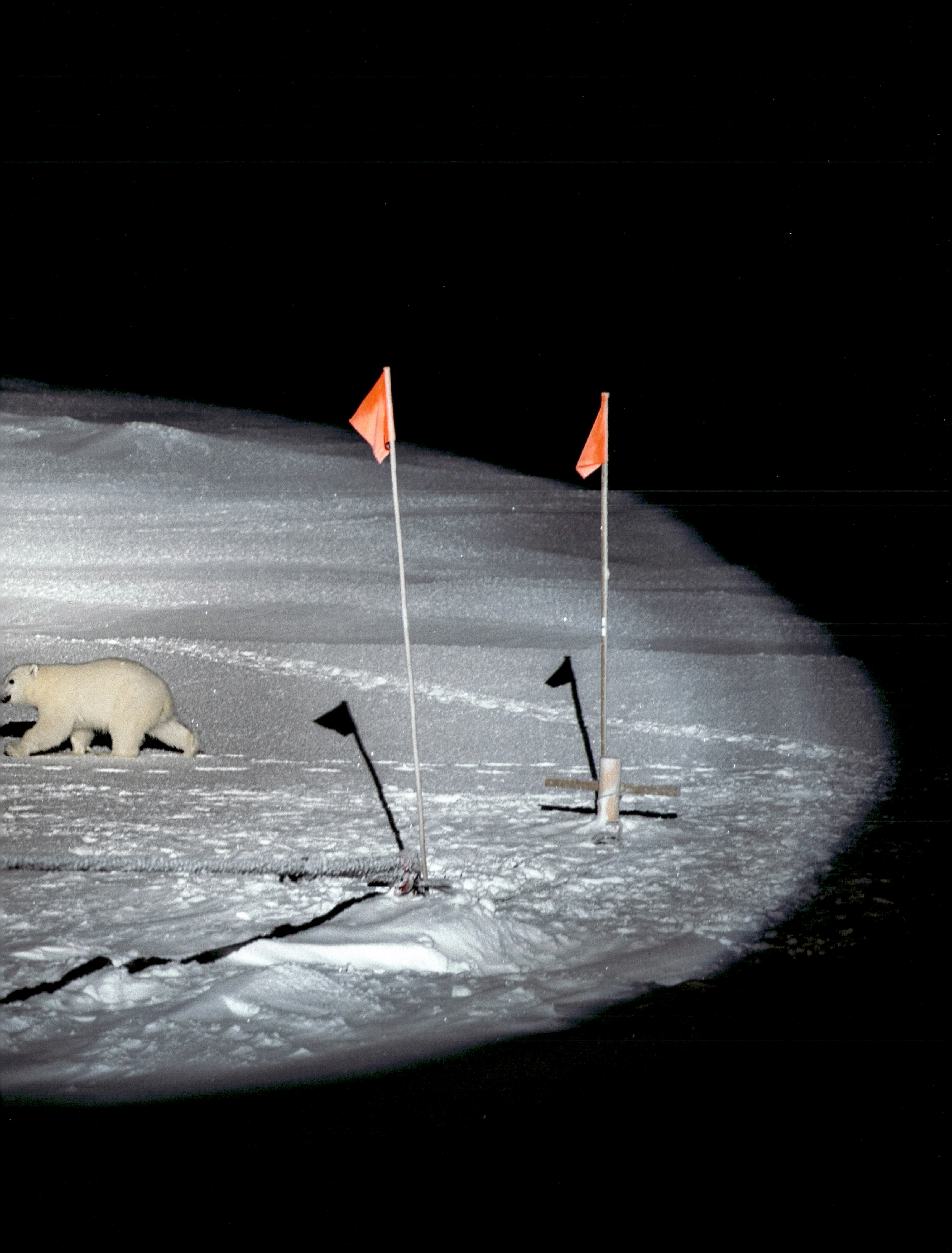

The research ship and expedition members look like a space vehicle and astronauts in the deep blackness of the polar night. While people wrap themselves in survival suits and thermal gear, the *Polarstern* is shrouded in the light of its three powerful searchlights. In their beams, the ice takes on its most bizarre shapes – and sometimes the true rulers of this hostile world, polar bears, make an appearance. Whether the ship's searchlight or a headlamp, only a limited cone of light ever illuminates the ice.

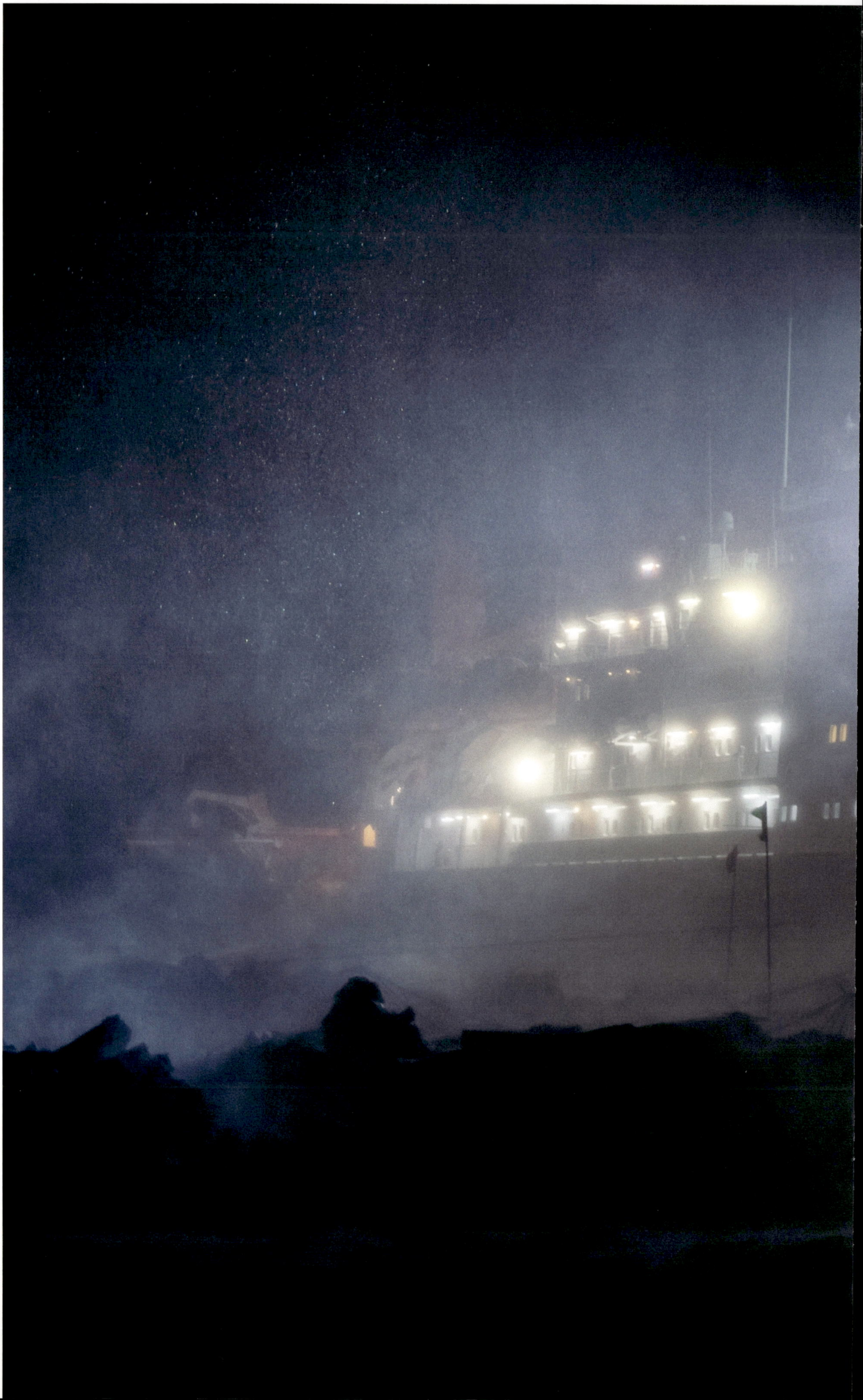

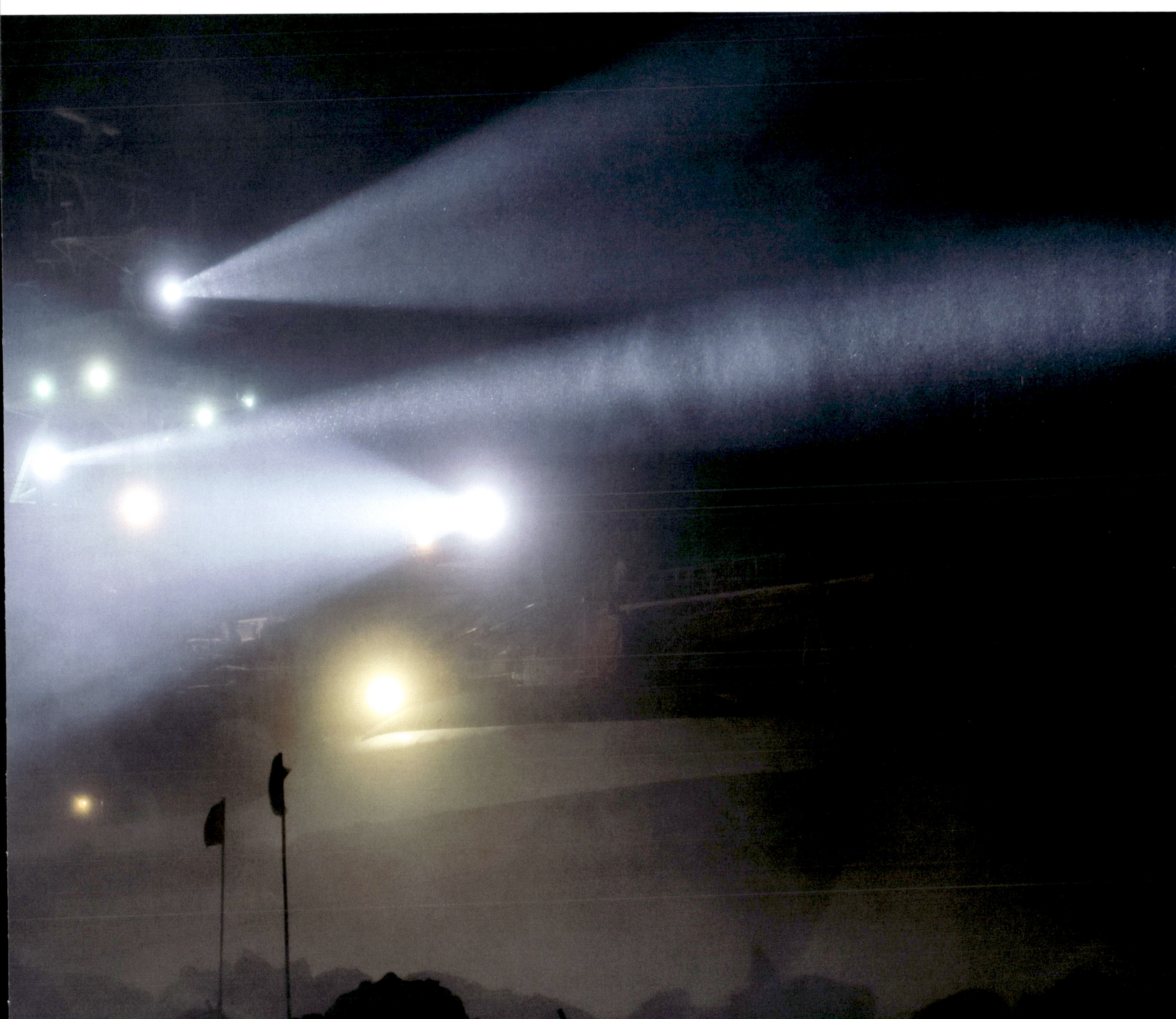

When a figure muffled against the cold to the point of unrecognizability enters the field of light, those on the ice often have to ask “Who are you?” to identify the person opposite them.

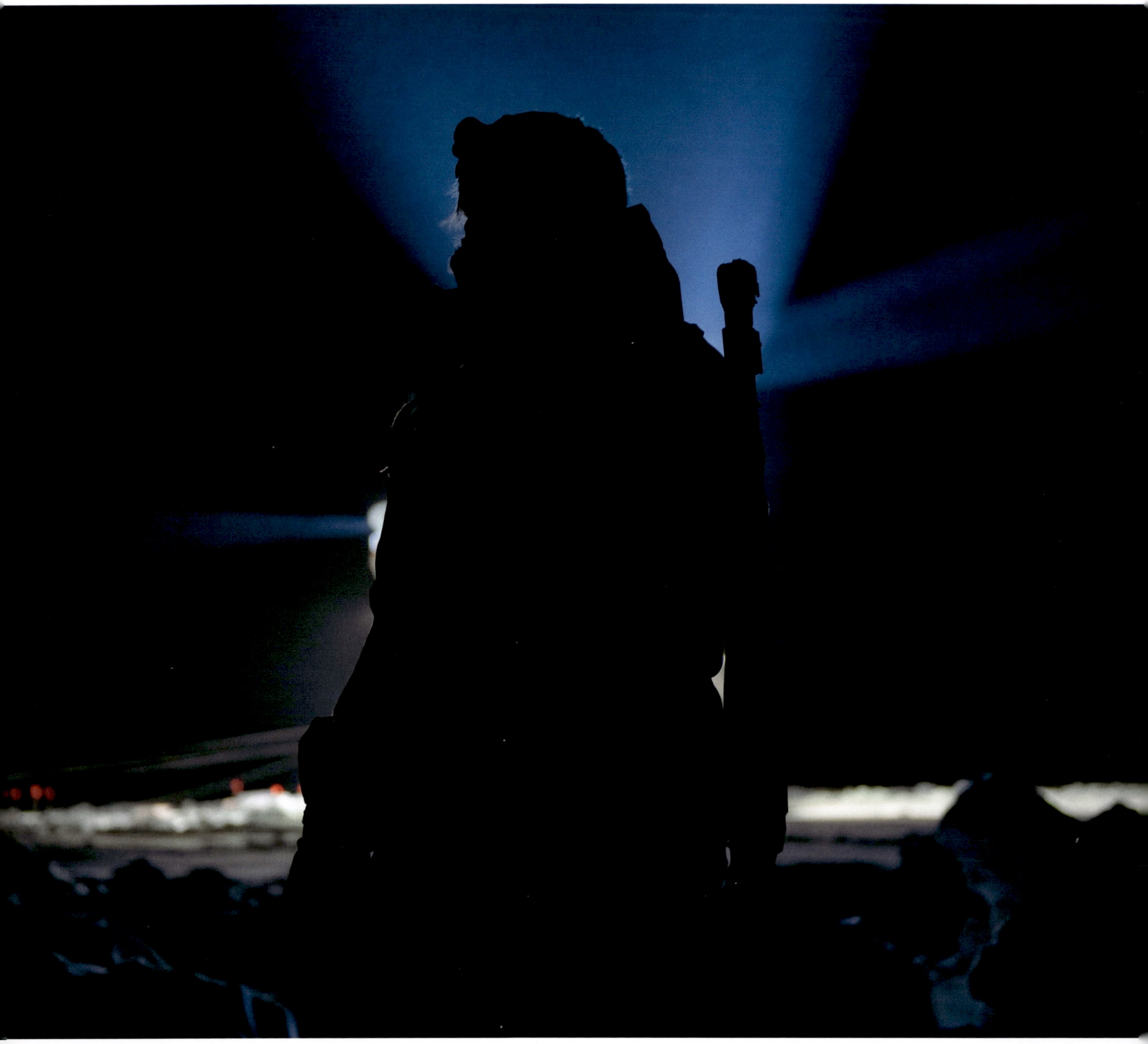

N

Night. Nacht. nat. nuit. notte. 夜. noche. ночь. yö. natt. nacht. 夜. noc. 밤. natt.

The people involved in the expedition come from all over the world, so among them there are many names for the permanent darkness that settles on the Ice Camp in mid-November. With a mixture of respect and scientific curiosity, they welcome the polar night in their own languages. They are aware that during the Arctic winter they will encounter an aspect of our planet that to date has been little investigated.

For six long months the polar night holds the North Pole in its icy grip. Once the sun has disappeared below the horizon, twilight likewise retreats further every day and finally takes its leave with dramatic shades of red in the sky. The phases of twilight start with what is known as civil twilight: the sun hides at up to six degrees below the horizon. It is still possible to read on deck without switching on a light. In the nautical twilight that follows, the sun sinks to between six and twelve degrees, but the horizon is still discernible. Finally, astronomical twilight sets in. The sun is then twelve to eighteen degrees below the horizon and no longer competes with the bright stars in the sky. Finally, by mid-November, even twilight no longer ventures above the horizon, and does not return to the North Pole until the end of January. In contrast to other places on Earth, where the nights are never in absolute darkness, if the sky is cloudy polar night immerses the Arctic in uncompromising blackness. At these times the perceptible world of the MOSAiC expedition is reduced to a microcosm, created by the powerful searchlights of the *Polarstern* which bathe the ship and the ice floe in its shining cone of light.

From time to time, however, another celestial body brings light to the Arctic: the moon. Then the sea ice reflects the moonlight in silver and grey, and at the same time the gravitational force of Earth's satellite makes the ice move. This is often announced by a sudden cracking and crashing, followed by banging and splitting, even screeching and groaning. Where seemingly solid planes of ice were at rest just a moment ago, within a few minutes cracks appear, then fields of ice break up and drift apart. In other places the force of the full moon pushes masses of ice together into huge, metres-high ridges of compressed ice. These movements of the ice thunder through the darkness for minutes on end, tugging and pushing at the hull of the *Polarstern* and making the ship hum. Then the Arctic relapses into icy silence again – and a world reshaped by the forces of nature is revealed to the researchers' eyes. After such nights and storms, it is often necessary to clear up the Ice Camp: to rescue power cables from ice ridges or relocate parts of the camp that have broken off and drifted away. Especially at the beginning and end of the polar night, parts of the floe break off and whole districts are shifted away. Then it is only possible to reach stations such as the ROV Oasis by kayak or helicopter in order to move equipment and protective tents to a new, more stable location.

Despite all the challenges, many expedition members fall in love with the sight of the Arctic night. In the blackness, artificial light sources shine with brilliant intensity. This world without light makes colours a precious rarity. The thermal overalls of people on the ice glow red when illuminated by their colleagues' headlamps – beams of light that allow a limited degree of vision and scope for activity on the ice. For the eyes, which have to be screened behind protective goggles at extreme temperatures so that they and the skin around them are not exposed to wind chill and the danger of frostbite, every period spent on the ice is a challenge – as it is for the rest of the body, which must be protected against the external cold but breaks into sweat during strenuous work under so many layers of clothing.

The darkness has a powerful effect, not only on the external world but also on the inner life of its human visitors. On winter days many of them sense the extent to which the human organism depends on light. Biorhythms are confused, and tiredness and lack of energy are a common result. In contrast, some passengers on the *Polarstern* struggle with insomnia as soon as they lie down in their bunks. Fridtjof Nansen's journal, published in English as *Farthest North*, written during the long days and nights of the *Fram* expedition, may well be the work most read during the weeks when the sun is absent. With vitamin D supplements and intensive sports programmes in the gym or the *Polarstern*'s swimming pool, as well as with daylight lamps, the team members counteract the influence of the night. Above all, however, chocolate is what many expedition members take in well-judged doses against the polar night blues. In the monotony of darkness, mealtimes provide much-needed orientation over the course of the day. When biorhythms fail, the galley team can be relied upon to proclaim noontime with exact regularity. The internal body clock no longer ticks on board according to the light, but it regularly chimes the hour for dining.

Winter temperatures also take the installed measuring devices to their limits. Wind chill causes even the best-prepared sensors to ice up, and several buoys in the many-branched autonomous network fall victim to the cold. Materials like plastic that are supposedly flexible simply break into two in the frosty air. A good deal of ingenuity and talent for improvisation are required to make equipment operative again. Ultimately, however, not even the adversities of the polar night can prevent the instruments from converting the previous enigmas and unknowns of the Arctic winter into processable data. A continuous stream of prospective insights flows into the researchers' databases. Even during the months of darkness it becomes clear that all of the efforts have paid off and that the expedition is abundantly harvesting the foundations for a fresh knowledge that will permit a better understanding of the Arctic and the global climate system.

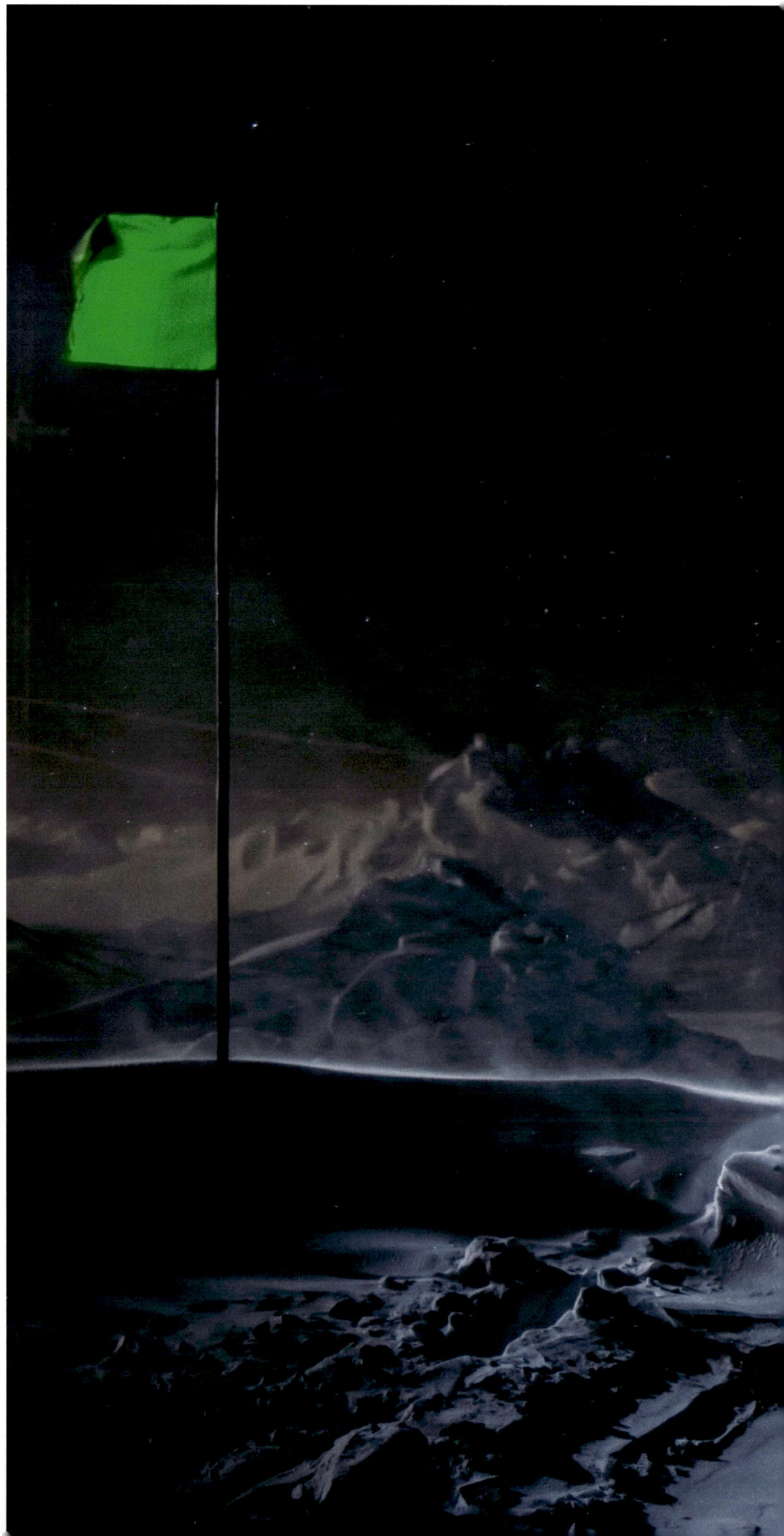

"During the polar night it became clear that we are only a very, very small dot in the expanse of the Arctic." Captain Stefan Schwarze

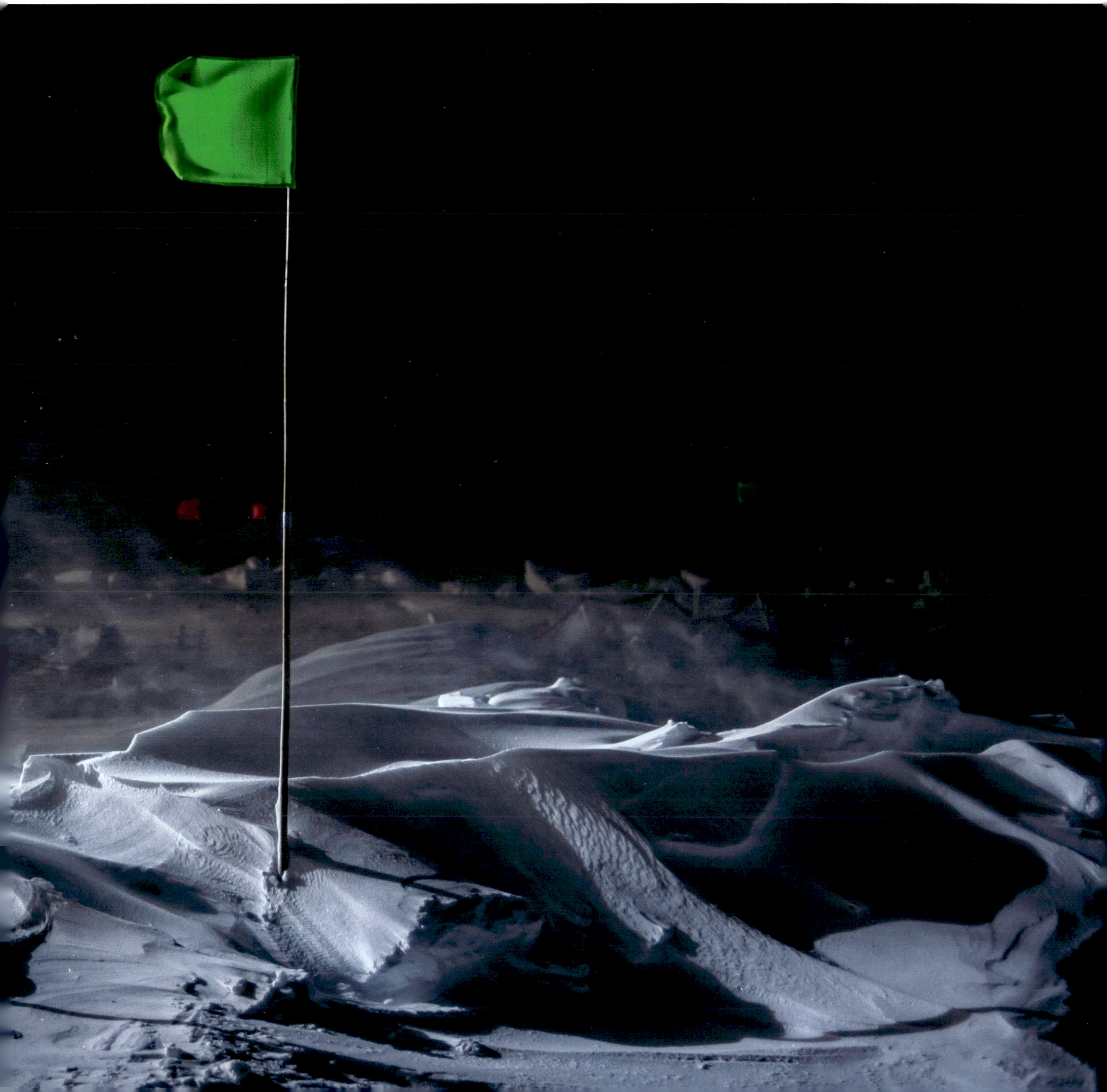

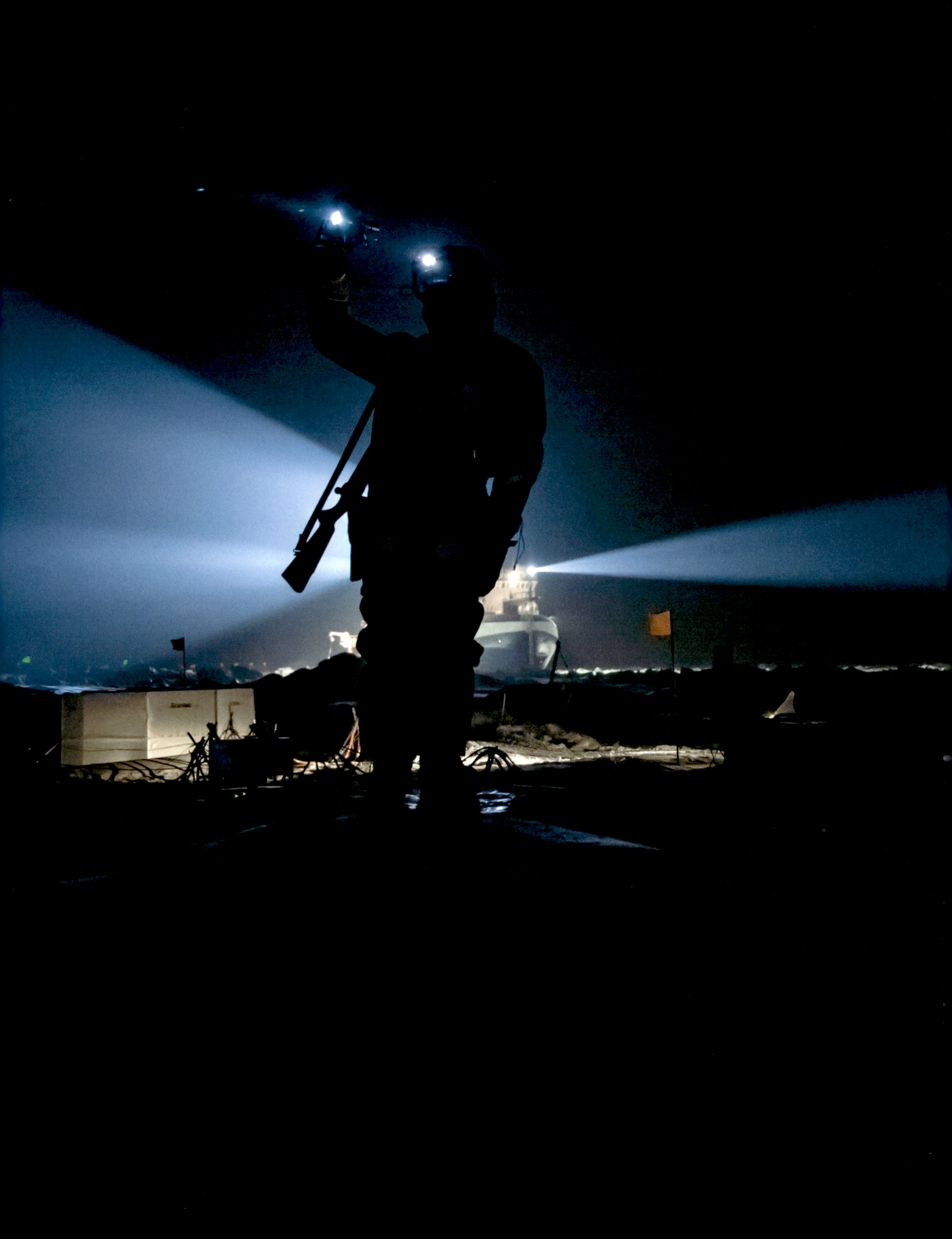

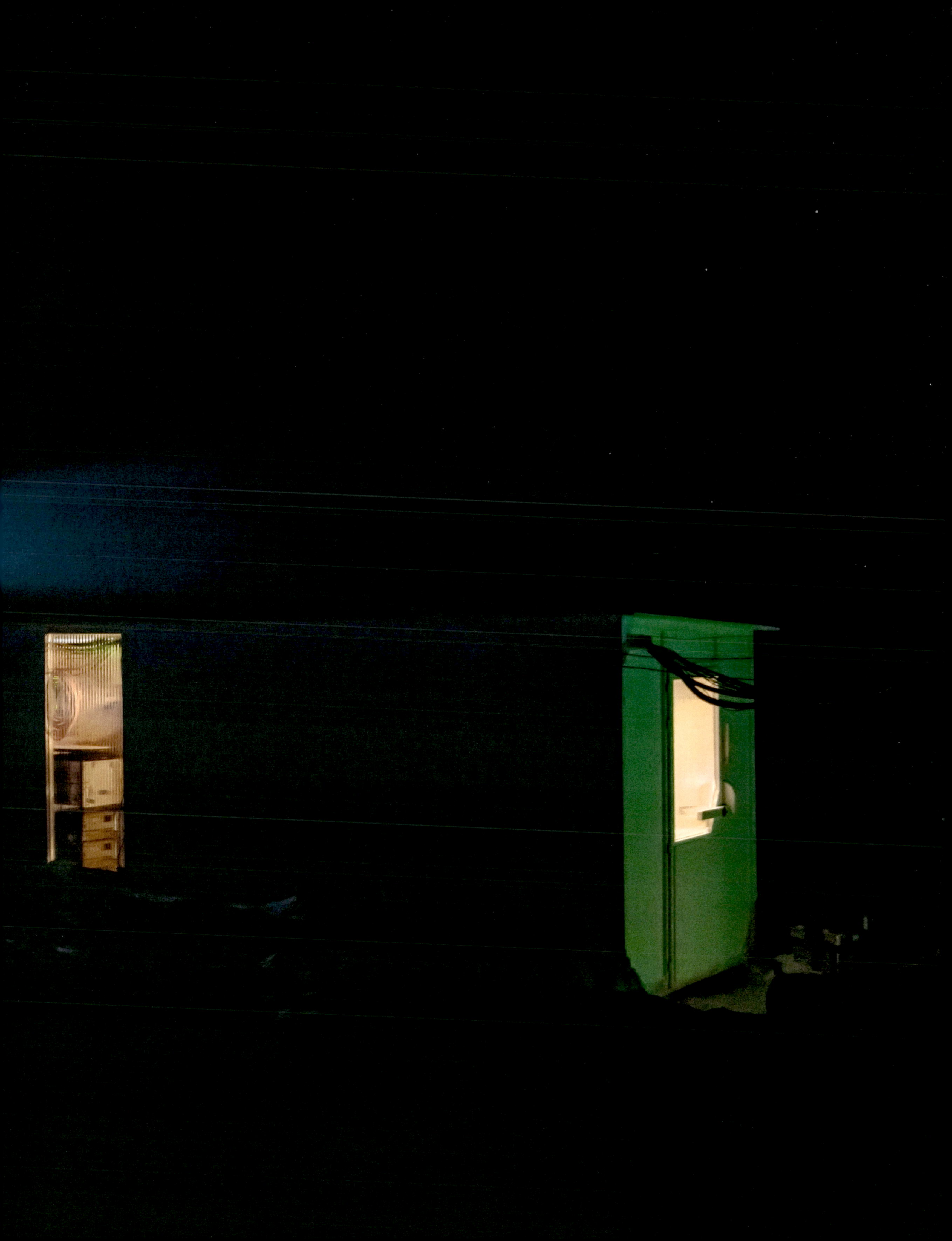

In mid-November, a severe storm for the first time took the expedition by surprise. This was a first highlight for the researchers, as they had never before had the opportunity to fully observe the effects of a storm on the Arctic climate system. The forces of nature make the ice move. Inventiveness is required for a safe crossing of the ice ridge that the storm has created: every step on freshly piled snow and ice holds the risk of stepping into the unknown. The wooden Nansen sledges prove to be reliable temporary bridges.

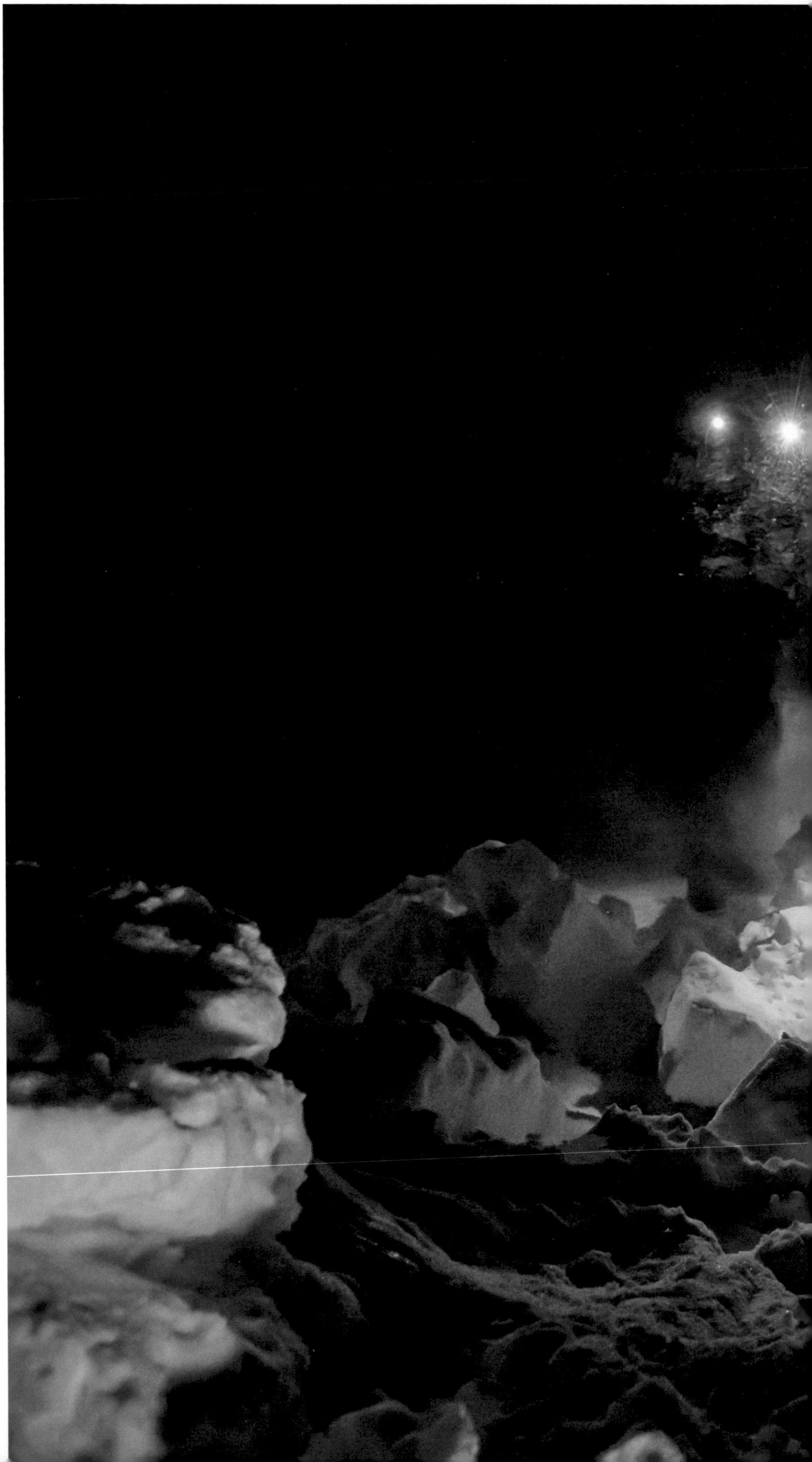

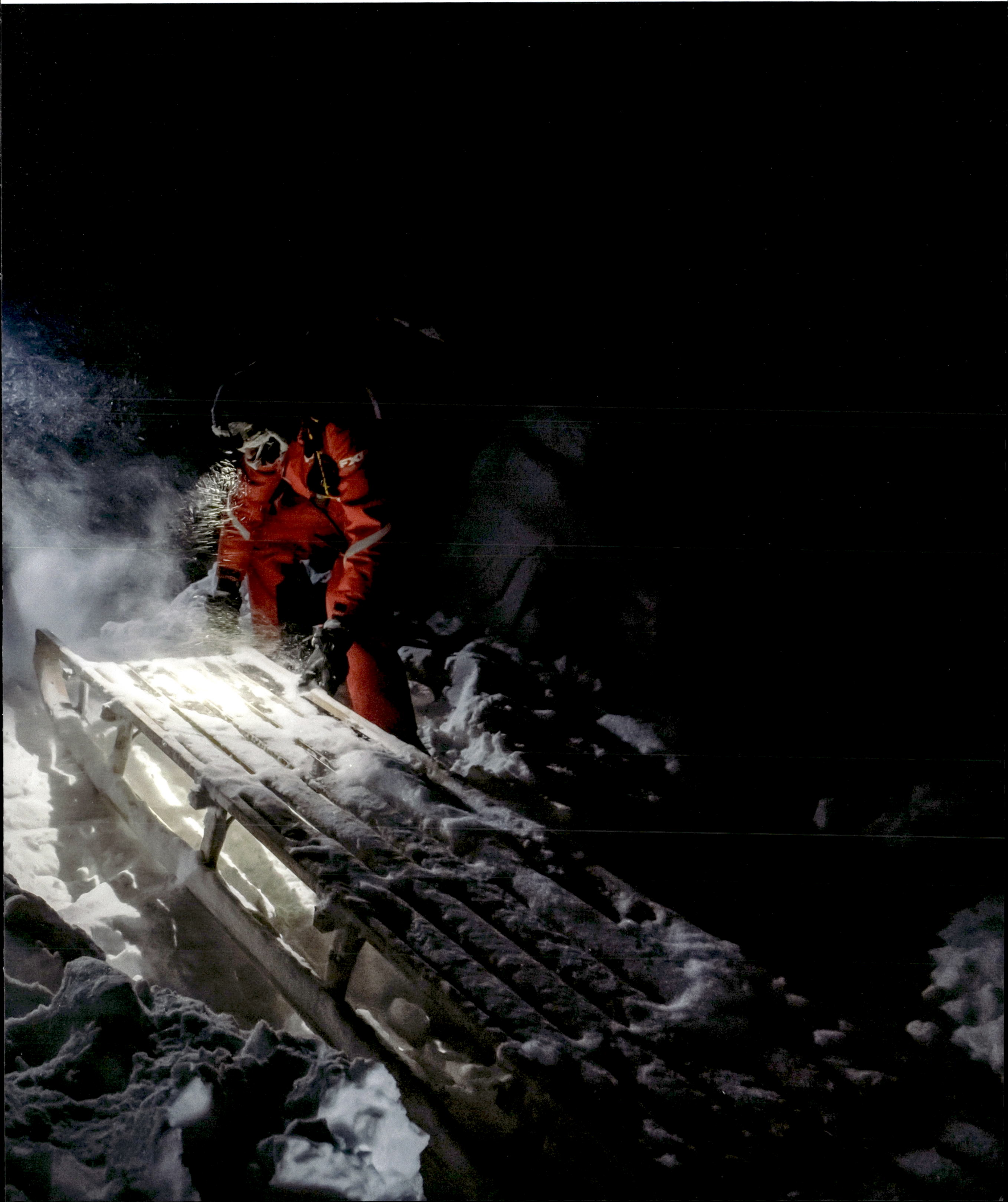

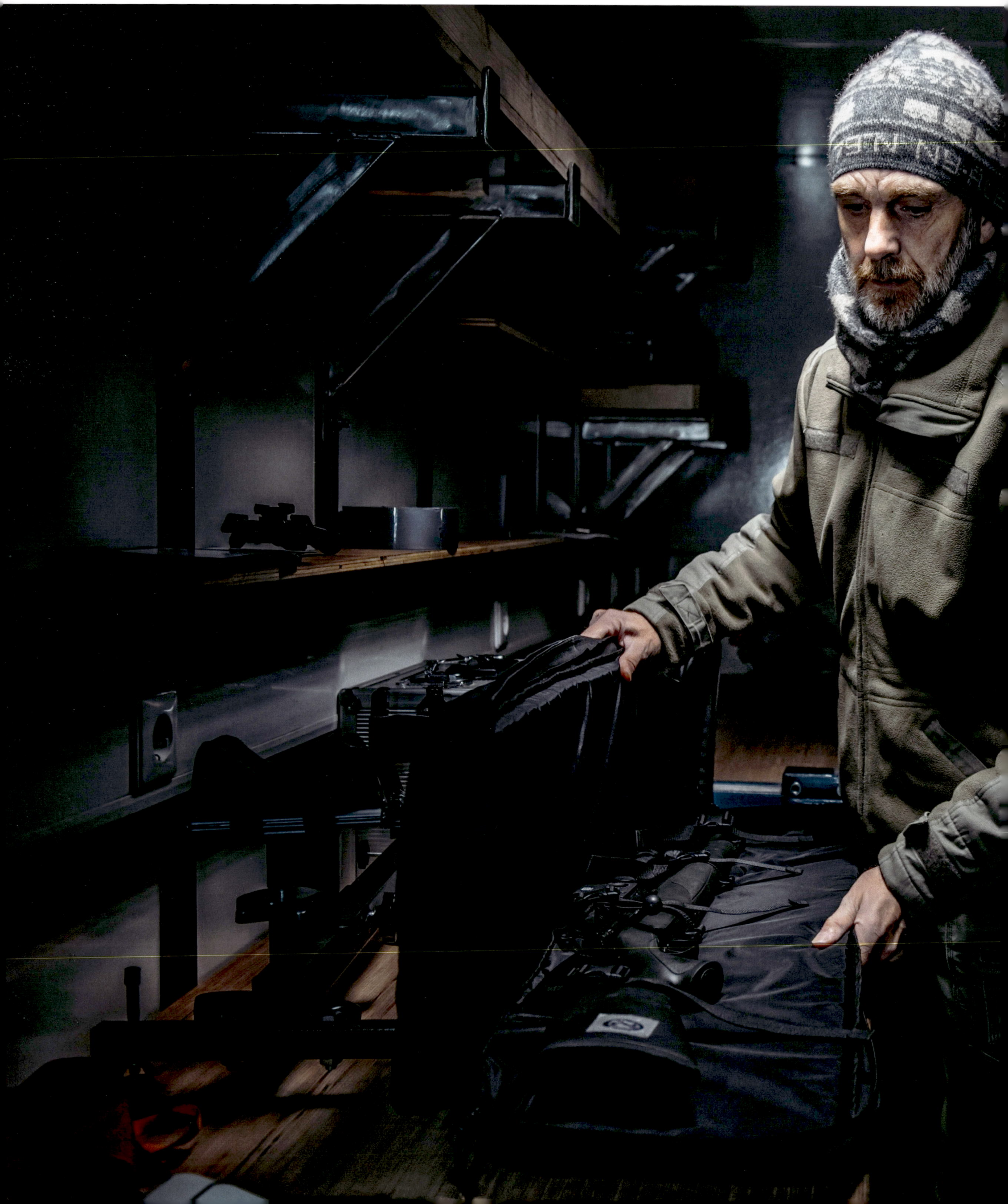

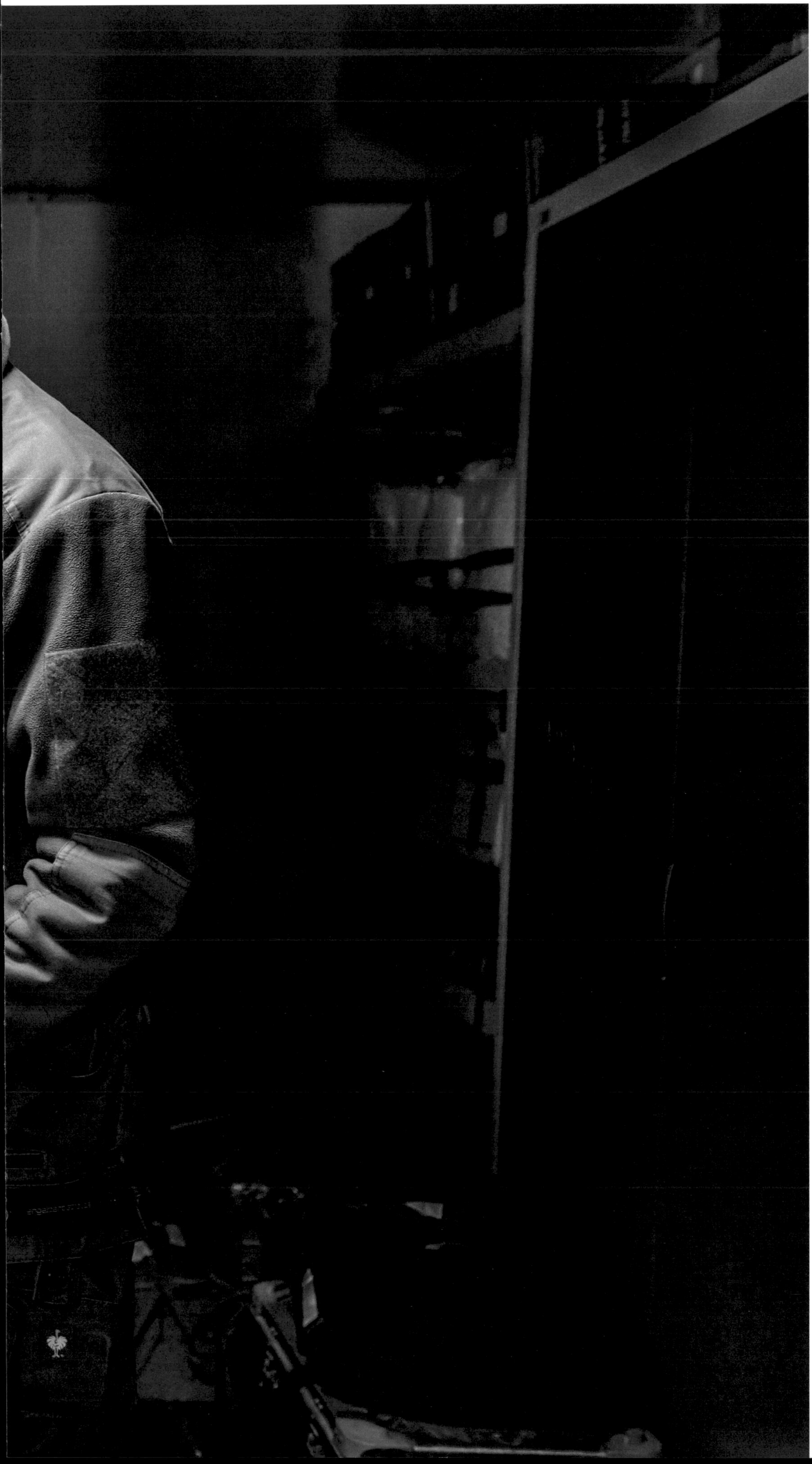

A strict safety regime reigns during the expedition. No one is allowed on the ice without the accompaniment of armed bear guards. In addition to flare guns, whose piercing and luminous projectiles can reliably drive the polar bears away, the guards also carry loaded firearms. Only when a human life is in danger may they be fired at an animal. To make sure that the firearms are fully functional, Olaf Stenzel checks the rifles in the well-secured armoury of the *Polarstern*. The decades-long experience of this police officer from Bremerhaven is invaluable in the hostile conditions of the Arctic.

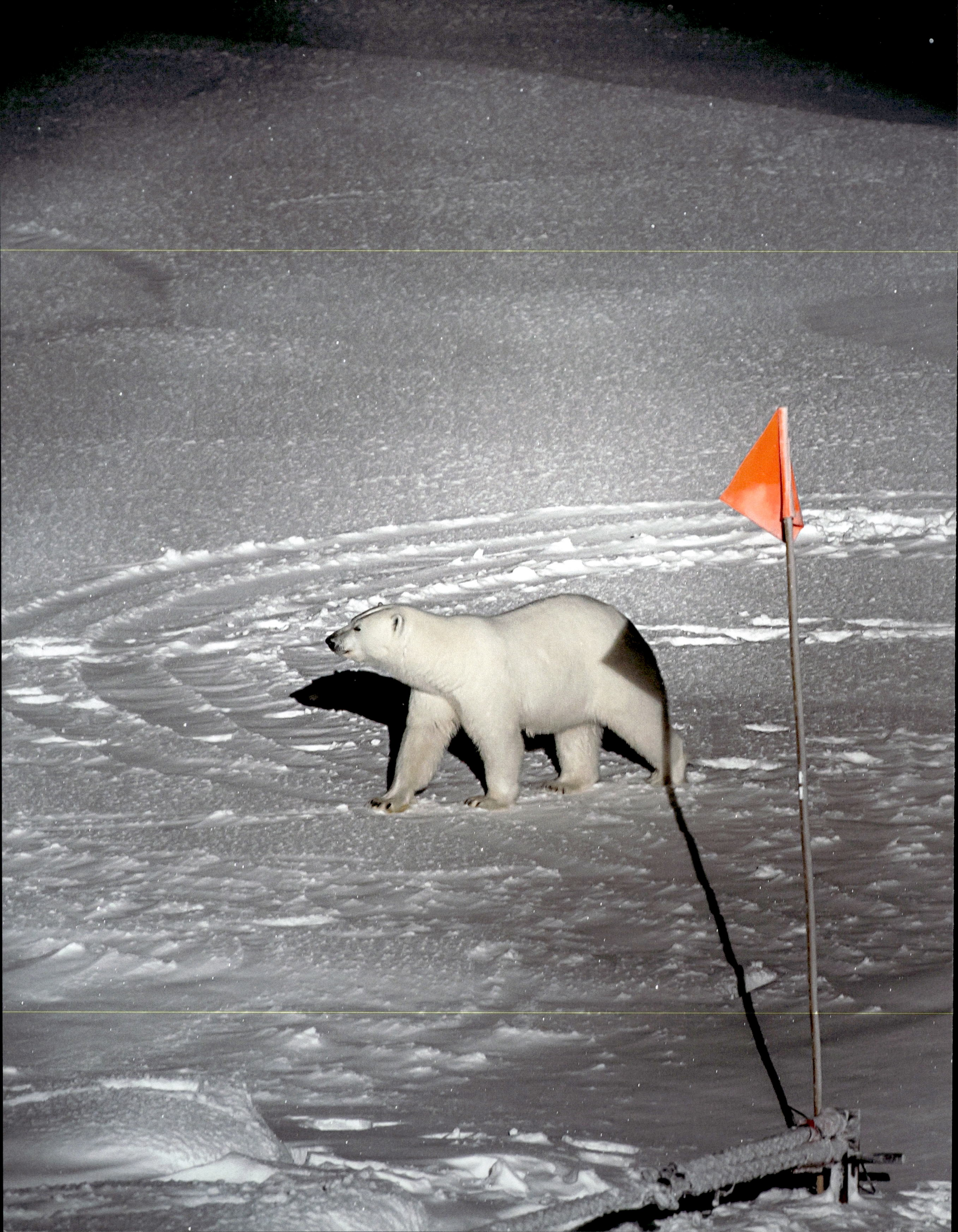

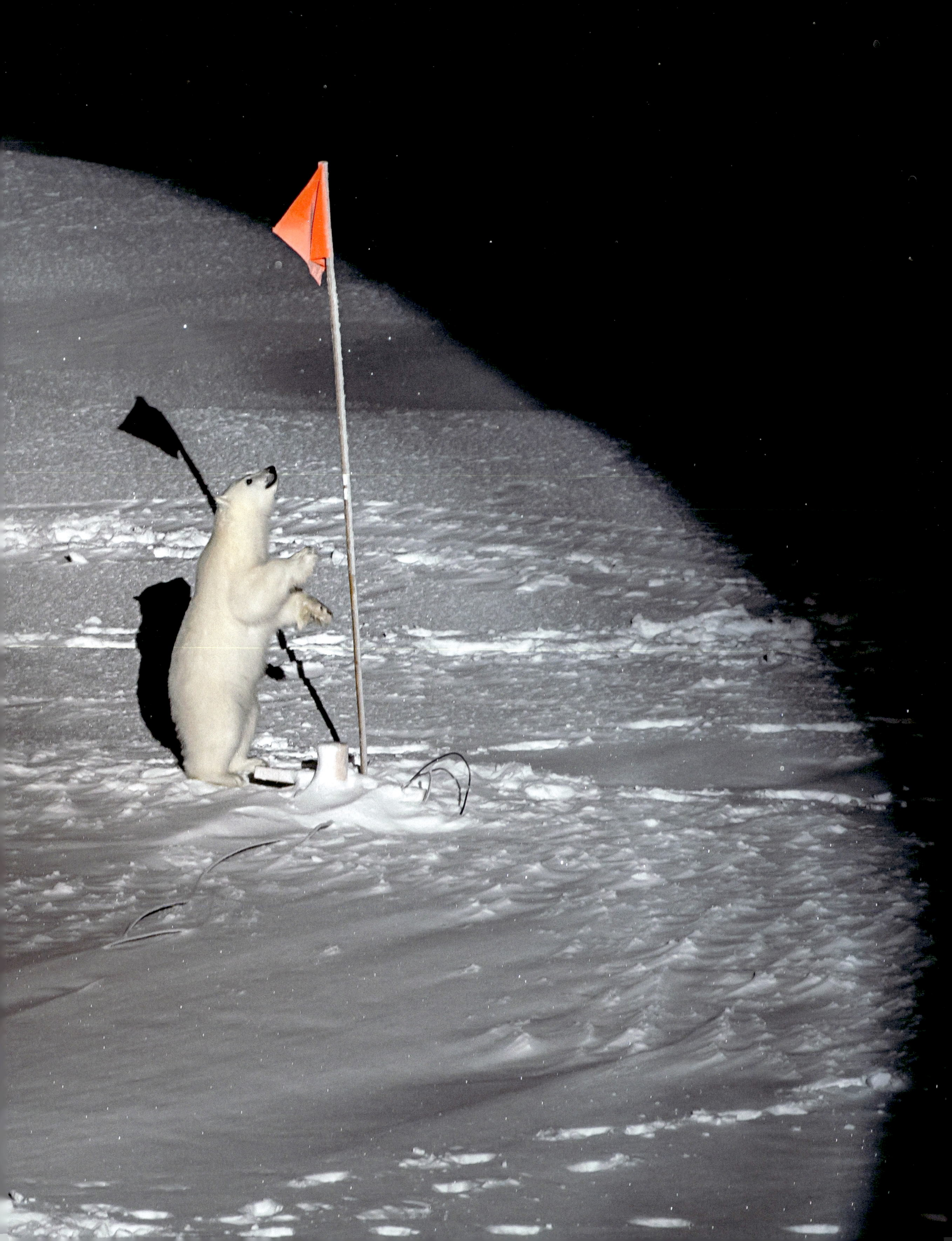

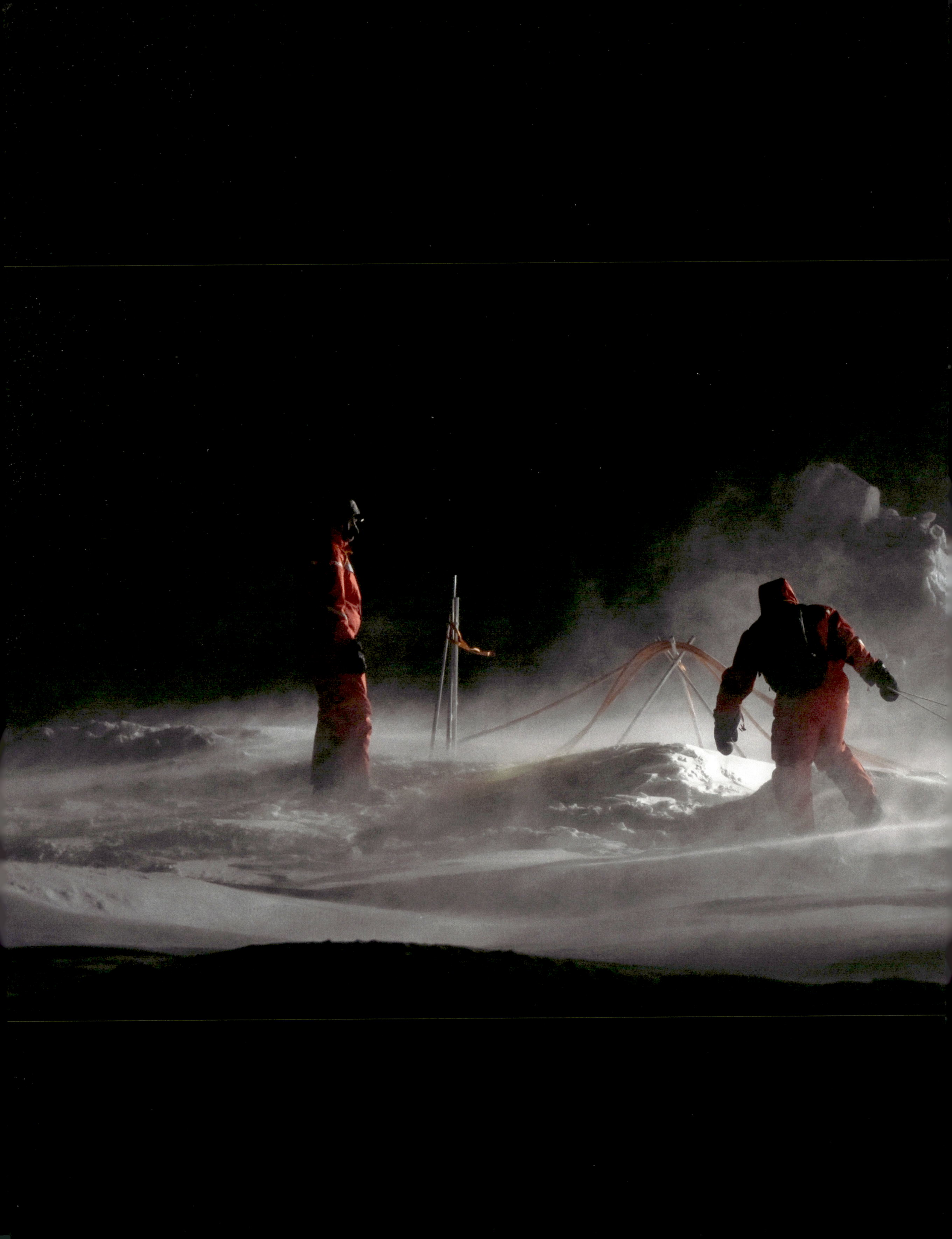

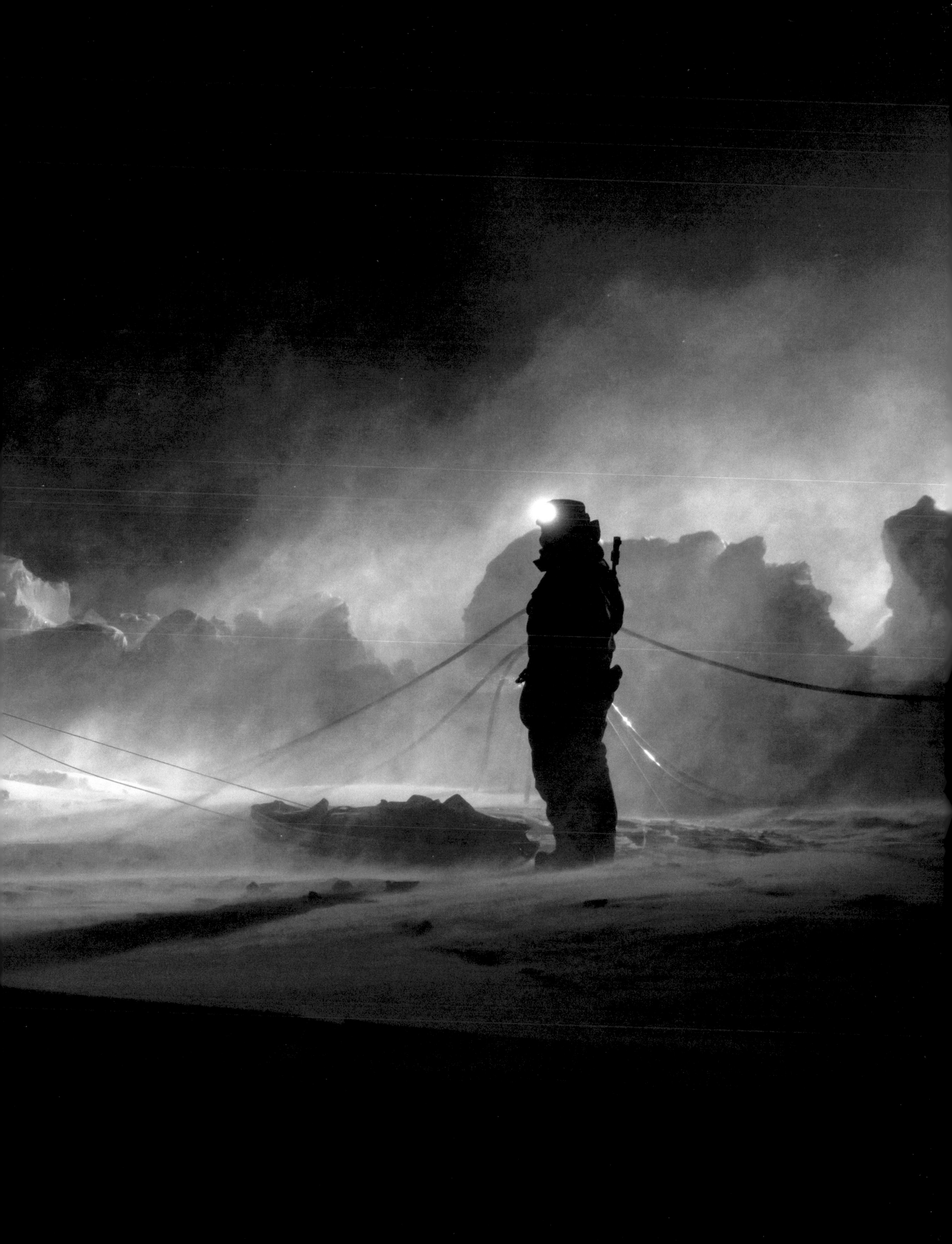

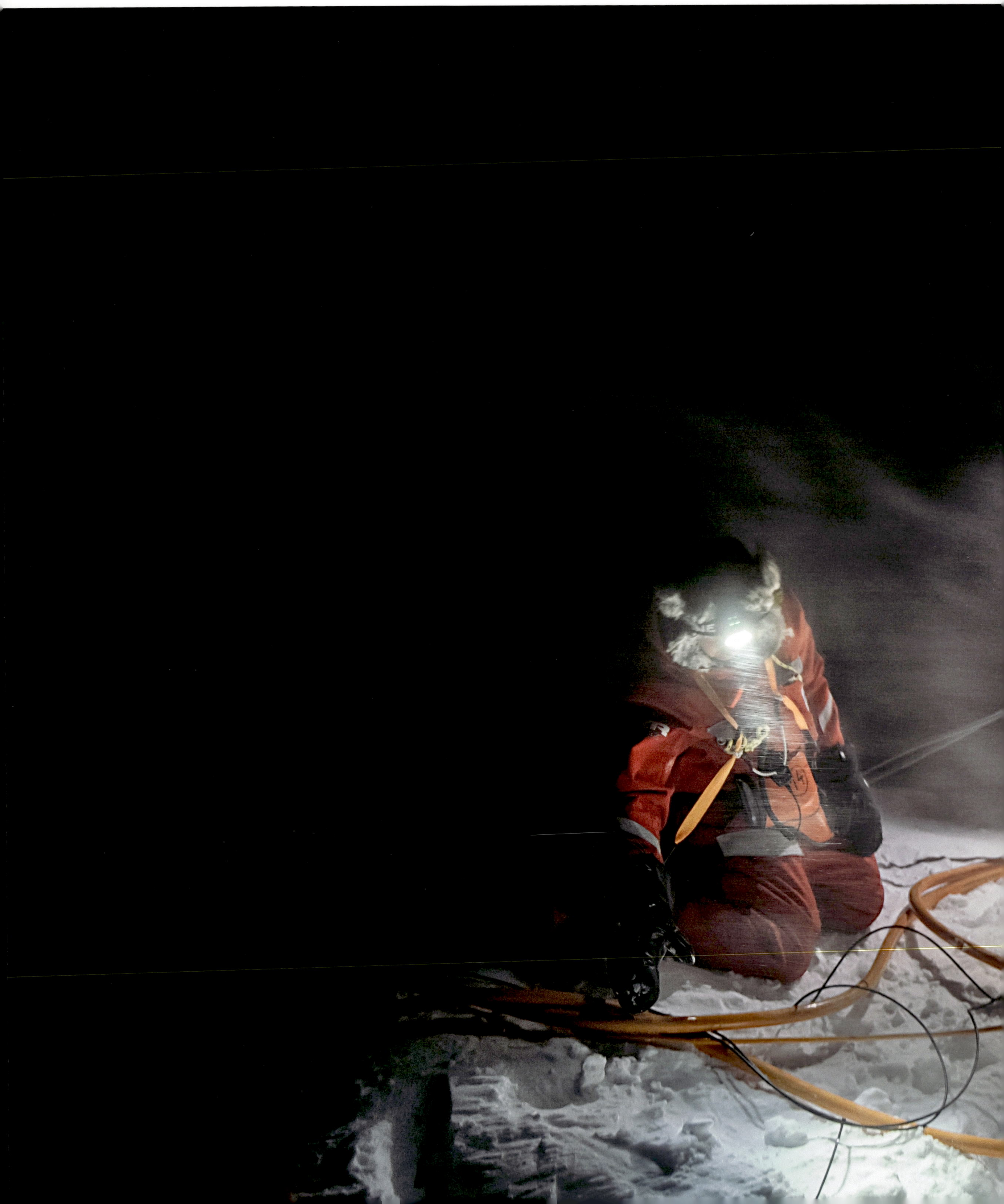

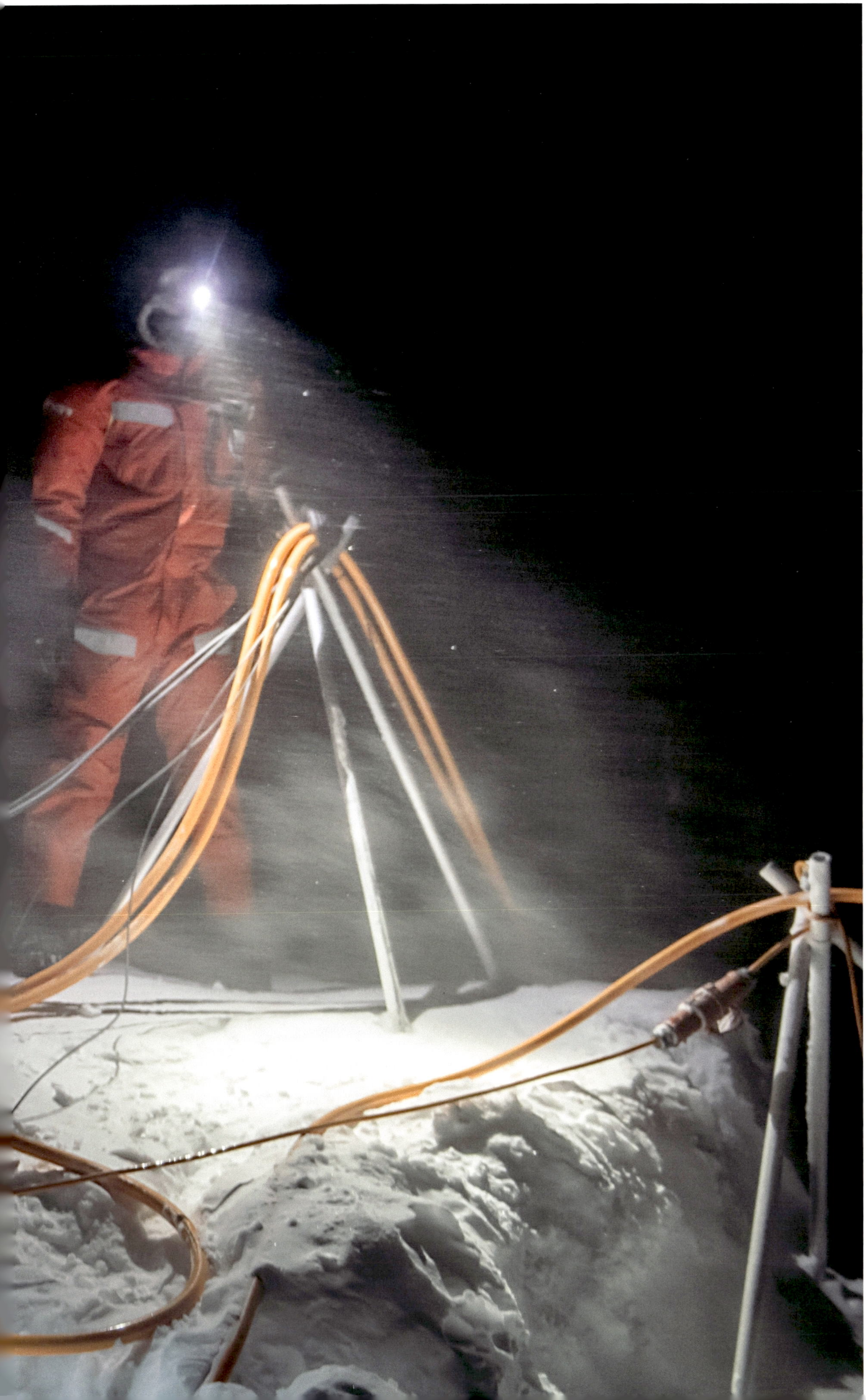

Today the wind is raging over the *Polarstern* and the ice camp at more than 100 kilometres per hour. Visibility in the blizzard is close to zero. Nevertheless, salvage teams need to go out repeatedly to save laboriously installed devices from the forces of nature. Verena Mohaupt (left) and Audun Tholfsen are cutting through an electrical cable to prevent the ice from snatching away not only the cable but also the precious instruments connected to it. Replacements would come, if at all, with the next changeover team at the earliest.

A wide crack in the ice has opened up at astonishing speed. It divides the floe, and thus the research camp, like a river. With a kayak two members of the team set out for the other bank, about 30 metres away. They have to check whether the camp's stations there have suffered in the storm.

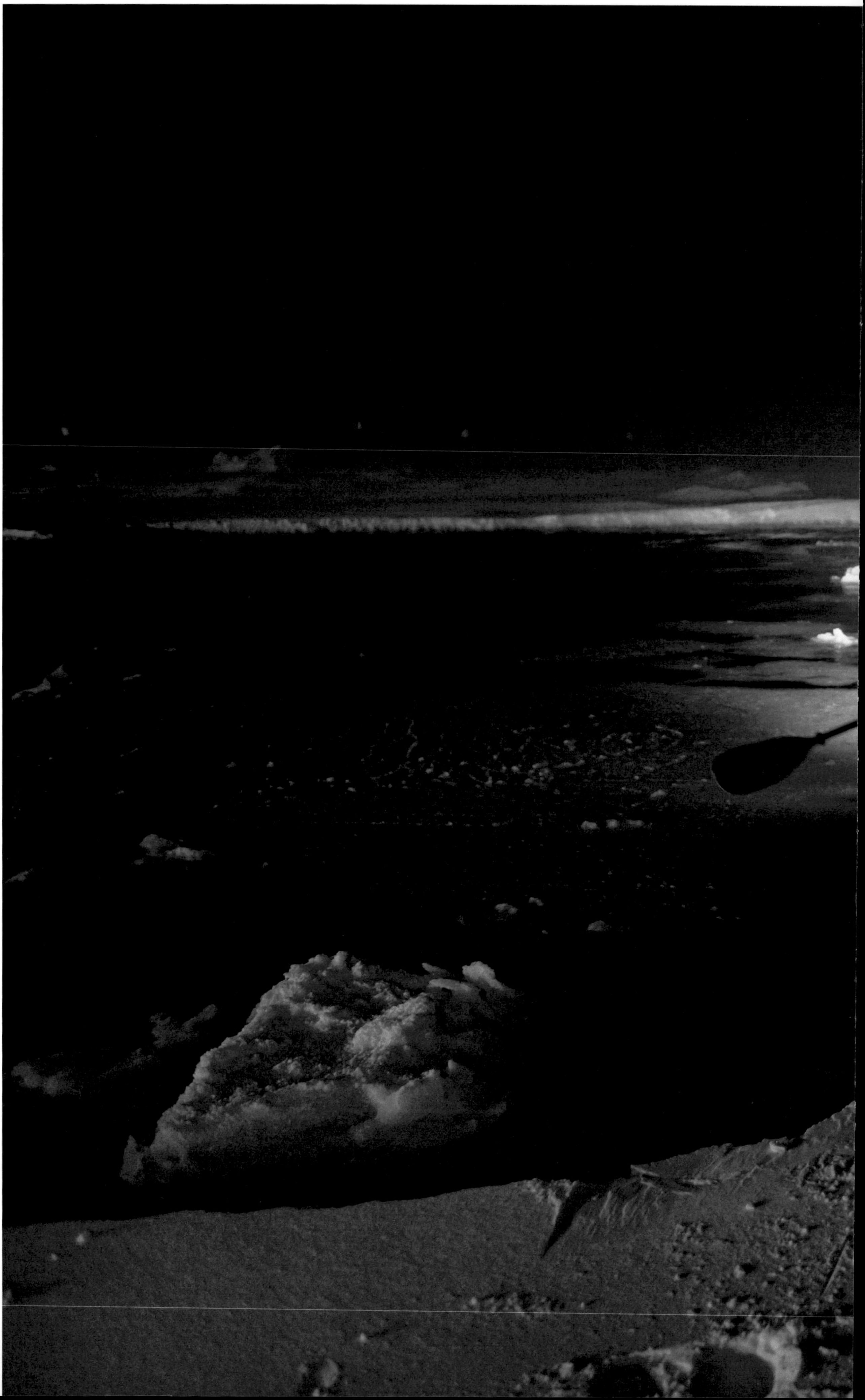

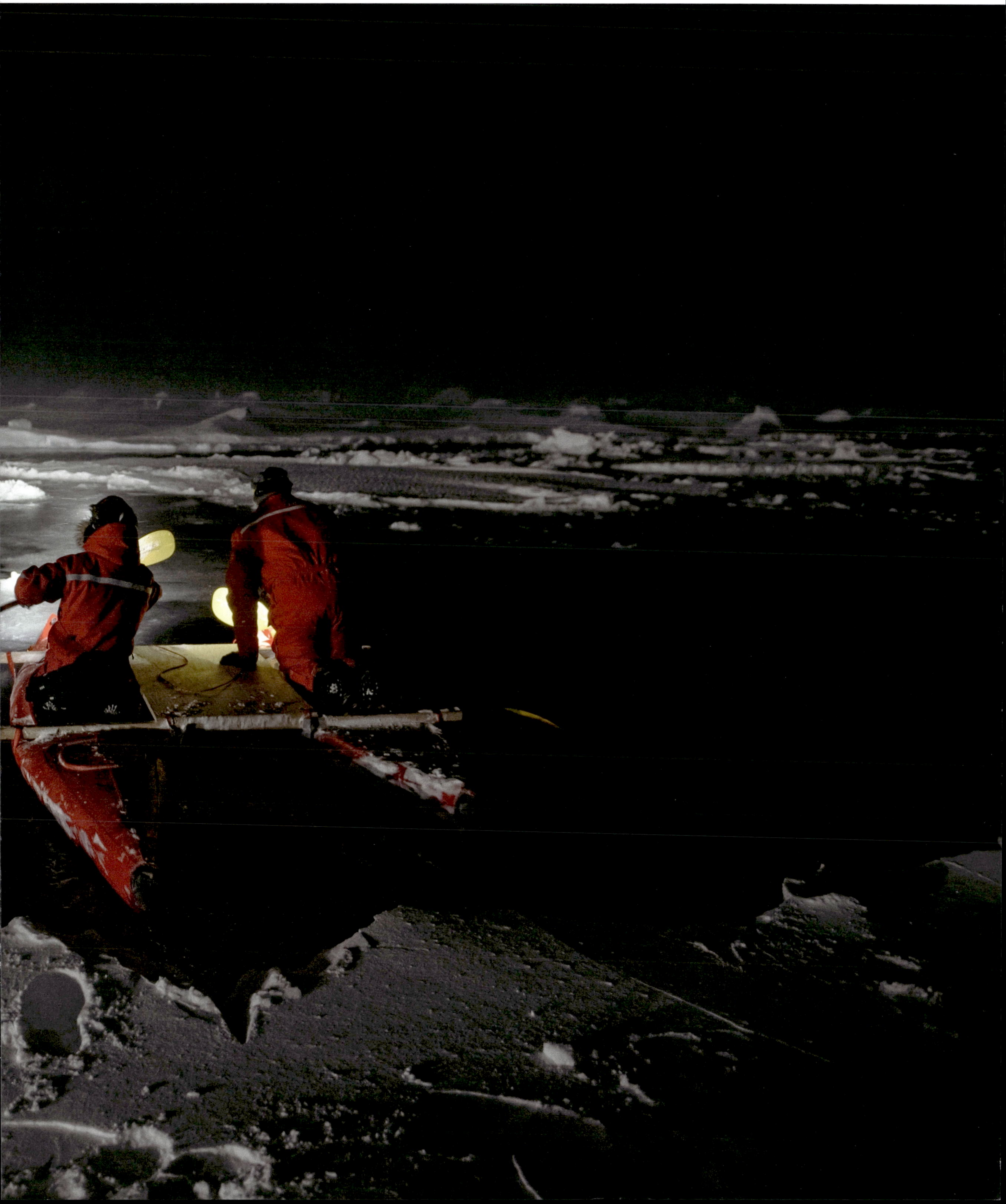

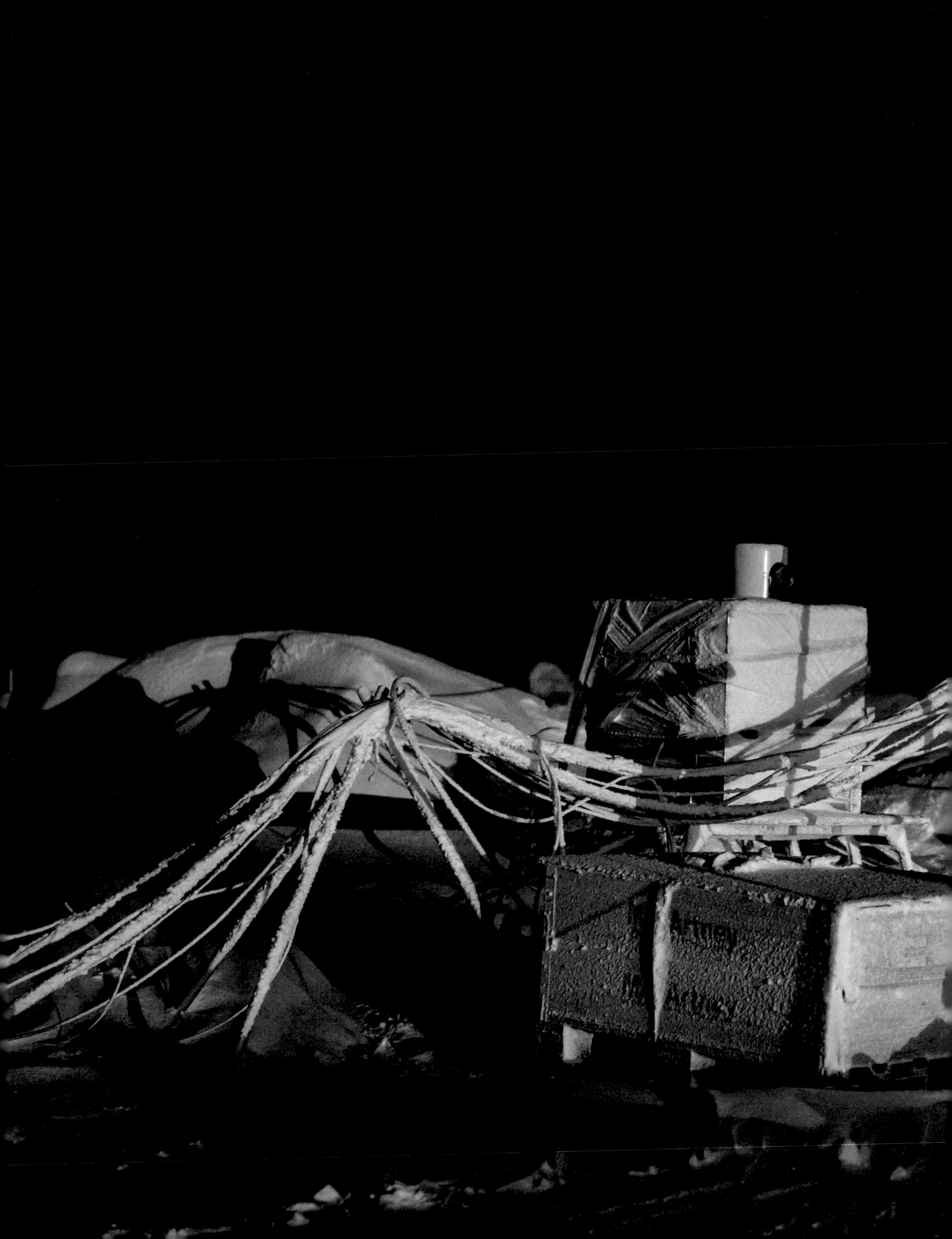

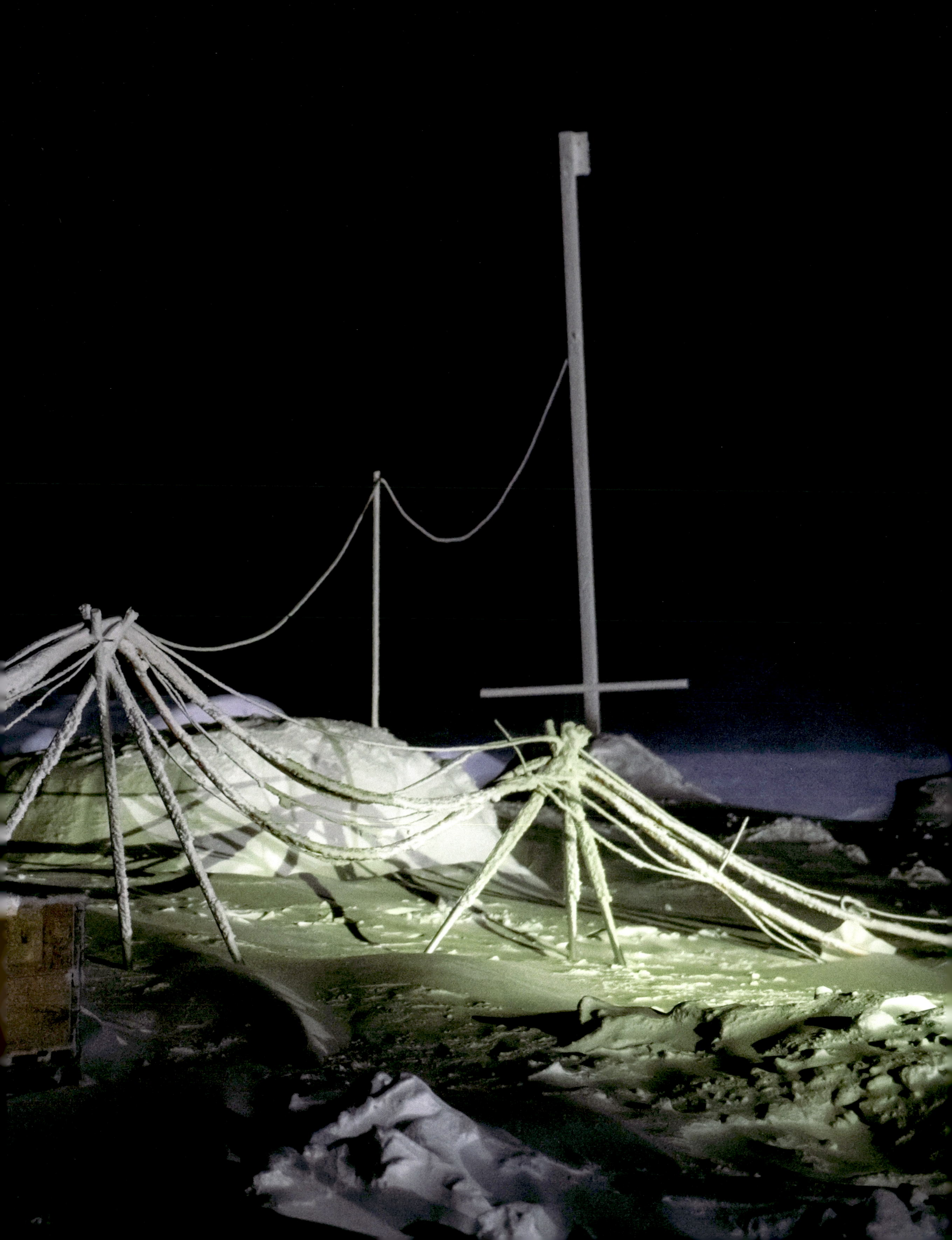

These days the ice moves without end. Ocean City with its round blue tent is increasingly threatened. The decision is finally taken to evacuate it and rescue all of the instruments so as not to lose them to the ice and sea. The German logistics expert Verena Mohaupt and the Norwegian bear guard Audun Tholfsen are a well-practised rescue team by now. It will not be the last time that the expedition members have to relocate entire research stations.

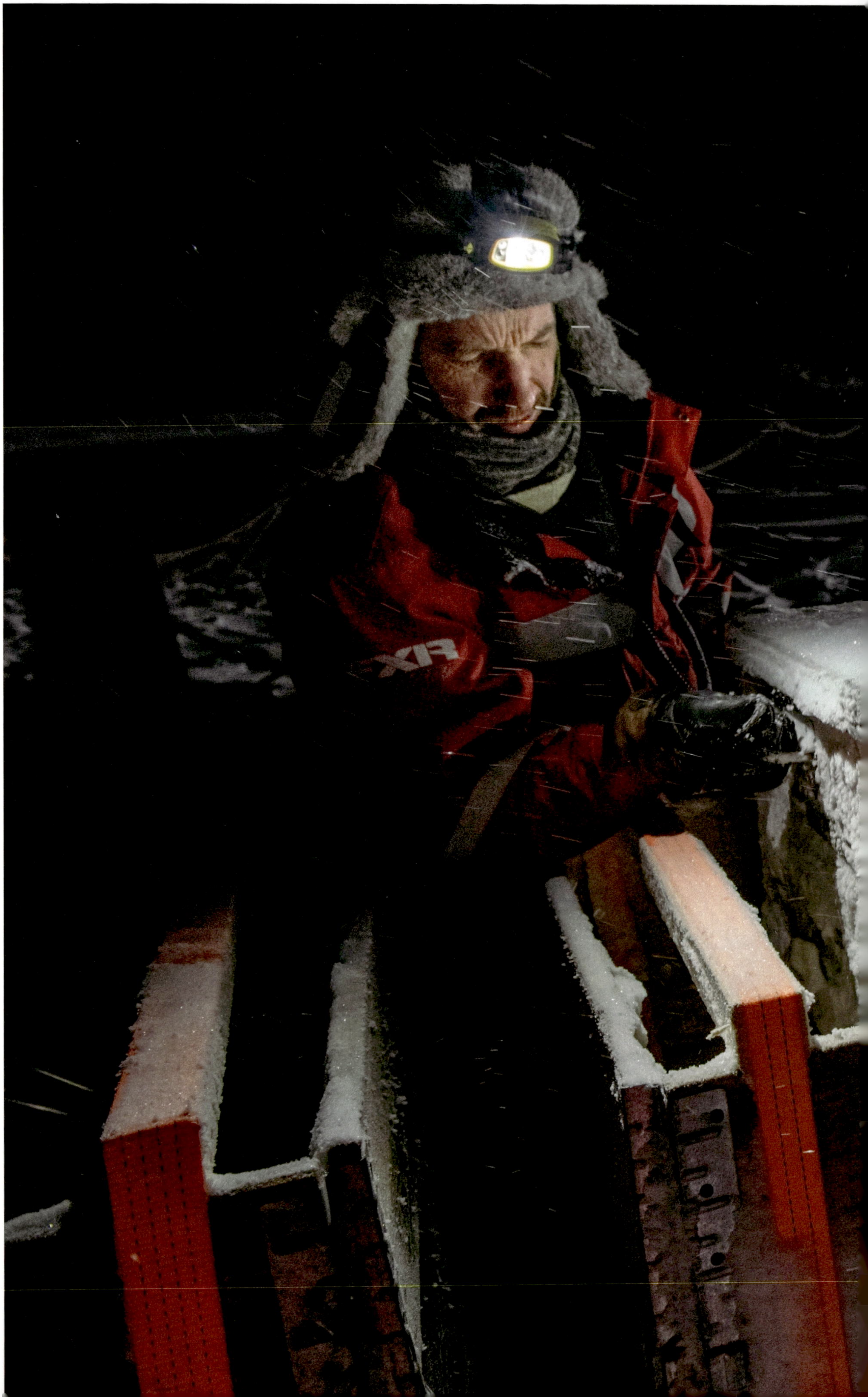

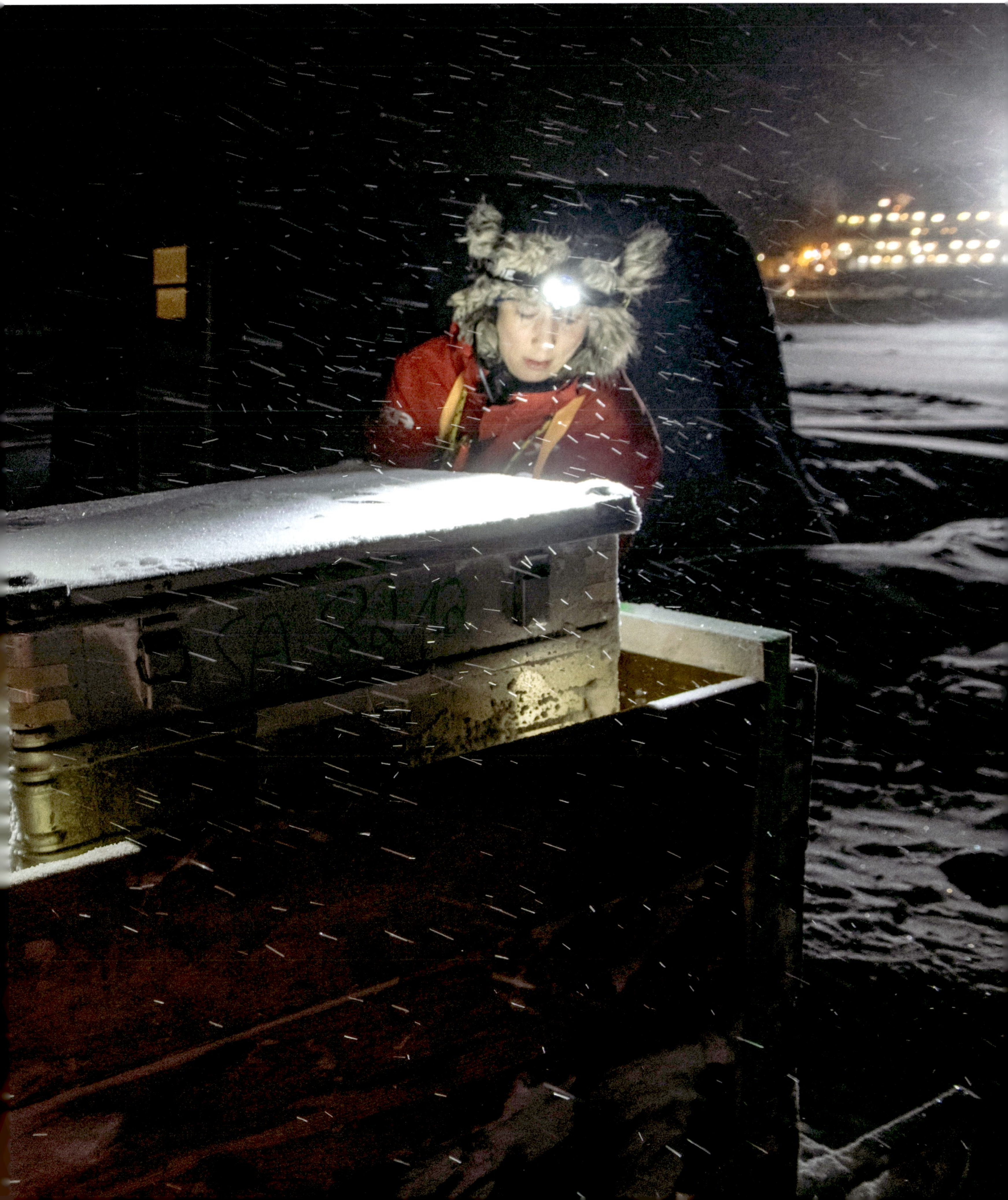

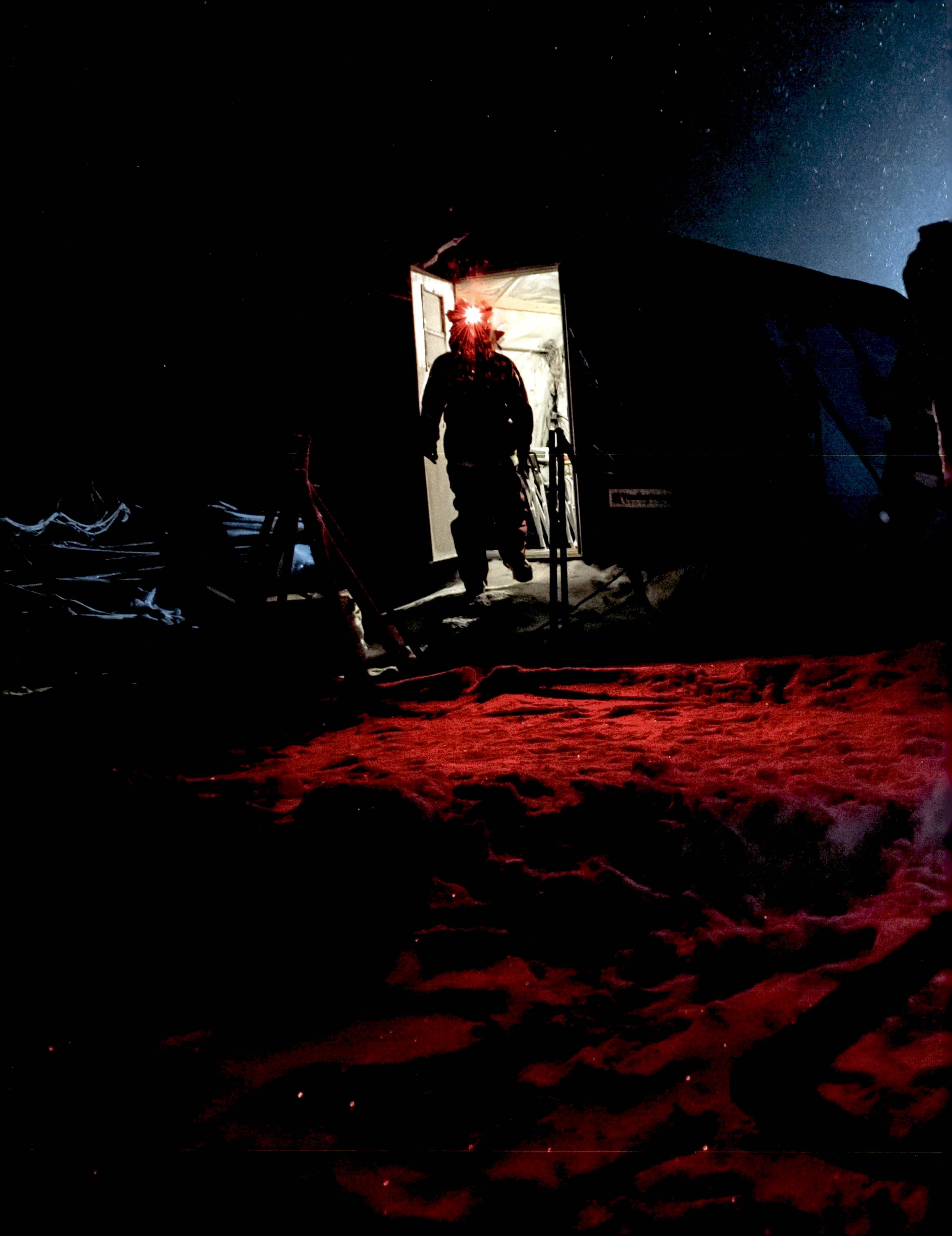

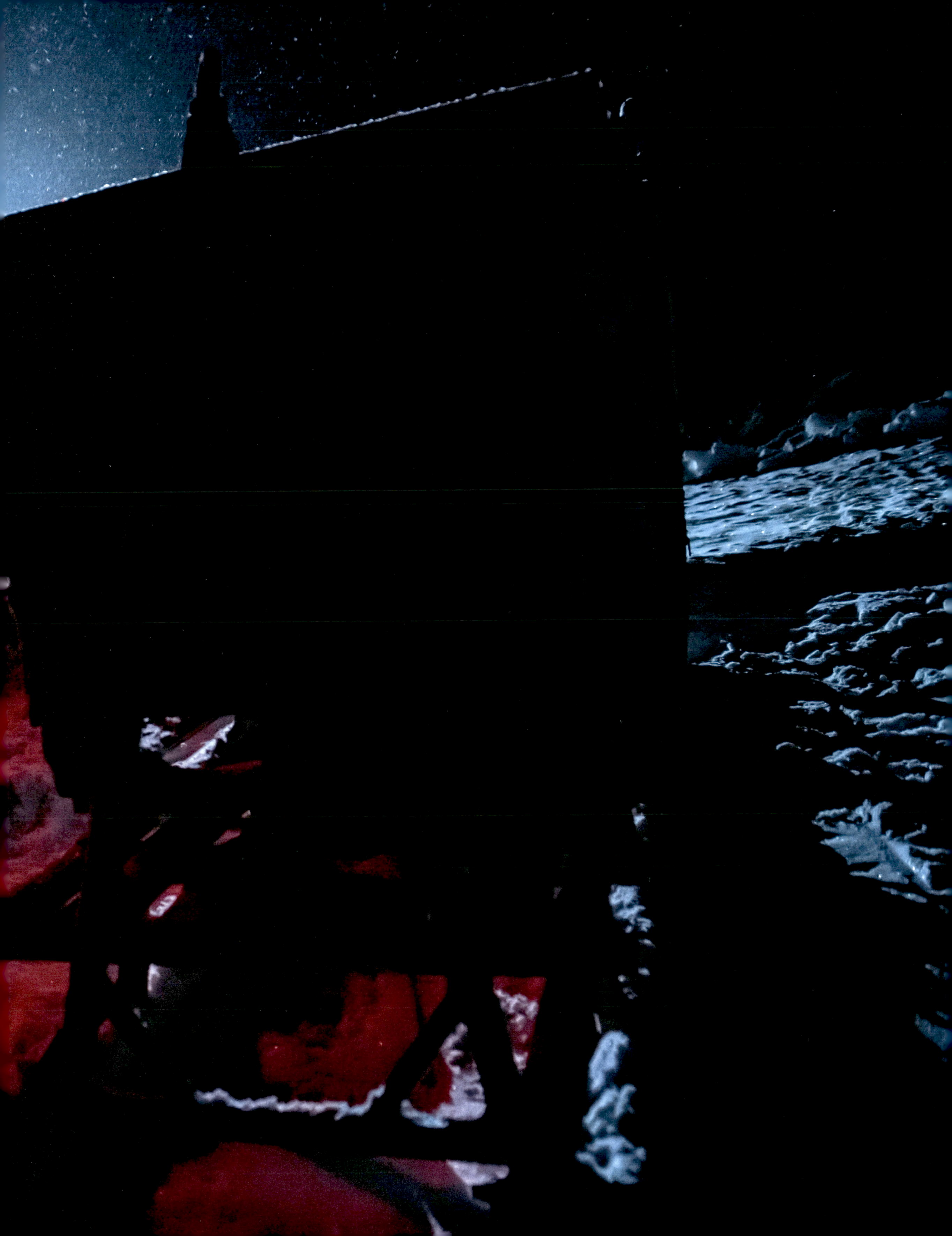

Top: The bright searchlights of the *Polarstern* bathe the surroundings of the ship in sharply contrasting light.

Below: Two participants dig a cable out of the snow that is at risk of being lost due to the movements of the floe.

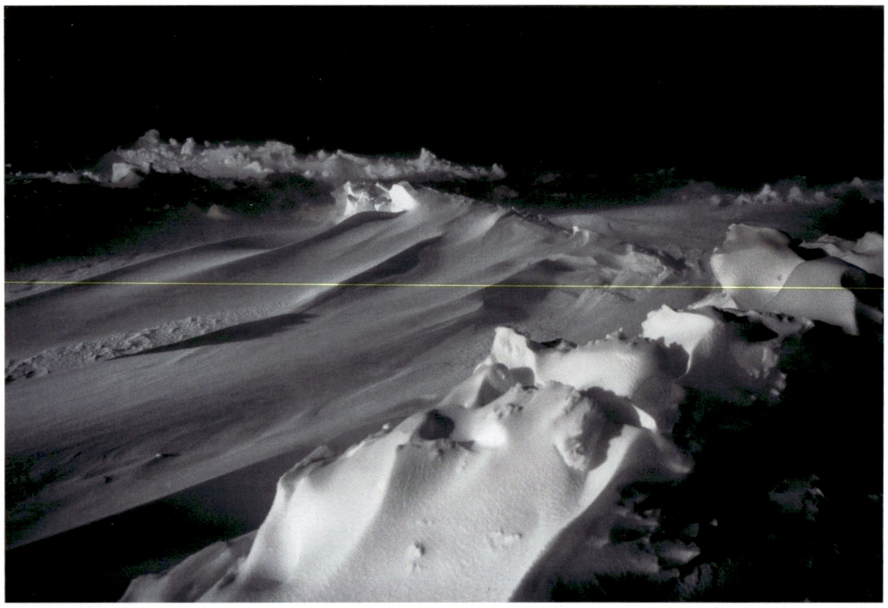

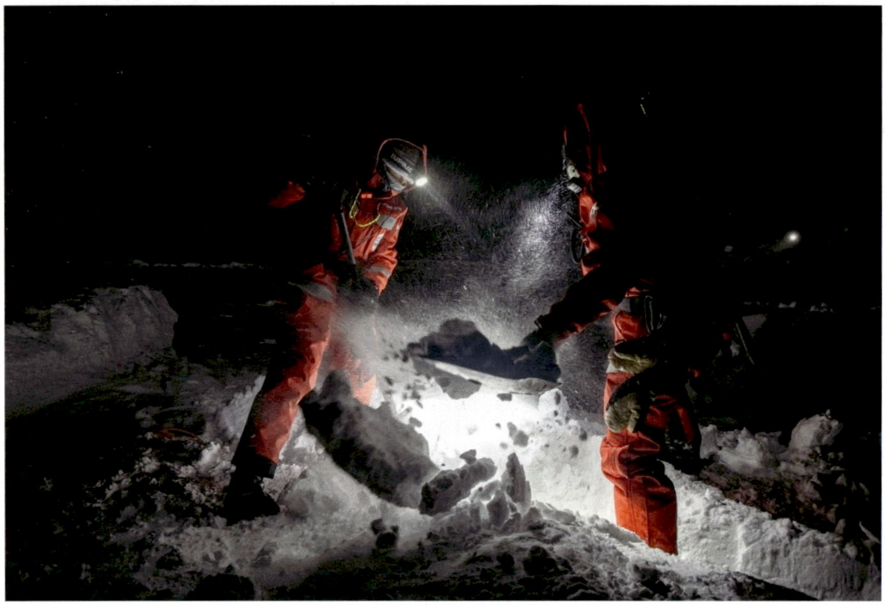

The bear guards often stand motionless for hours on the floe, their concentrated gaze directed to the open spaces. Even the best equipment cannot permanently keep out the biting Arctic wind, and then the cold creeps into their bones. The bear guard Gaute Hermansen therefore constructs a windbreak from frozen blocks of snow. This protective wall makes it much easier to hold out during his long, cold watch.

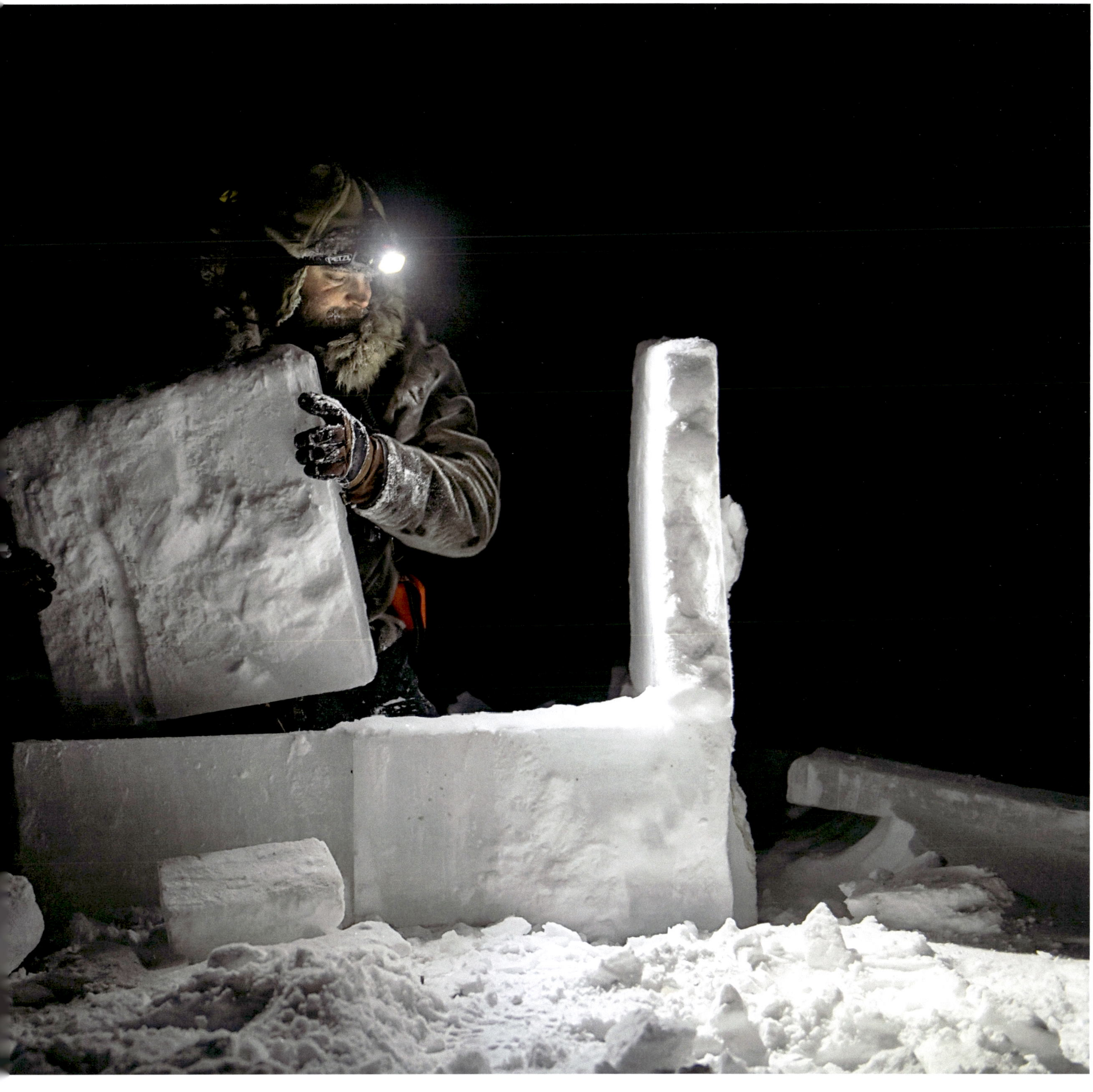

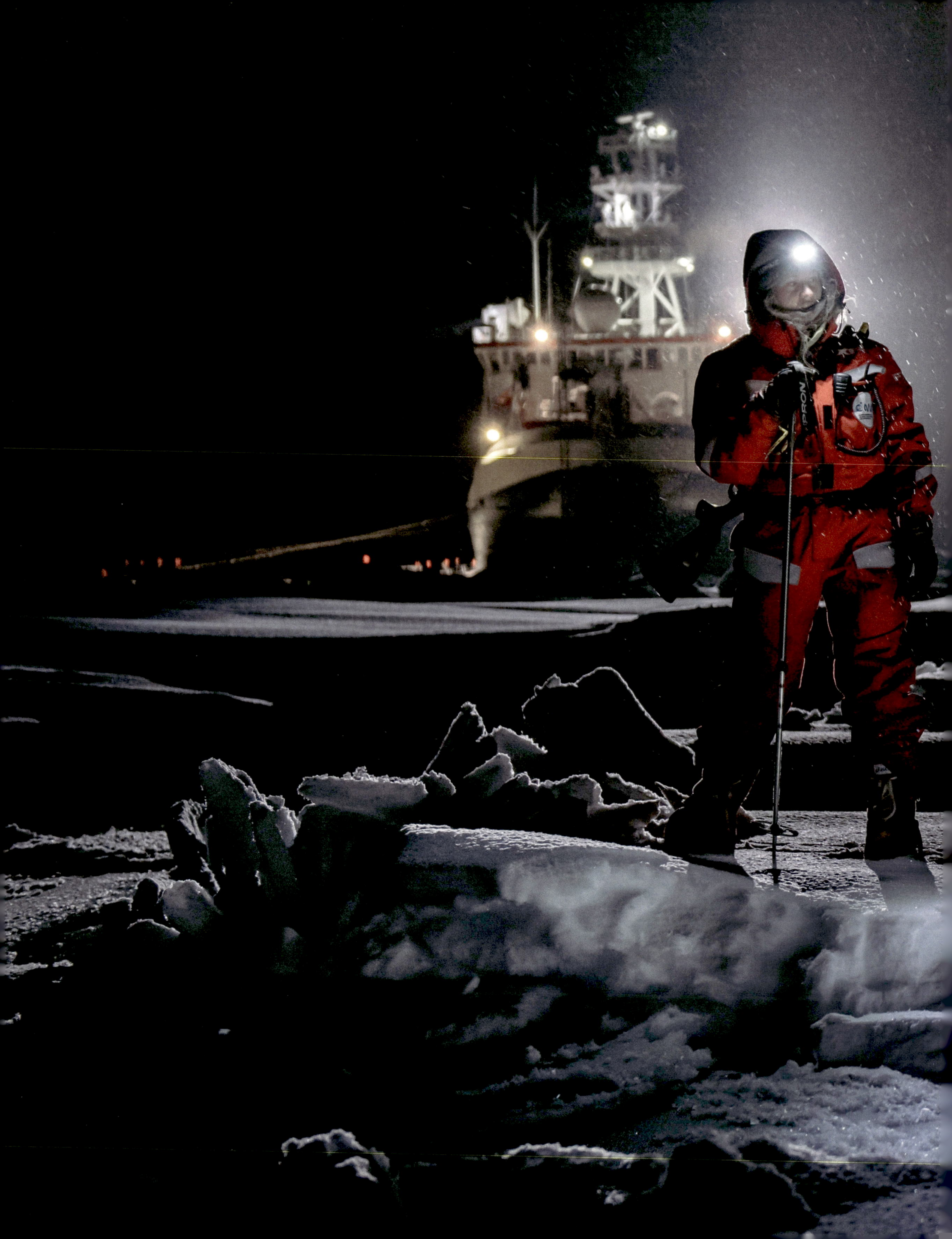

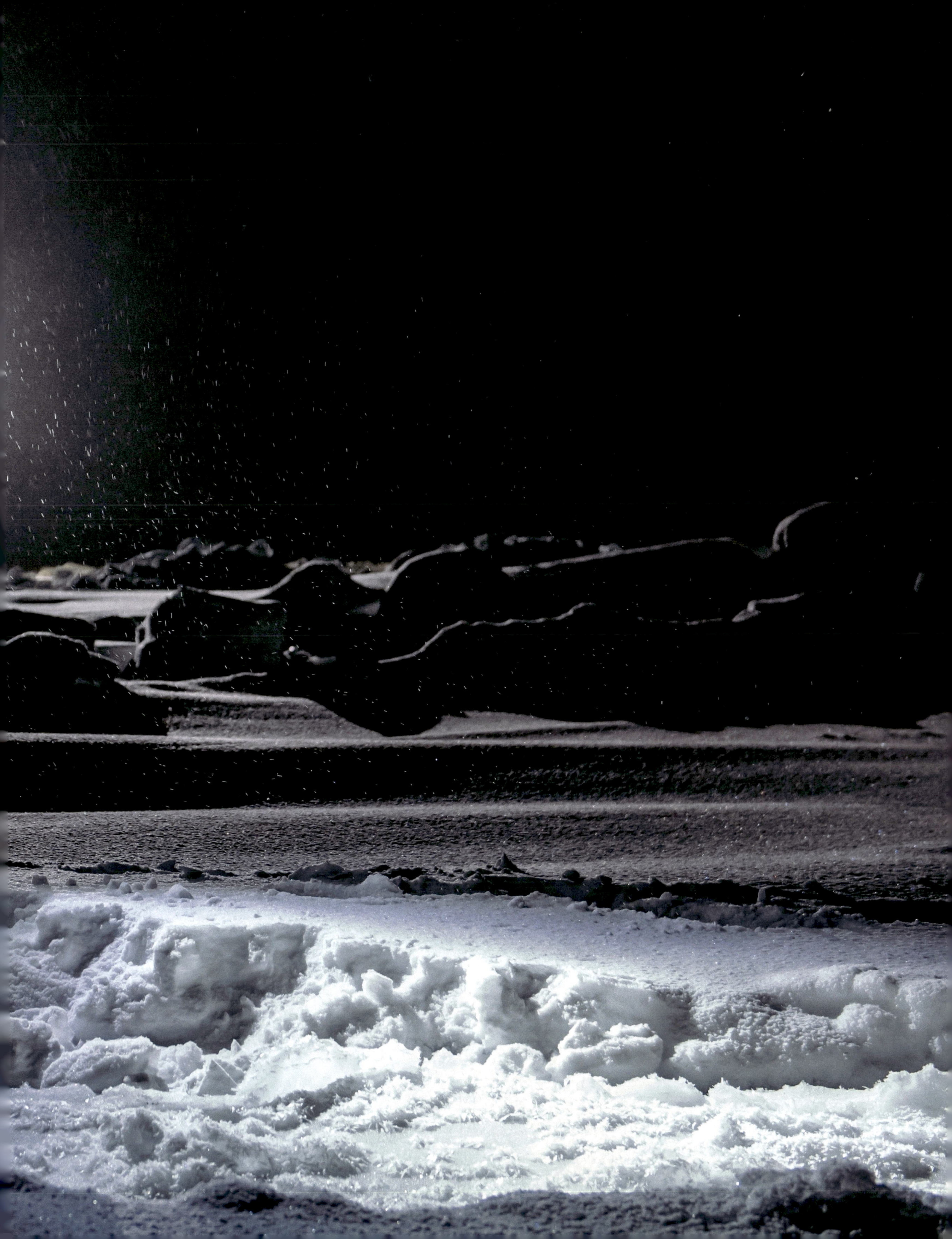

Ocean Research and Climate Change

Looking into the Depths

Deep black, a window opens into the depths next to the *Polarstern*. The scientists can lower measuring devices to a depth of around 4,000 metres – right down to the ocean floor – and collect water samples through this hole drilled in the ice. During the *Fram* expedition, Fridtjof Nansen was the first to measure the depth of the northern polar sea, and later expeditions gleaned large amounts of data from the Arctic Ocean, especially in the summer months. Until the MOSAiC expedition set off into the ice, the ocean during the winter was a great enigma to science, almost as black and deep as the seawater itself. Team Ocean is now harvesting invaluable data through the hole next to the *Polarstern*. In addition, there are other measuring instruments in Ocean City, the camp district of Team Ocean, and numerous drifting buoys, widely scattered around the *Polarstern*: a network of stations that assist the researchers to gain insights into the highly complex world of the Arctic Ocean. In collaboration with other expedition teams they are investigating the interactions between the ocean, sea ice and the atmosphere, including the influence of the ecosystem and the exchange of gases. They are researching currents in the depths and the diverse events in the layer of surface water from which the sea ice is born. They are producing profiles of currents, measuring the temperature and salinity, and turbulent vortices of water that transport nutrients and also warmer water through the sea.

The connection between the sea in the central Arctic and the climate and weather in mid-latitudes is closer than one might generally assume. The Arctic Ocean, sometimes also known as the North Polar Sea, is the midget among the world's oceans, although with an extent of 14 million square kilometres it is still much larger than Europe. Nevertheless, its influence on the earth's climate system is enormous. In the Arctic Ocean and its Antarctic counterpart, the earth's sea currents near the surface are cooled, allowing this denser water to descend to the bottom of the ocean, and back toward the equator. These currents transport not only large volumes of water across the planet but also huge quantities of heat energy. For the regions north of the equator, this near-surface ocean circulation acts as a kind of heat supply. The best example is the Gulf Stream, to which Europe owes its pleasant temperatures, even though we live at a latitude that should have a much cooler climate. If the Gulf Stream did not carry heat from equatorial regions to the north, Europe would be on average 5 to 10 °C (9 to 18 °F) colder and considerably more barren. Across the globe, the circulation of the oceans considerably mitigates the temperature difference between the equator and the poles. This means that the ocean currents are especially important factors in moderating the earth's climate system.

In recent years, ocean research has documented some of the major effects of anthropogenic climate change. At the same time, it has been discovered that climate change would have had a much greater impact on the earth without the oceans, as they have a massive ability to store heat and carbon in the surface of our planet. They absorb more than 93 per cent of the heat energy generated by the greenhouse effect caused by humans. Since the 1970s, the global ocean temperature has been carefully recorded, and thus today a kind of fever curve can be traced for the water in the seas: in the upper 75 metres, the global trend for this period is a gain of 0.11 °C (0.2 °F) per decade. This does not seem like much at first, and this temperature rise does not cause fish and other marine creatures to literally break out in sweat. However, the metabolism of many marine animals is greatly stressed by this warming, with threatening consequences for certain species, such as corals. A local rise in temperature can also be considerably higher than this global average. When the water grows warmer, the biodiversity of the seas is endangered.

Ocean warming also causes a well-known problem that is affecting some 680 million residents of coastal regions and 65 million inhabitants of small, often poor island nations: the rise in sea level has accelerated strongly since the early 2000s. Warmer water not only expands and rises up the coastlines but it also reinforces melting of the massive Greenland and Antarctica ice caps, leading to even higher sea levels. An example of this effect is apparent in the Fram Strait, towards which the *Polarstern* is drifting during the year of the expedition: this sea passage between Greenland and Svalbard is one

There is a large hole in the ice not only in Ocean City but also right next to the *Polarstern*. Excellent cooperation between scientists and the ship's crew was required to make this hole. Huge blocks of ice were removed from the floe with the help of the ship's crane. Here a large CTD rosette is lowered into the water in order to explore the ocean down to the sea floor at a depth of about 4,000 metres.

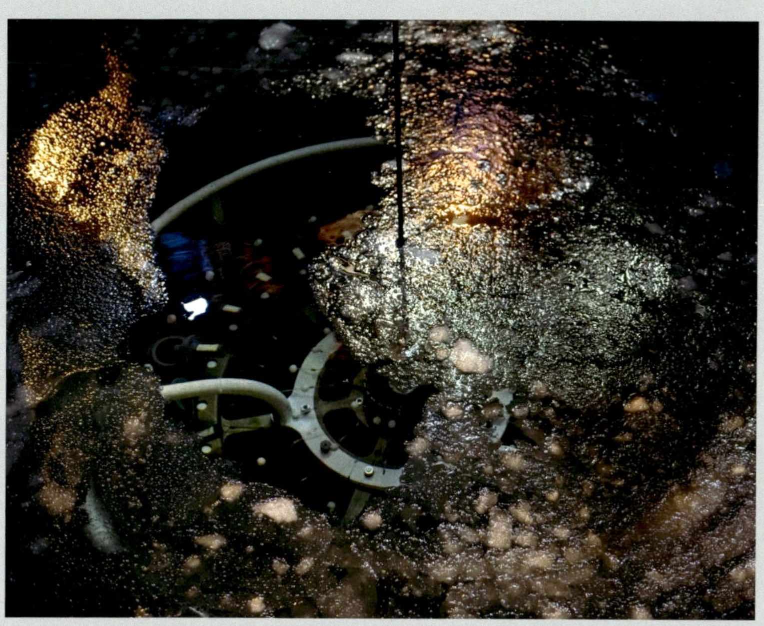

of the main arteries of ocean currents into and out of the Arctic. The temperature of the water that flows northwards has risen here by about 1°C (1.8°F) since the late 1990s. These warmer masses of water meet the Greenland glaciers, in effect melting away their bases. If humanity does not drastically curtail greenhouse gas emissions and thus the consequences of climate change, forecasts by the Intergovernmental Panel on Climate Change expect a rise in the global sea level of 60 to 110 centimetres on average by the end of this century.

But it is not only heat that the seas take up. The oceans absorb more than one-third of the carbon dioxide that humans pump into the atmosphere as a result of transport and industry. This, too, has so far spared us from even more severe effects of climate change. But at a high price: these increases of carbon dioxide alter the chemical balance of seawater. As the level of carbon dioxide rises, the pH value falls: the oceans acidify. To date the acidity of seawater has already risen by 30 per cent since the beginning of the Industrial Revolution, and forecasts project it could rise by as much as 100 to 150 per cent by the year 2100. The acidification of the oceans has long been considered a twin of global warming – and in the medium term no less troubling, as it threatens the life and the food web in the sea. As cold water absorbs carbon very effectively, the polar seas are particularly impacted. Marine life forms that clad themselves in shells are particularly at risk, as it is increasingly difficult to form the necessary lime. Among the endangered species are pteropods: sea snails. These tiny creatures which float elegantly in the Arctic water are part of the zooplankton and thus at the centre of the marine food web. In the Arctic, the fish and seabirds that survive on zooplankton feed seals, polar bears, whales and, ultimately, humans. Increasing acidification makes the delicate shells of pteropods thin. This can even result in holes appearing in their protective casing.

Creatures like pteropods are ideally adapted to the extreme environmental conditions of the Arctic. But this fragile ecosystem also depends on the nutrients that are carried by marine currents from neighbouring oceans. The researchers of the MOSAiC expedition are therefore closely examining the sea currents and what they carry with them, as well as the thermal and saline conditions. The complex heat flows in the Arctic are still open research questions. But one thing is certain: the masses of warm water that flow deep into the Arctic Ocean from the Atlantic and Pacific are capable of entirely melting the Arctic sea ice. However, this heat does not typically make its way from the depths up to the ice, because the increases of seawater salinity and density with depth largely prevent mixing of this deeper, trapped heat. The researchers are particularly interested in vertical mixing in the upper ocean, as this is where the fascinating interplay of ocean and sea ice takes place to the rhythm of the seasons: a perpetual cycle of changing aggregate states with a huge impact on the Arctic climate system.

In contact with the frosty atmosphere, surface water cools down to −1.8°C (28.8°F). At first its salinity prevents it from freezing, but if it gets even colder, eventually even seawater turns to ice – and expels its salt during the process. Sea ice is therefore much less salty than the ocean itself, and a particularly saline layer of water forms under the ice layer. Weighed down by the salt, it sinks to greater depths and thus generates movement in the upper water layer. But when the sea ice melts, a lighter layer of fresher water covers the Arctic Ocean. As a result of climate change, the salt content of the northern polar seas has been reduced, especially near the surface – partly because more and more sea ice is melting, but also because Siberian rivers are flushing great quantities of fresh water into the Arctic, fed by increasingly frequent and increasingly strong precipitation. And it is precisely this layer of fresher water, according to a hypothesis of the researchers, that might in turn play a part in making the sea ice melt even more, and more quickly, in summer. As a consequence, in the late summer the highly reflective sea ice layer separating the atmosphere from the ocean disappears, and in the months-long Arctic summer the sun's rays shine directly on the open sea. The dark water of the ocean almost completely absorbs the energy of the light and further drives the warming of the Arctic Ocean and the further melting of the Arctic sea ice – with serious consequences for the earth's climate system.

Change of Shift

Handing over at the North Pole

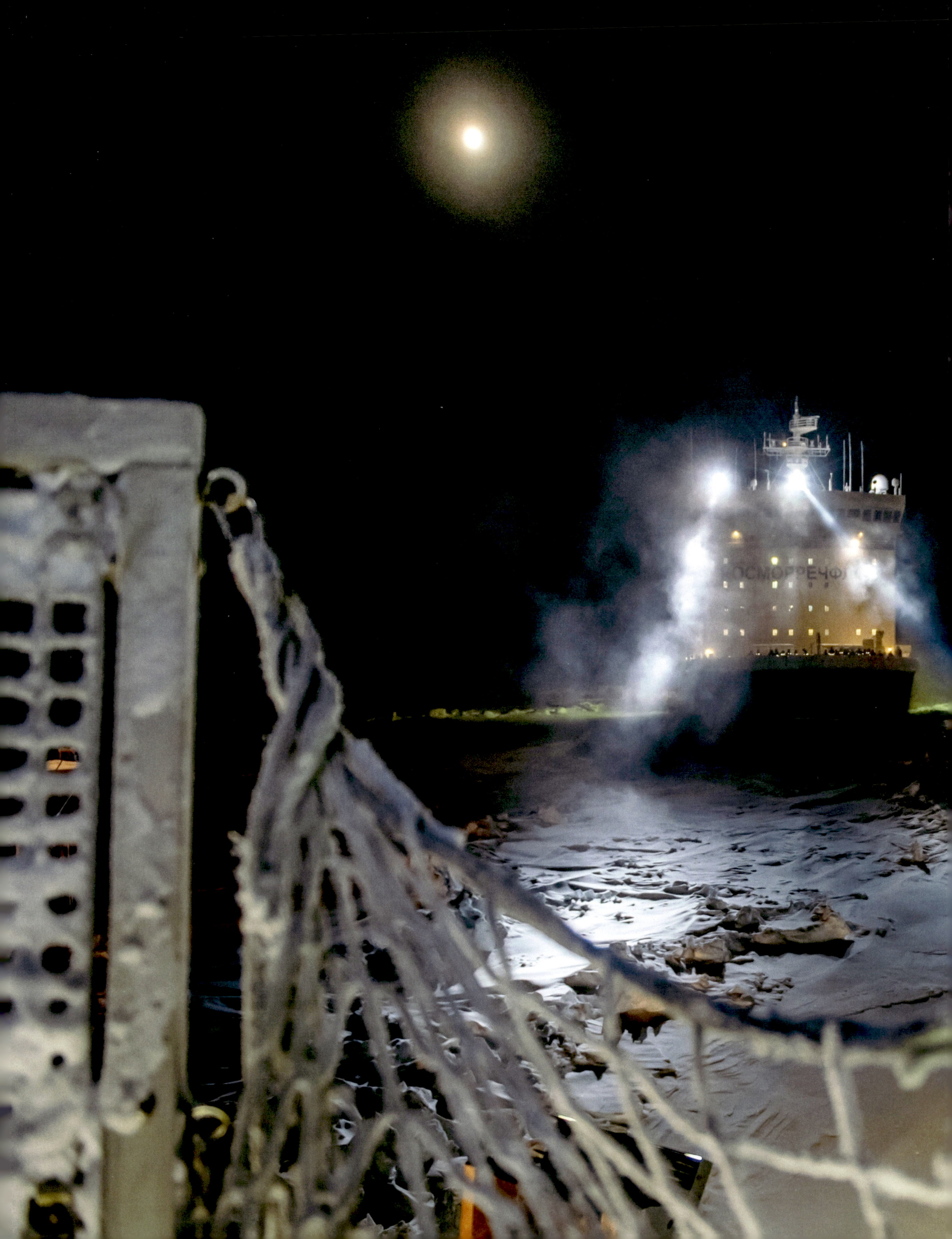

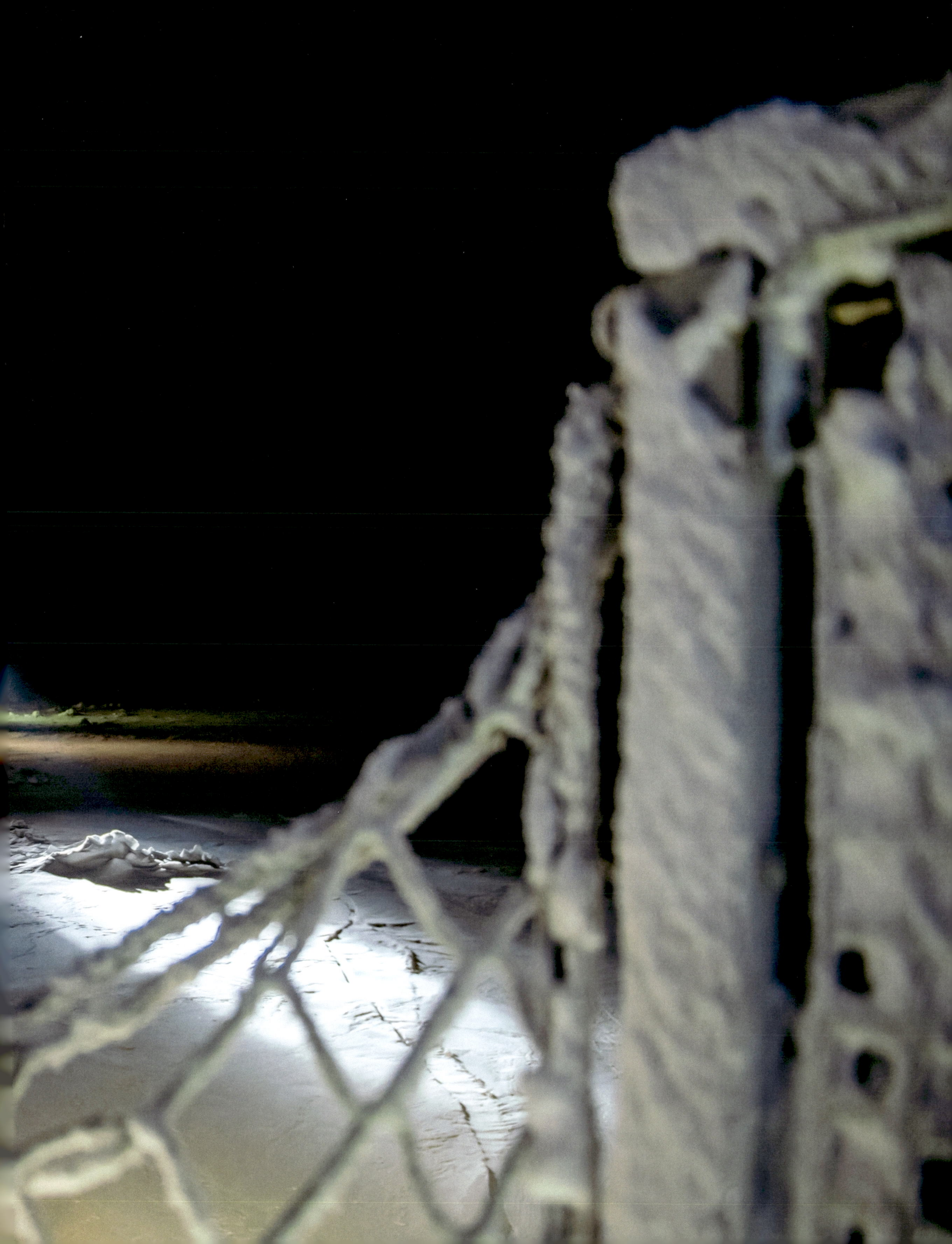

The first changeover of the expedition team is scheduled for mid-December 2019. Will the exchange of about one hundred people and freight in the middle of the sea ice take place successfully? It is a tense time for the crews on board the *Polarstern* and the Russian supply icebreaker *Kapitan Dranitsyn*, but the exchange goes ahead according to plan. A few days later, the team of the first leg begins the journey home. The mood fluctuates between longing for home and melancholy. For the team, which has grown close during months in the ice, a voyage of sixteen days remains until they finally disembark. In this period the *Kapitan Dranitsyn* presses ahead slowly through night and ice, with a monotonous motion and rhythmic sounds from the engines. The scientist María Josefa Verdugo observes this hypnotic monotony from the dark bridge of the *Kapitan Dranitsyn*.

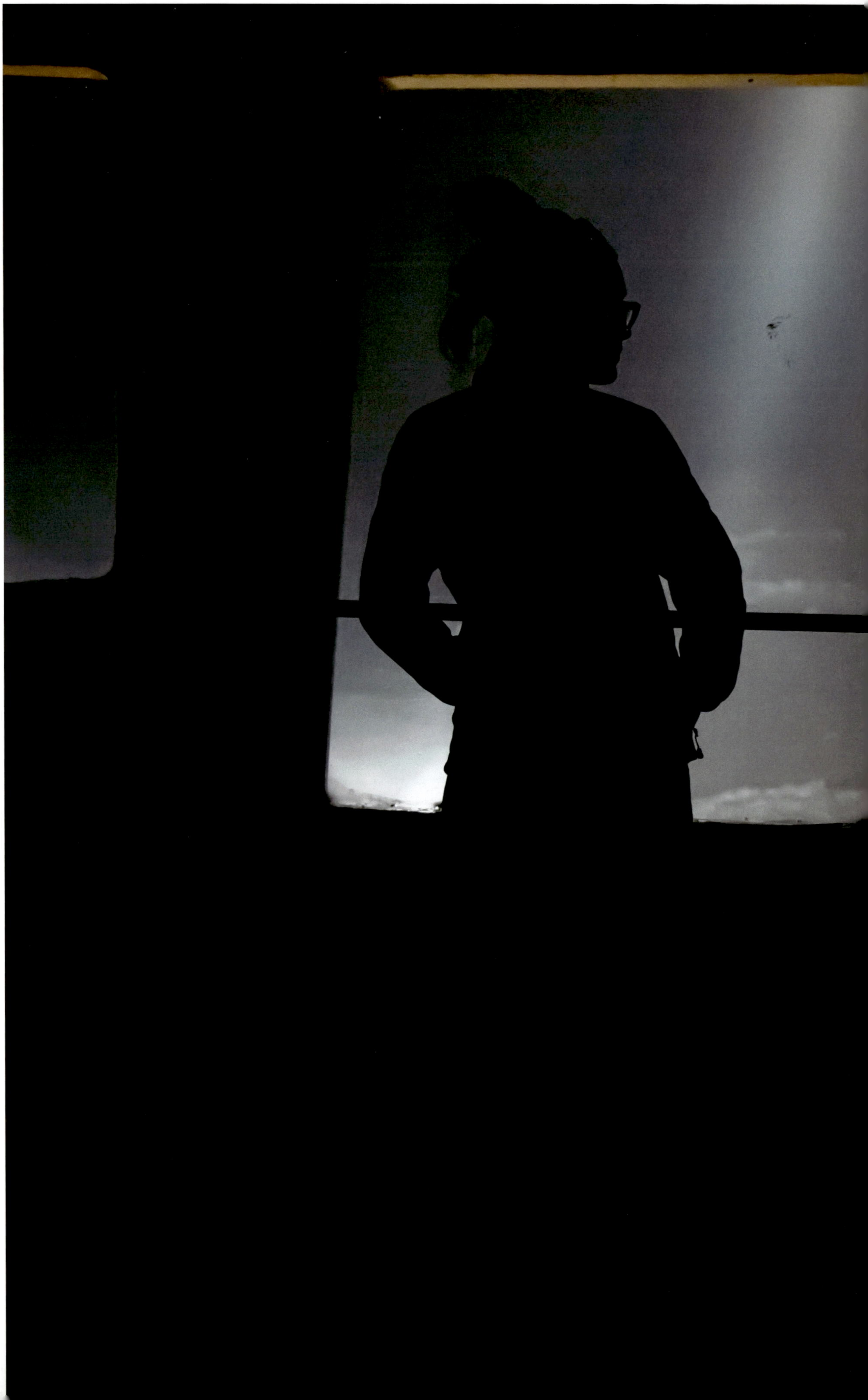

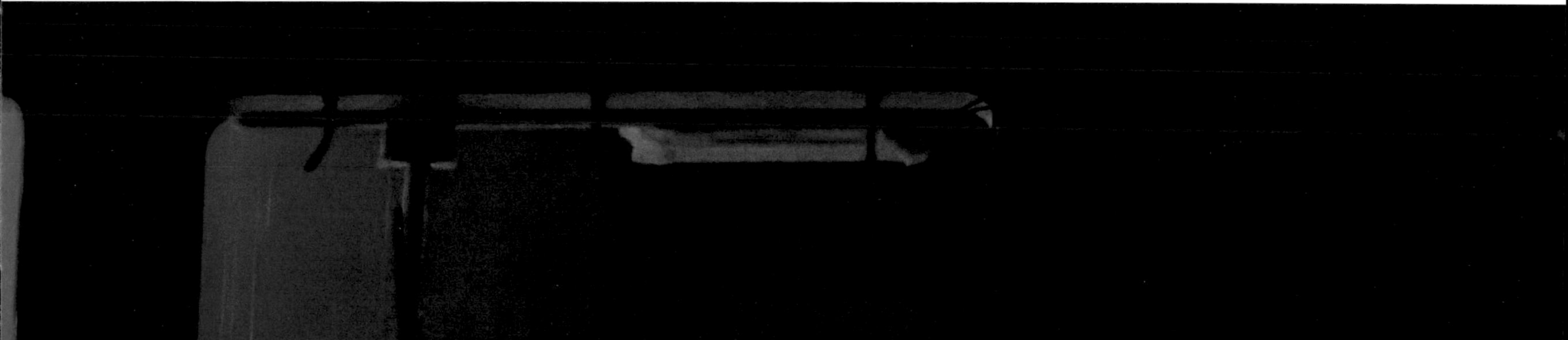

As the team of leg 1, now on the supply vessel *Kapitan Dranitsyn*, celebrates the Christmas holiday on its return journey to the south, only the *Polarstern* itself is closer to the home of Santa Claus. At Christmas, lots are drawn – with gifts that started their journey to the Arctic back in September and now finally have their day. The mood is good, even though, as is sometimes the case with Secret Santa, not every gift suits the taste of its recipient. For this occasion, Santa Claus personally pays a call to the supply icebreaker.

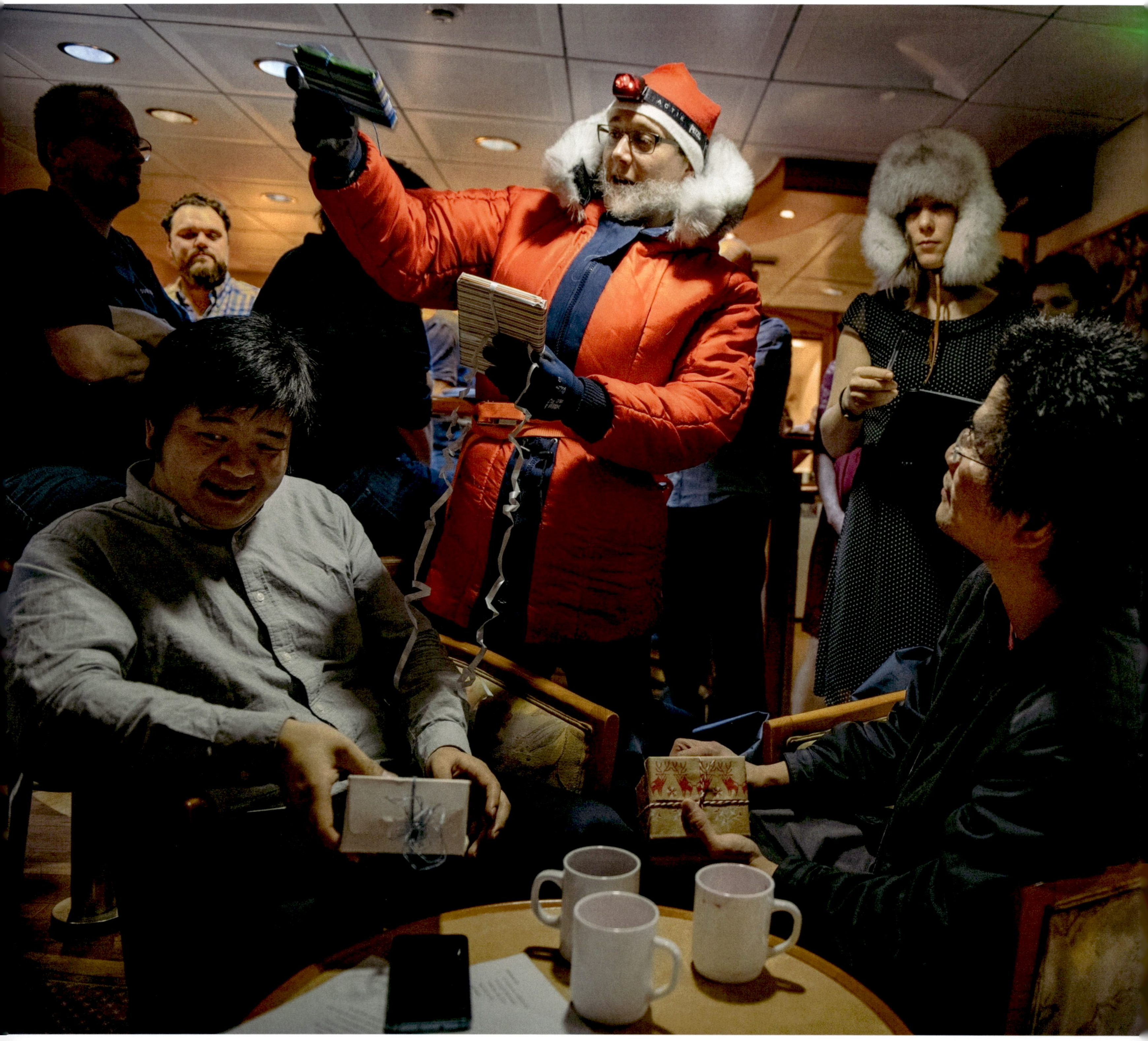

I

It's December 2019 and the Russian icebreaker *Kapitan Dranitsyn* is resolutely battling ahead through the ice. Its goal is the *Polarstern*, which by this time has already been drifting with the sea ice for more than two months.

The cargo hold of the *Kapitan Dranitsyn* stores food: items that keep such as flour and preserves, fresh fruit and vegetables, but also potato crisps, chocolate and beer, which are important mainly for the mental well-being of the expedition members. The icebreaker has also taken scientific instruments on board and materials that the team on the ice floe has ordered for repairs. Above all, there are about one hundred additional members of the expedition on board: the new team that will be passed the baton on the *Polarstern* – and that will also take on the forthcoming "night shift". The team for the second leg of the voyage, which begins in mid-December 2019, is starting out in the darkest and coldest phase of the expedition. The *Kapitan Dranitsyn* is taking them to the Arctic winter.

The drift of the *Polarstern* takes it to regions so remote that even satellite links between the ship and the land only function to a limited extent. The regularly planned supply and exchange trips – to changeover the team of scientists, logistics specialists and ship's crew at intervals of a few months – are thus only possible by means of sophisticated logistical choreography. The MOSAiC expedition has had a decade of meticulous and complex planning. Various alternative plans are held in reserve to guarantee provisioning and exchange in all circumstances, however adverse. Nevertheless, during the MOSAiC expedition it becomes clear that all polar exploration is dependent upon the forces of nature and that even the best planning of such a large-scale mission has to measure up to this force majeure. So on every supply trip the question comes up: Will the exchange of personnel and goods go according to plan?

The *Kapitan Dranitsyn* finds itself exposed to the forces of nature on its very first supply trip. A hurricane-force depression is raging over the Barents Sea, so the ship spends a few days in a fjord waiting for better weather – a delay for which the schedules of the polar expedition are prepared. Finally, after some two weeks at sea, the team on the *Kapitan Dranitsyn* spot the bright lights of the *Polarstern* on the horizon. Carefully, so as not to damage the ice floe and ice camp, the Russian supply icebreaker manoeuvres alongside the *Polarstern*.

The five days following are a busy time again in the Arctic: tonnes of cargo are carried, partly by muscle power across the ice, partly hoisted directly from ship to ship by massive cranes. The temperature during these operations is −30 °C (−22 °F), so it is important to hurry when moving goods that must not freeze. At the same time, members of the new team are instructed in work plans and safety measures by their colleagues. Which instrument is located where? What is the course of the roads on the floe? What about experiences with visiting polar bears? In contrast to the first team, the new one has never had the opportunity to explore the floe in daylight. The newcomers have to learn how to find their way around in the darkness of the polar night. For this they cannot rely on their eyes alone; technical aids such as laser scanners and infrared cameras allow them to make a precise map of the terrain through a series of carefully calibrated helicopter flights over the floe landscape.

When the scientists and logistics specialists, bear guards and crew members who had set off for the Arctic months ago finally board the *Kapitan Dranitsyn*, many of them feel a mixture of longing and melancholy. Though the joy of returning home certainly grows, it is hard for many of them to leave behind the *Polarstern* with its carefully maintained measuring instruments in the ice. At home, quite a few will find the routine of civilization too much: the endless choice of supermarkets and crowded city centres will be overwhelming. Some will long for the inhospitable vastness of the Arctic.

The first changeover goes very well. As time passes, however, the expedition will face ever greater challenges that will make extreme demands on the organizers, requiring them to append completely new considerations to the existing alternative plans – within a very short time.

In February 2020, when the *Kapitan Dranitsyn* once again sets off into the ice for a changeover, it encounters ice conditions that are considerably more difficult: the sea ice that is progressively disappearing as a result of climate change is precisely what slows the journey of the Russian icebreaker towards the North Pole. While global warming increasingly prevents the formation of thicker, years-old sea ice – the massive pack ice that once enclosed and destroyed many an expeditionary ship – in the new Arctic the wind stacks up and forces together the thinner and smaller ice floes to make thick pressure ridges. Where this ice protrudes up to 4 metres out of the water, the *Kapitan Dranitsyn* is hardly able to pass, as the masses of ice often go down to depths of 20 or even 30 metres. These floating mountain ranges repeatedly block the path of the ship, which laboriously struggles onwards through continually shifting channels in the ice.

Finally, the Russian captain Alexandr Erpulev achieves a masterly feat of seamanship. On 28 February 2020 he guides the *Kapitan Dranitsyn* to within almost a kilometre of the *Polarstern*, thus passing very close to the North Pole in the Arctic winter. In reaching a position at 88°28′ N on its voyage, the Russian icebreaker not only successfully smashes through the difficult ice conditions of the new Arctic, it also smashes a record, as no ship has ever before penetrated so far north so early in the year under its own power. The *Polarstern* has, in the meantime, also set a record: it has drifted to a position at 88°36′ N, and thus only 156 kilometres from the North Pole. Never before has a ship been so far north in the Arctic winter.

The distance between the ships means that the snowmobiles and large snowcats are put to use for exchanging crew and materials. Including the wind-chill factor, the temperature sinks to −58°C (−72.4°F), making the switchover challenging both physically and in terms of planning: heated containers are required to transport longed-for fresh fruit and vegetables undamaged to the *Polarstern*. The wide channel that the *Kapitan Dranitsyn* has cut into the ice freezes over again within just one day.

While the exchange in the Arctic is still in full swing, in Russia a further icebreaker, the *Admiral Makarov*, sets off and sails towards the *Kapitan Dranitsyn*, whose enormous efforts to break through to the *Polarstern* have consumed its reserves of fuel. While still in the ice, the two icebreakers rendezvous and the *Kapitan Dranitsyn* is refuelled so that it can continue on its course for Norway. When it finally puts in at Tromsø on 31 March 2020, delayed by more than three weeks, the expedition members return not only to daylight after months of darkness but also to a world that has unexpectedly changed...

> "During the MOSAiC expedition, I experienced the complete darkness of the central Arctic, the isolation from civilization and frigid subzero temperatures, all which can be challenging to work and live in. But, at the same time, it is one of the most beautiful places I have ever been and certainly one of my most memorable experiences. To know that I am one of a handful of people in the entire world who has set foot on the rapidly evolving Arctic sea ice is a profound honor."
>
> **Jessie Creamean, scientist**

The team of the second leg of the expedition spend Christmas on the drifting *Polarstern*. Over the holiday the men and women of the first leg, by contrast, are already on their way to Norway on the Russian icebreaker *Kapitan Dranitsyn*, sailing into the year 2020. Christmas decorations in the mess of *Kapitan Dranitsyn* create a festive mood.

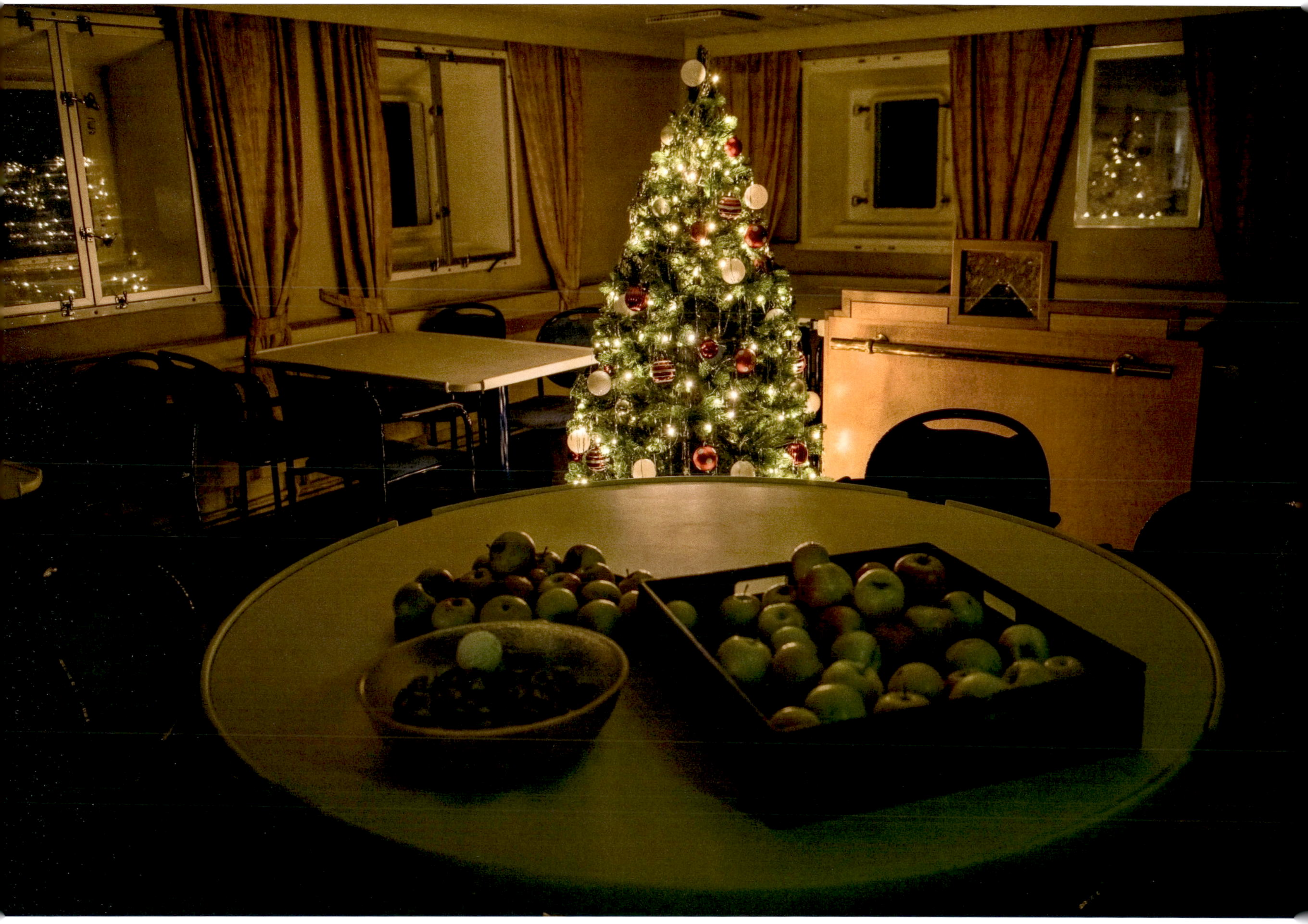

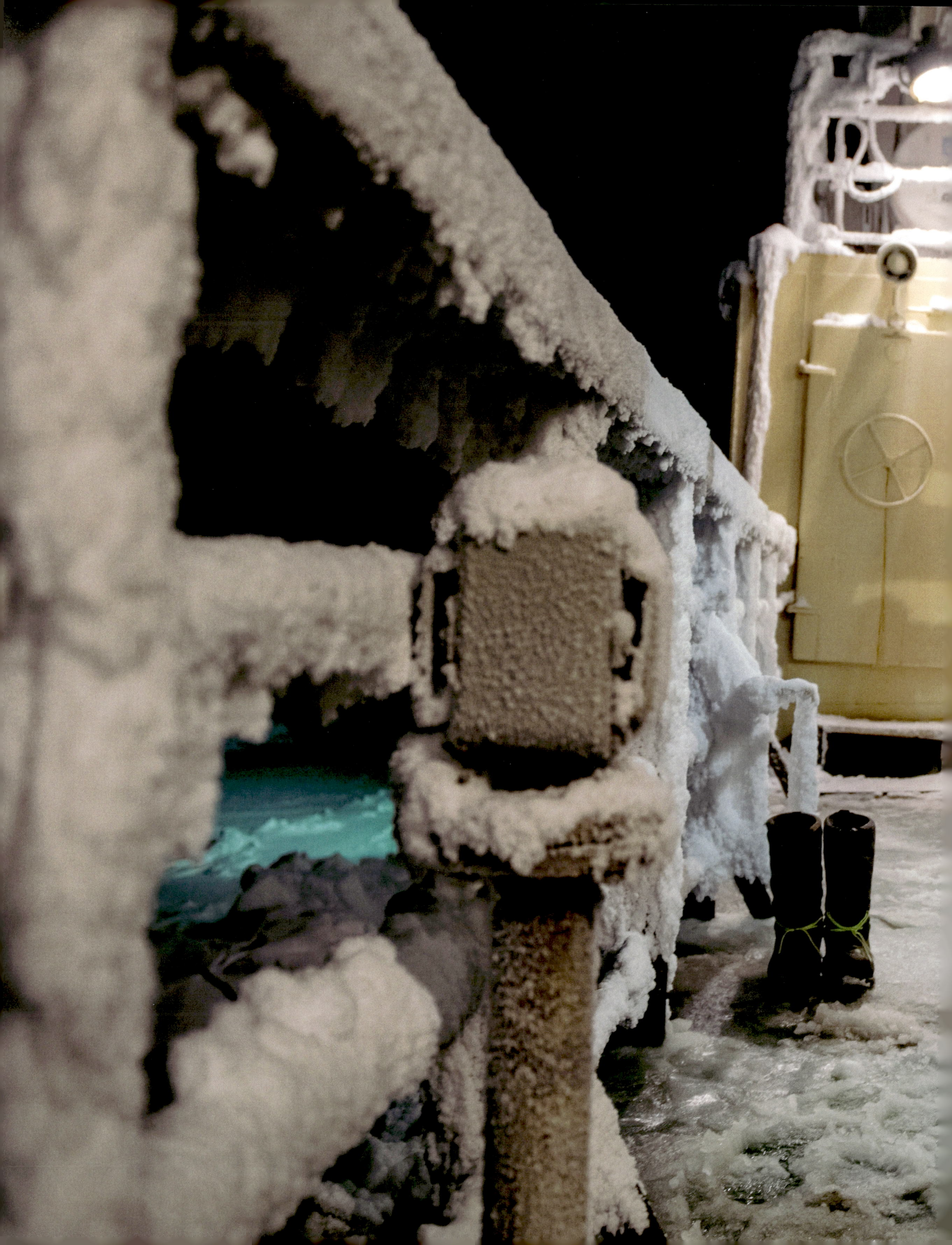

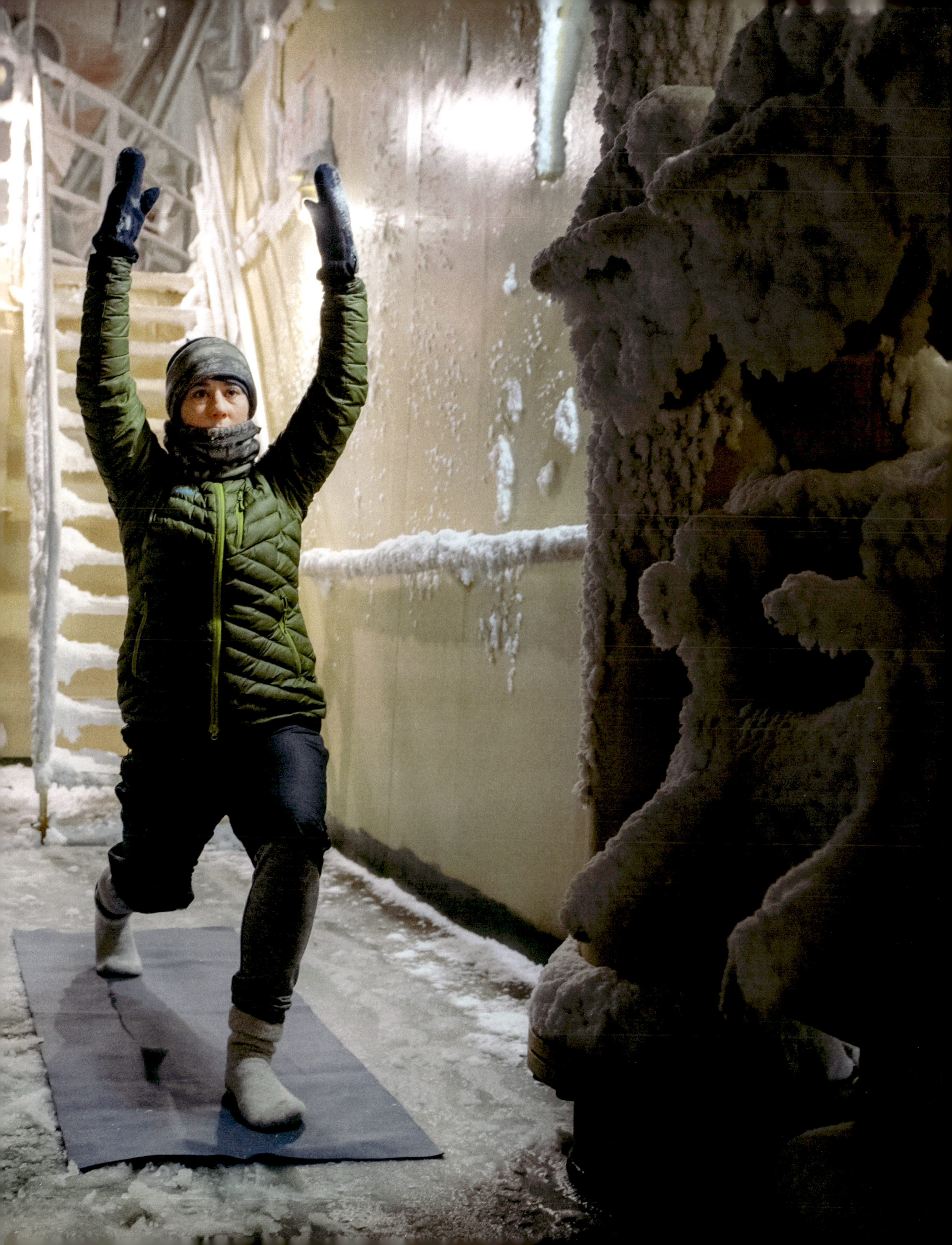

Participants from all over the world have the opportunity on board to experience some typical and charming peculiarities of different cultures and countries. The ice researcher Jari Haapala stays true to Finnish tradition by taking a daily sauna. In this way the others learn the Finnish sauna greeting "Hyviä löylyjä!", which can be translated as "Pleasant steam!"

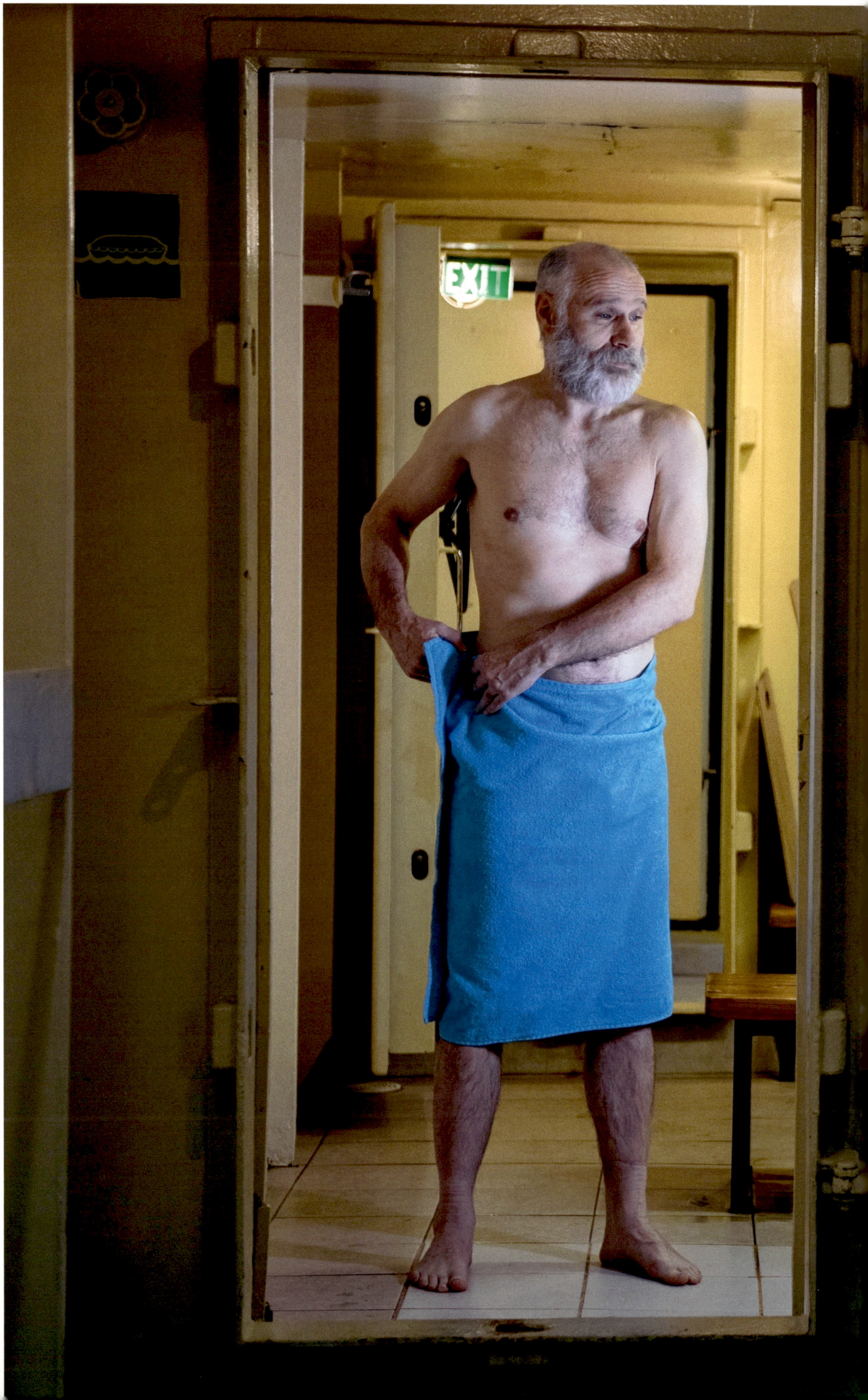

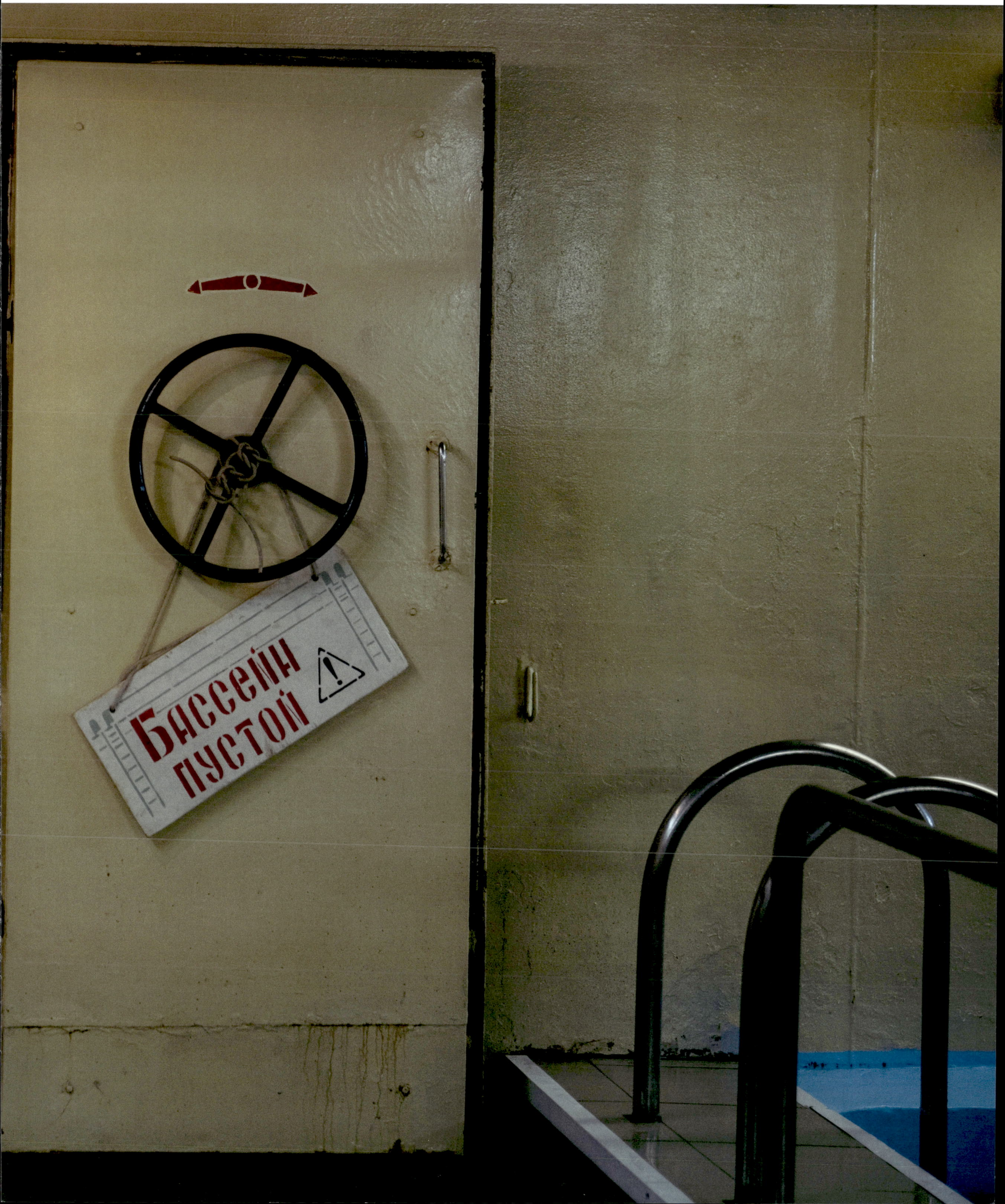
Бассейн
пустой

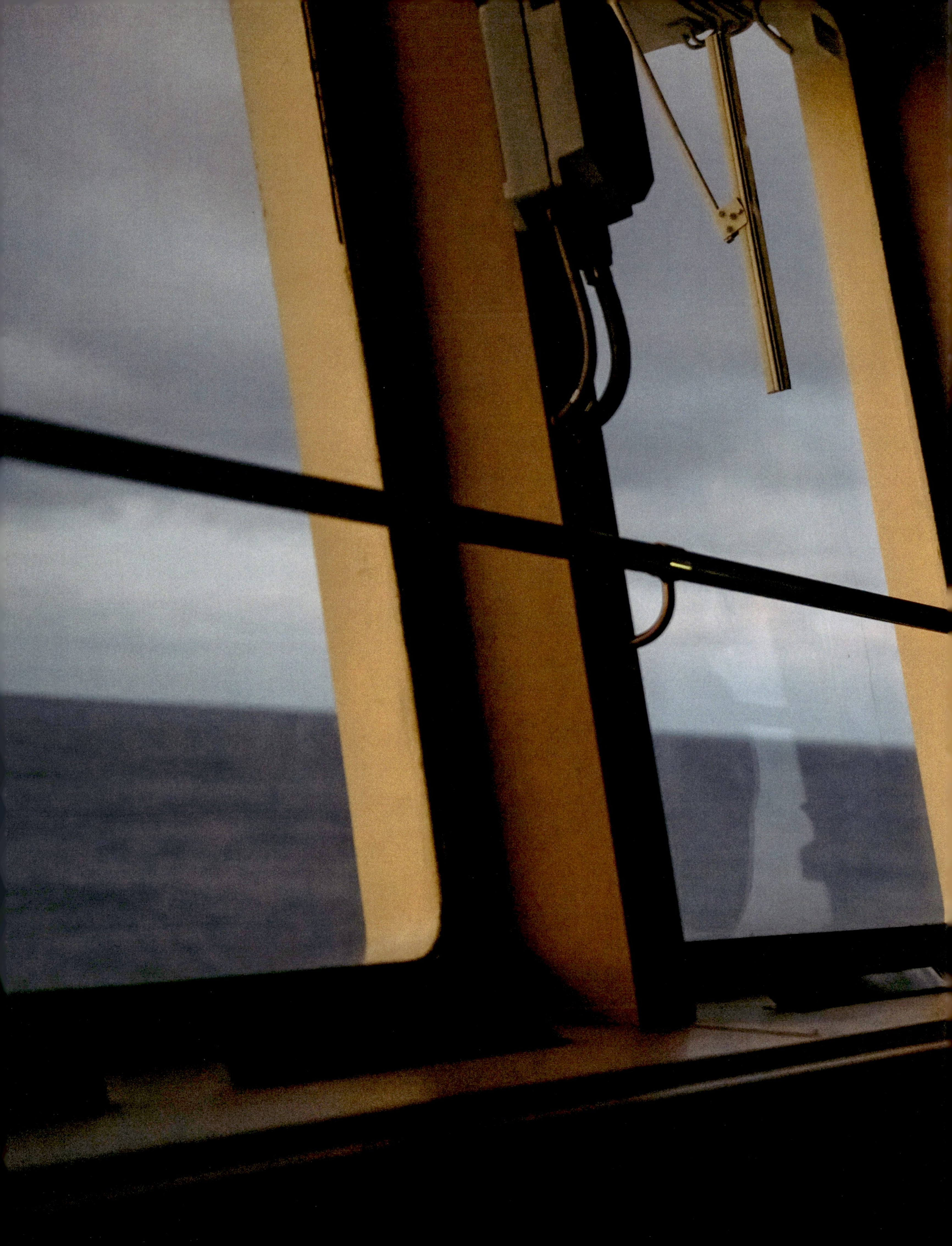

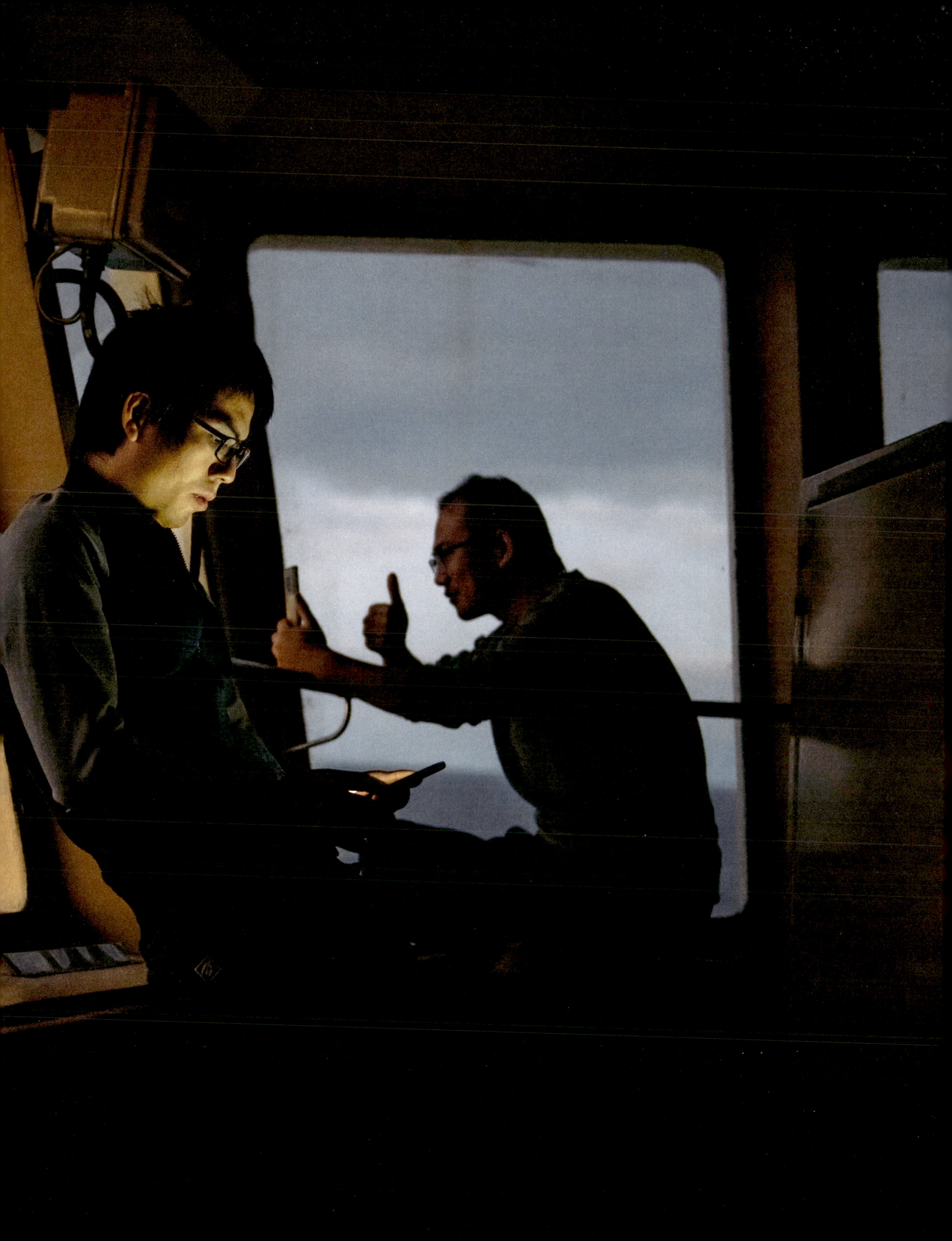

The darkness of the polar night is endlessly fascinating

Esther Horvath spent several months as the expedition photographer on the icebreaker *Polarstern* near the North Pole. Hardly any region is as difficult to capture photographically. Every day photographer and camera were exposed to extreme cold. An interview about the unique Arctic light, a helicopter pilot who sews, and visits from polar bears in the long Arctic night.

How would you describe the MOSAiC expedition in an image?
There is a scene that always comes to mind when I think of the expedition: one evening three scientists are returning to the *Polarstern* through a storm after taking measurements. It is almost completely dark, and the only light comes from the ship. Outside the beam of light there is nothing but darkness. This darkness, combined with storms, describes very well our daily life on the ice during the polar night.

You spent almost four months in the Arctic. How did you prepare for the expedition?
Every expedition needs good preparation. I have been to the polar regions several times, but I have never prepared as intensely as for the MOSAiC expedition. For almost half a year, everything I was doing revolved around it. Like the other participants, I did preparatory training. That gave me a lot of security. If anything happened, I would have known how to react, for example if someone fell into the water. Thanks to the preparations we were highly sensitized to potential dangers during the expedition. During the training I was also able to prepare very well mentally for the expedition.

> "I fell in love with that light. It is not a monotonous white landscape. When the sun and clouds move, the colours in the ice often change." Esther Horvath

MOSAiC was not your first expedition to the Arctic. From a photographic point of view, what was different for you this time?
On my earlier expeditions to the Arctic, it was often light for twenty-four hours a day. I had never experienced the polar night before. I love taking photographs in the dark and playing with light and shadow. The Arctic winter provides a unique visual mood. For me as a photographer, the opportunity to get close to the North Pole in the polar night is something you only get once in a lifetime. I was aware of that every day. That is why I took every opportunity to go out on to the sea ice.

What photographic equipment did you take?
I was shooting with three full-frame cameras, and always had two of them with me on the ice. So when a camera stopped working for a while due to the cold, I could switch quickly. I also had a few special LED lamps with me to provide light. Because batteries drain quickly in the cold, I had to take plenty of replacements. And of course I had a lot of memory cards and hard drives to back up the photos with.

What were the greatest challenges for you as a photographer during the expedition?
Cold hands were the worst. The camera is made of metal and it conducts temperature extremely well. I suffered from that almost every day. Sometimes my hands hurt so much that tears were running down my cheeks. During the last few weeks I found a solution to this problem: I modified the camera a bit and wrapped it in a kind of coat using a special adhesive tape. Because my gloves were also quite bulky due to the hand warmers, I also had to alter the camera shutter with sticky tape so that it was easier to use. I had to work

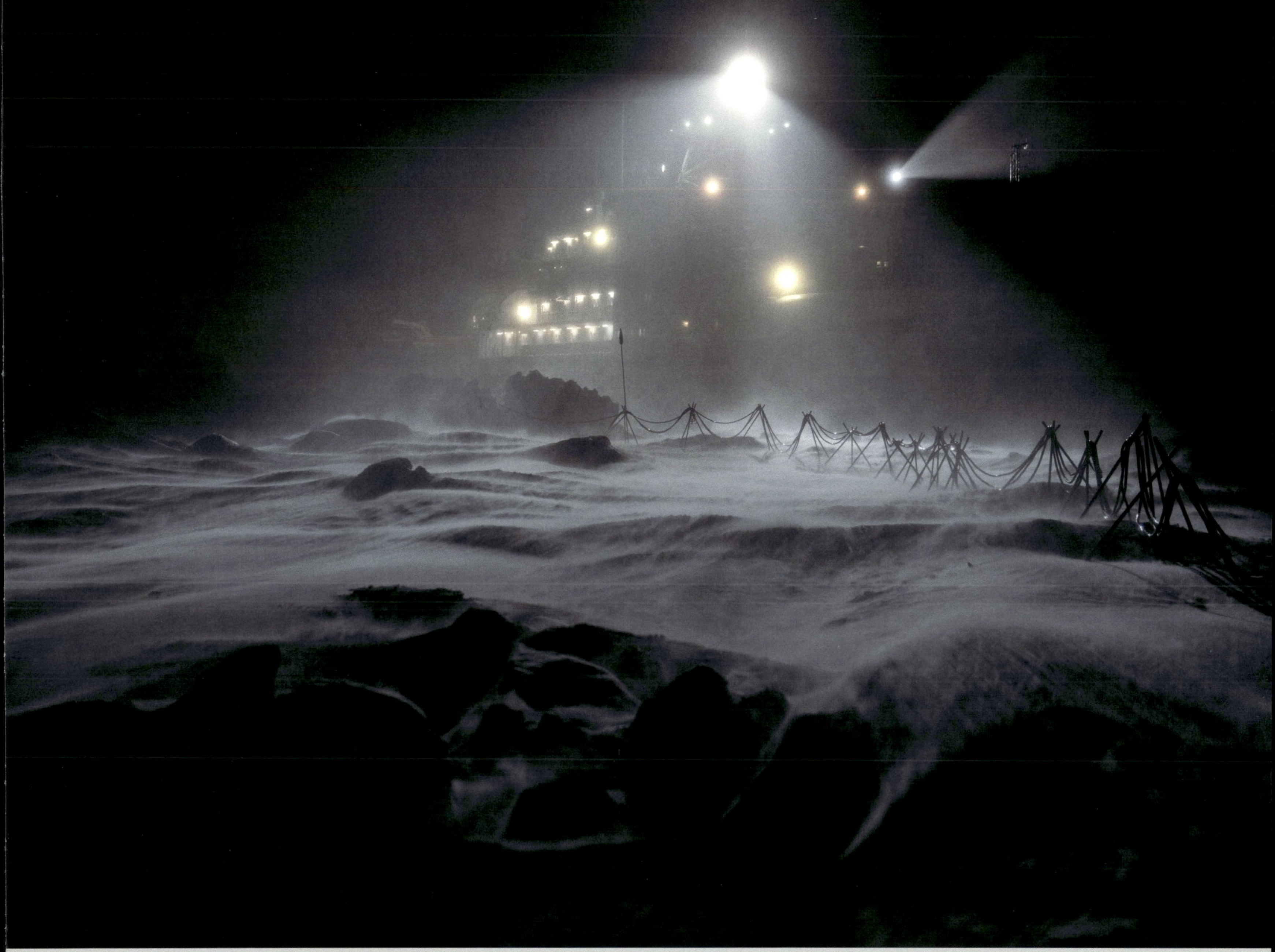

out a quick routine for changing the battery in a few seconds. The frequent snowstorms were another challenge. Sometimes the wind speed would reach 100 kilometres per hour. Without snow goggles you can't open your eyes, but I could barely see what was in focus in the viewfinder through the goggles. On top of that it was dark, of course, and snow whirled up into the air. I often only had a rough idea of where the focus was.

The expedition reached the Arctic before the polar night began. What was the light like at this time?

I fell in love with that light. It is not a monotonous white landscape. When the sun and clouds move, the colours in the ice often change. The reflections of the sun on water and ice are also fascinating. Especially when the sun is low, there are beautiful and diverse colours... orange, blue, purple. You won't find a light like this anywhere else. But the weather can change extremely quickly. Fog suddenly emerges and creates a ghostly mood. You feel as if you are in a film.

What was it like to take your first steps on the ice?

I was allowed to go out with the very first group to explore the ice. It felt like a historic moment, because no one had ever been there before. It reminded me of a moon landing. The idea that this floe would probably be our new home was very exciting. At that time the ice was still almost entirely flat. But here and there you could discover snow-covered melt ponds and pressure ridges. Later the landscape constantly changed. The ice cracked or piled up into gigantic pressure ridges.

What were the most exciting moments of the expedition from a photographic point of view?
Photographically this expedition was a dream come true. I am extremely grateful to have had the chance to photograph this magical scenery. Even though at first glance the Arctic landscape seems rather homogenous during the polar night, I felt I was constantly discovering something new. Life on board, too, was fascinating. We had no internet, no television. It was interesting to see that in isolation on the ship the expedition members spent their time together in a quite different way than at home. Taking photographs of this was very intimate.
How do you proceed when taking photographs?
For every image, I first of all have a story in mind. I do my research beforehand and talk to people a lot. I know when and why I press the shutter release and what I want to say in this way. I usually take photographs with a wide angle of 24 or 35 mm, because I also want to capture all the details in the background. I usually also set up my own light. Then I make no more adjustments, but sometimes I have to wait forever for the moment that I have as an image in my mind, and then I can press the shutter.
What kind of stories do you tell with the camera?
One of the helicopter pilots told me that he sews his own flags to mark the landing sites on the ice. He had a sewing machine in his cabin for this purpose. I liked this story so much that I asked him if I could be present the next time. For this photo it was important for me to also show his cabin. What is it like to live and work in a few square metres? There is a story like that behind every image in this book.
As a photographer, what was your everyday life on board like?
Although every day was different, I had a fixed routine on the expedition and had to organize my time well. I always got up at eight o'clock in the morning. That is relatively late for life on board, and I often missed breakfast. Then I drank coffee and got ready. At nine o'clock I was standing on the work deck, and usually went on to the ice with the first group. Then I took photographs there until lunch. At one o'clock in the afternoon I went out for the second time and stayed until the evening meal. After that we had various discussions to plan the next day. Late in the evening I often sorted and edited the photographs.
For one of your photographs you received the World Press Photo Award in the Environment category. It shows a polar bear mother and a cub walking inquisitively through the research camp. How did this photo come about?
It was shortly after the evening meal and I had gone out on deck again to take pictures. Suddenly I noticed that many of the expedition members were observing the ice floe from the bridge, because two polar bears were approaching the ship. Then I went right to the front, to the bows. To see two polar bears in the dark wilderness and at the same time to be safely on the ship myself – this was a one-off opportunity. I am not a wildlife photographer, but I wanted to record how the polar bears react to our research camp. After all, we were not alone in the Arctic, but only guests in the realm of the polar bears.
Was there anything on the expedition that could not be expressed through photography?
Yes! It is very difficult to convey photographically our limited field of vision. We could only see what lay within the light of the ship's searchlights or in the beam of our headlamps. And when it was also windy, you couldn't hear anything either. In this extreme region, the senses work quite differently than usual. The darkness of the polar night is endlessly fascinating. Every day out there had me spellbound again, standing in the middle of the sea ice in this blackness. And every day I wondered whether this alien and at the same time magical world could be real at all, or whether I was looking at a film set.
What personal experiences have you taken away from the expedition?
As a photographer I learned a great deal. In my time there I gained a lot of experience with photography in extreme cold and darkness. And I also got an increasing understanding of the work of the scientists. When I look at my photos today, I can see how I developed. But most of all it is the unique experience of standing on the sea ice and being aware that below us there is an ocean several kilometres deep. That is such an incredible feeling. It is almost indescribable.

The tracks of the snowmobile stand out in stark contrast from a landscape reminiscent of the surface of the moon. Some participants in the expedition admire this view with amazement, and many share a feeling that the blue Earth must be about to rise from behind the colourless horizon.

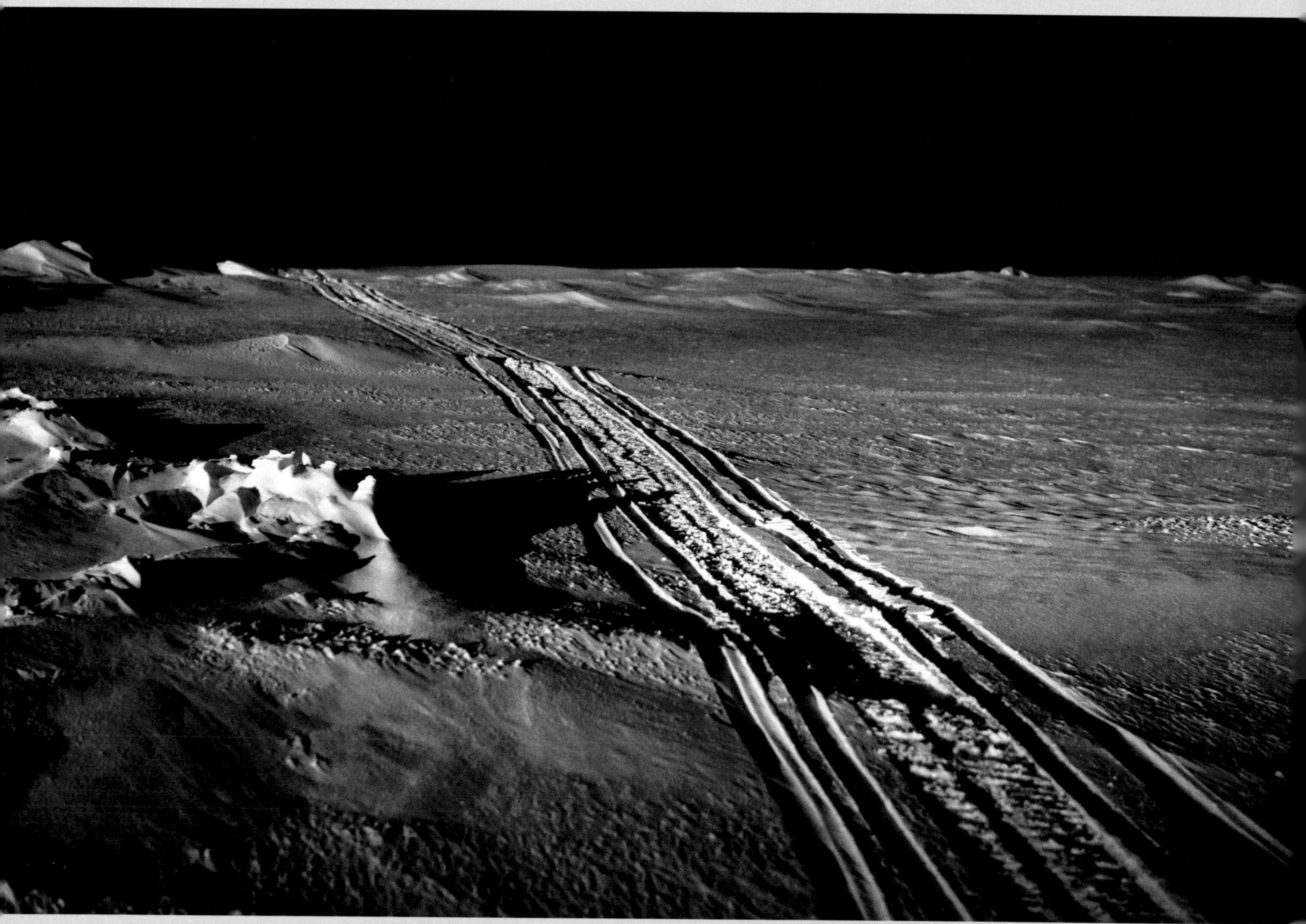

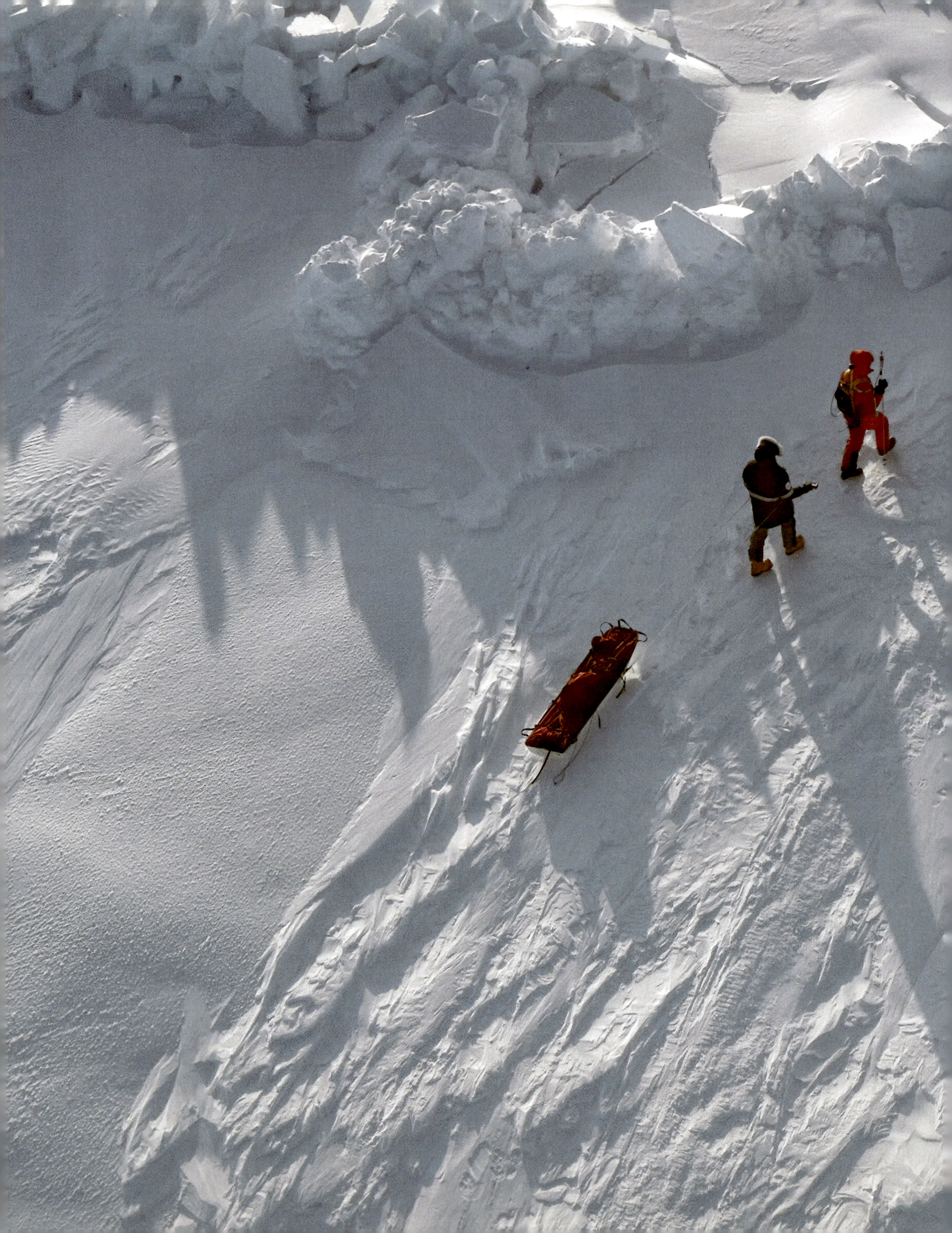

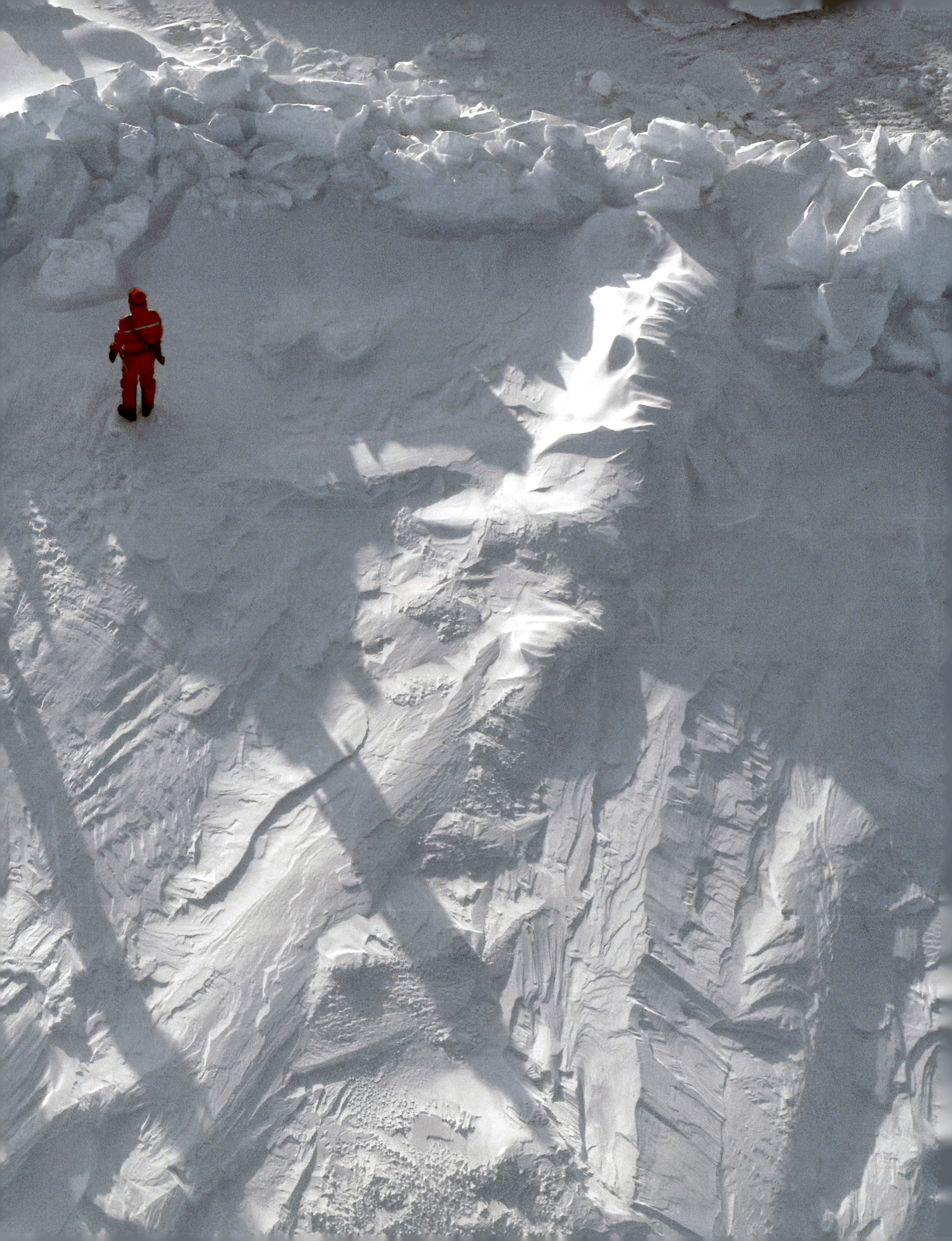

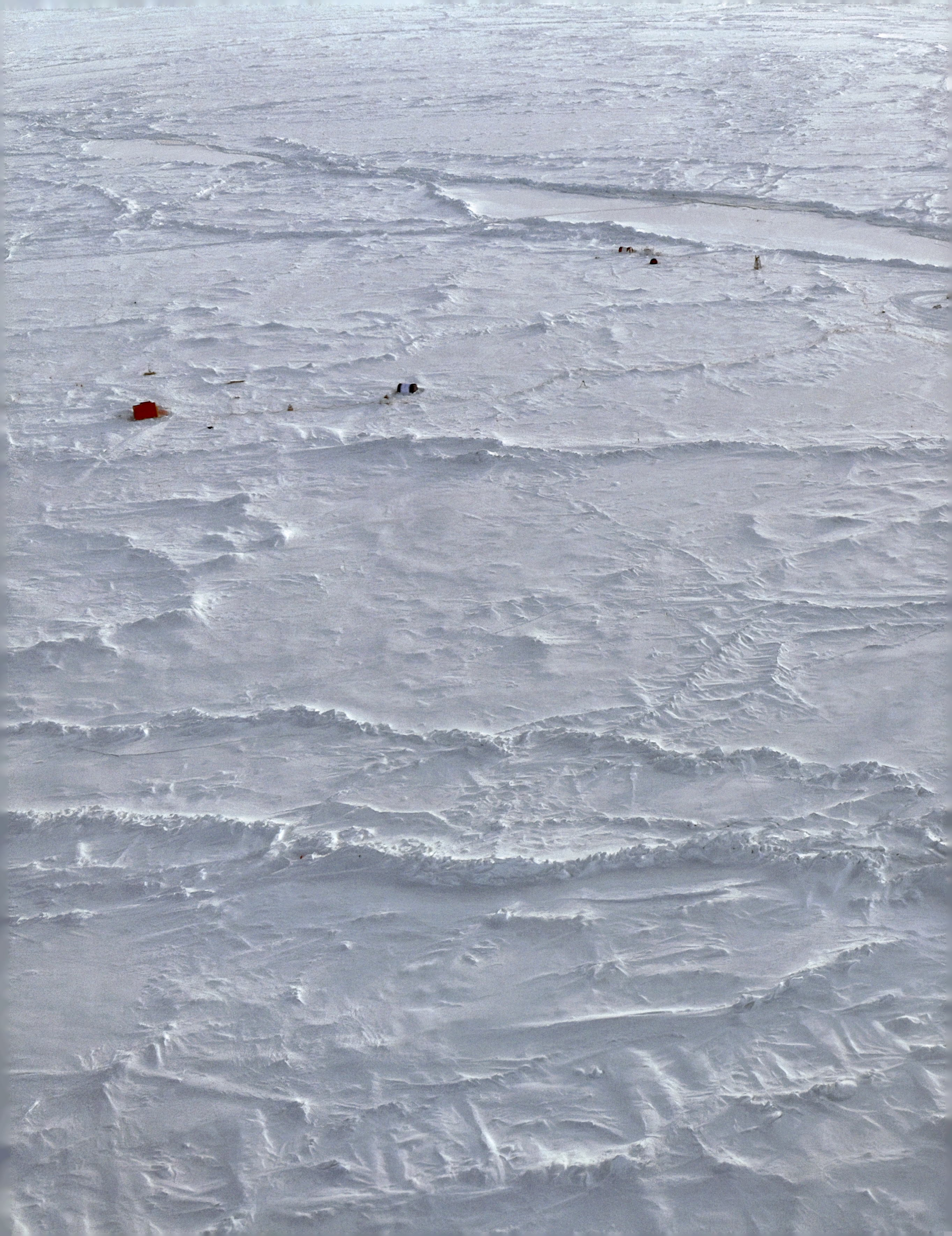

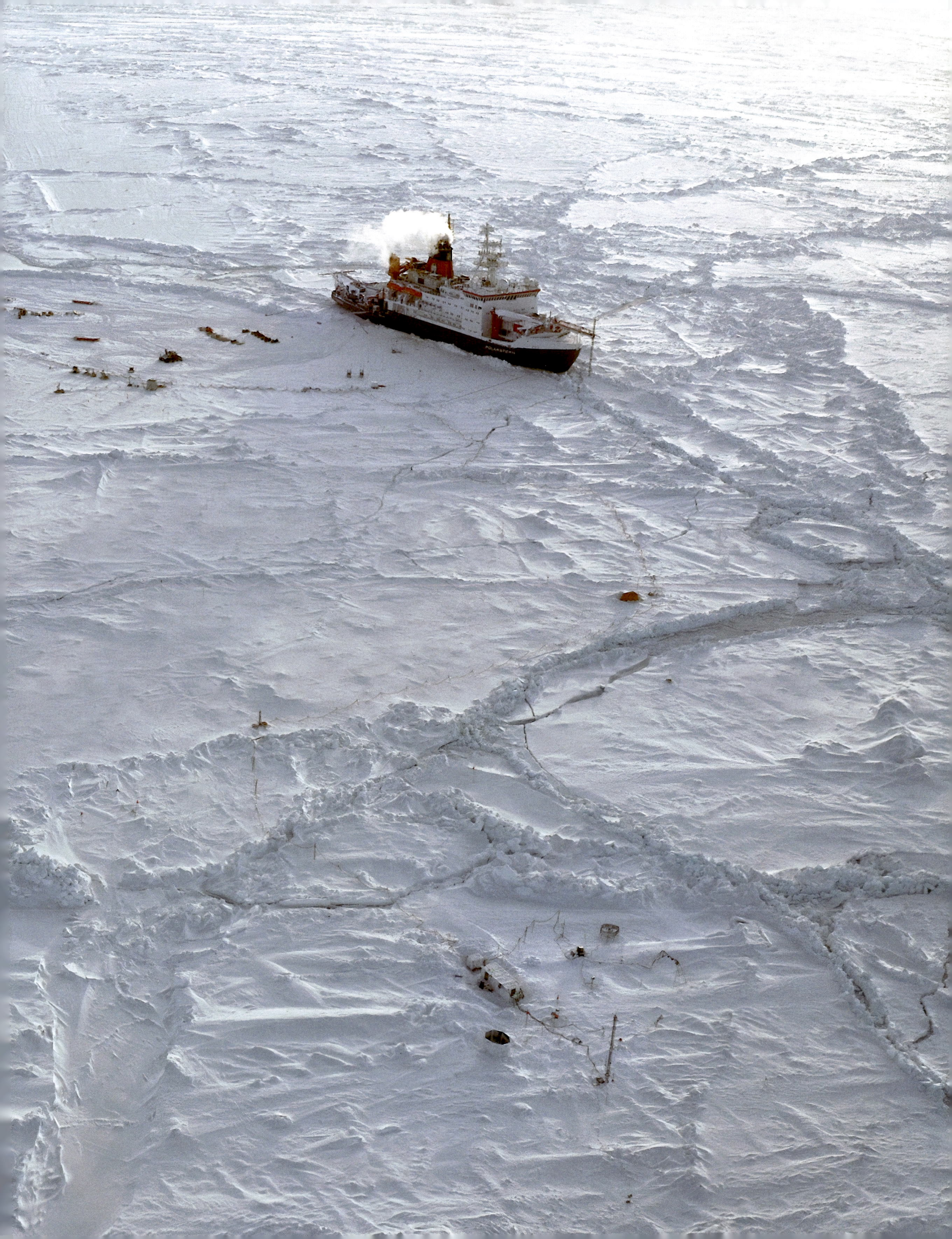

In early March the sun reappears over the horizon for the first time. It heralds the polar day, which begins soon after. Daylight remains round the clock. Thus in a short time the scientists' working environment changes from one extreme to the other – from constant night to permanent day.

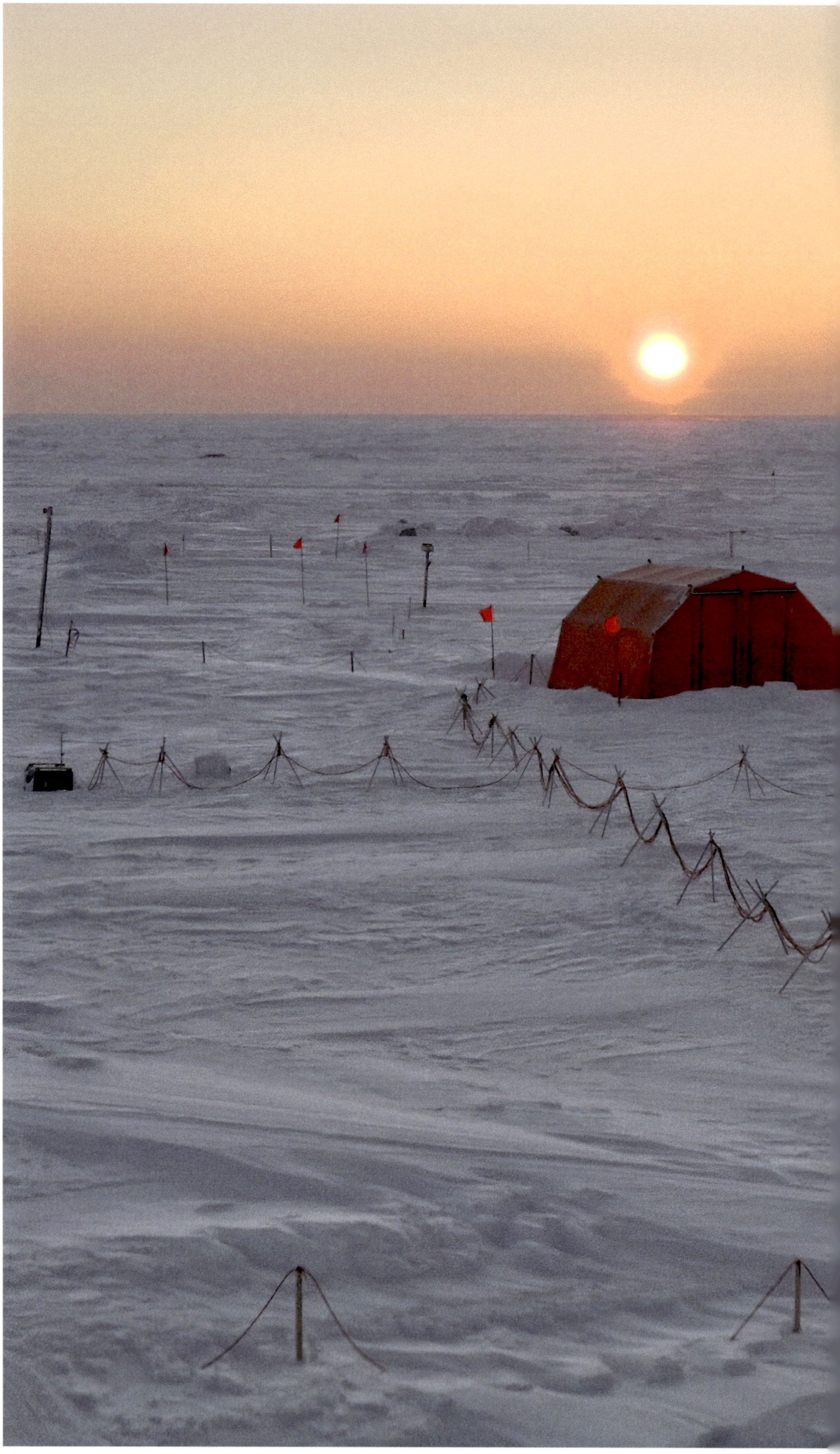

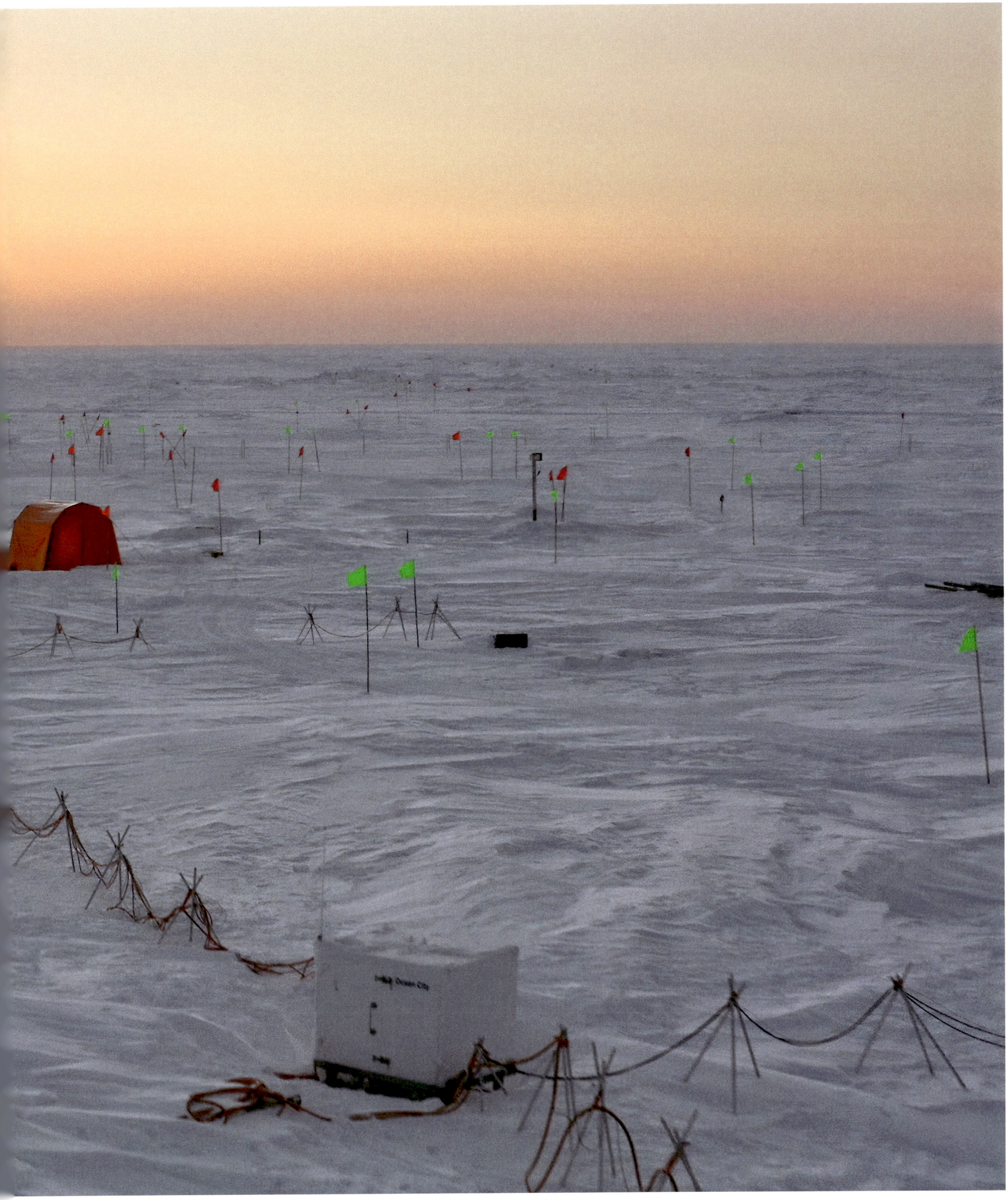

I

In March 2020, as the team from the second leg were setting their feet back on terra firma after months at sea, the team for the third leg was already drifting through the Arctic Ocean. The expedition was progressing as planned, but on land great changes were on the horizon.

Even before the third team set off in January 2020, news of a new type of lung disease in China had swept the world. But the news then had yet to suggest that the coronavirus pandemic would soon have the entire world it its grip. Only two months later, countries around the world were closing their borders in the face of the pandemic. Public life and international traffic had come to an almost complete standstill – and the coronavirus pandemic became a threat to the continuation of the MOSAiC expedition, putting into doubt whether new participants and supply ships from all over the world could travel to the *Polarstern*.

The team for the fourth leg on the ice floe was to have been rotated in by aircraft taking off from the Svalbard archipelago, as in spring the sea ice is at its thickest and represents an almost insuperable obstacle even for icebreakers. But the coronavirus had undermined all expedition planning. To protect Svalbard, the Norwegian authorities sealed off the archipelago. The alternative route via the Russian archipelago Severnaya Zemlya was also closed. One after another, various alternatives proved to be unfeasible. Although the team was adequately supplied with food and fuel, the *Polarstern* was suddenly cut off from the rest of the world.

The MOSAiC expedition nearly came to an end only halfway through its projected duration, but thanks to great resolution and commitment, the team refused to give up and within weeks developed an entirely new alternative plan by which the research mission could continue despite the global standstill. For those then on board the *Polarstern*, this meant remaining on the expedition almost twice as long as originally planned. The participants were to spend more than four months in the Arctic and – with the world watching – maintain their position in the ice.

In the meantime, the light returns and the long polar night gives way to the equally long polar day. The sun rises ever higher over the icy horizon until at last it no longer sets. Sunglasses become a vital piece of gear, and the headlamps disappear into the bottom of duffel bags. The icy landscape, previously reduced to the reach of searchlights, now reveals its full extension before the expedition members. Their extended stay may have been involuntary, but the researchers take advantage of their extra time in the ice. They investigate the ice floe under the new springtime conditions from every conceivable perspective and document how the life of the Arctic becomes ever more active under the light of the sun. At the same time, and with ever greater intensity, the floe demonstrates the dynamic nature of sea ice, and the work of the research team is anything but easy. Ridges of compressed ice push upwards and cracks both large and small appear in the ice. The expedition is now living up to its name. On the increasingly fragmented floe mosaic, the different stations of the research camp begin to move. Permanent routes are no longer guaranteed: the meteorological measuring tower in the ice camp can at times only be reached by kayak. For a while the ice is so unstable that the team even discusses "parking" the *Polarstern* in a new site, having it frozen into the ice again somewhere else.

Meanwhile, strong winds drive the *Polarstern* away from the North Pole and towards the Fram Strait – and therefore close to Svalbard earlier than expected. This is advantageous from a logistical point of view for the newly adopted plan: two German research ships, the *Sonne* and the *Maria S. Merian*, are en route to Svalbard to take the next team to the Arctic. To avoid transmission of the coronavirus to the *Polarstern*, the newly arrived members undergo two weeks' quarantine in Bremerhaven. However, the two ships are

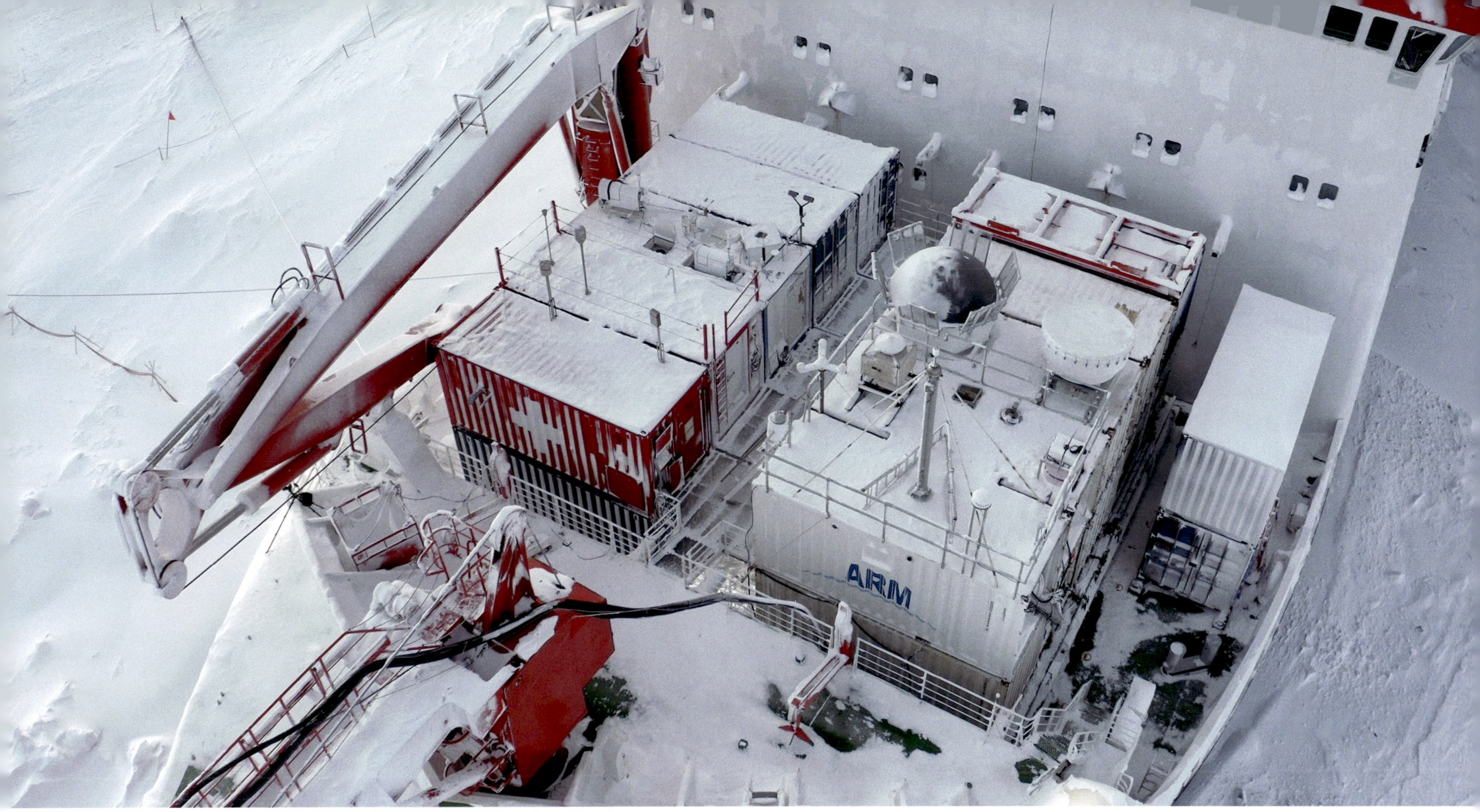

not icebreakers and can only take the team and cargo as far as the waters around the archipelago, in other words, to the edge of the sea ice. The new plan thus means that the *Polarstern* itself has to sail out: after seven months and twelve days, the research icebreaker interrupts its drift. And even as the *Polarstern*, after a tremendous effort, sets off for the edge of the sea ice, the researchers still keep an eye on their home floe: with autonomous measuring instruments they continue to gather important data now that the melt season has begun. More and more often the temperature rises to a level around the freezing point. The snow cover disappears from the sea ice. On the way to Svalbard the team notice how ice floes that were previously a radiant white are slowly acquiring a bluish colour and how melt pools are expanding. Nevertheless, the sea ice does not make life easy for the *Polarstern*. The changing, melting surface conceals solid two-year-old ice which finally, only after a fierce struggle, releases the *Polarstern*.

But then it's done. After nineteen days the icebreaker reaches the other two ships in the Adventfjorden off Svalbard. The teams carry out the exchange of provisions between the three ships in calm waters. As the newly arrived crew members settle in on board the research vessel, the crew from the third leg make their way home at last.

The *Polarstern* finally sets off again. In the approaching summer months the scientists will continue their search for missing pieces of the Arctic climate puzzle. They will take back tonnes of samples of water, snow and ice. Huge quantities of zooplankton and phytoplankton will end up in laboratories all around the world. Hundreds of terabytes of data will be fed into model calculations. It is data and knowledge about a changing Arctic – about a region that reveals our idea of eternal ice to be an illusion. Climate models are already predicting with high probability that before 2050 the Arctic Ocean will be largely ice-free in summer. As an ice-free ocean completely changes the face of the Arctic, this new North Pole has long been of interest to industry and a subject of political ambitions: where the sea ice no longer stands in the way, there are prospects of large deposits of raw materials, new fishing grounds and long-coveted shipping routes – for cargo and even cruise ships. This outlook represents a great risk to the Arctic environment and its sensitive ecosystem, thus making it all the more necessary to have a profound understanding of the region. With this knowledge it will be possible to establish a clearer framework for sustainably managing the Arctic.

The data the *Polarstern* brings home from its drift expedition are thus of inestimable value. In the years to come, the work of ana lysing this data will occupy scientists across the globe and provide urgently needed knowledge about the Arctic and the global climate system. Only then will we have a better understanding of the role that it plays in the earth's climate and of what it means for the whole world to lose the old Arctic.

The North Pole is far, far away. Yet what happens there does not just stay in the Arctic, and, conversely, the behaviour of people around the world has an impact on this distant region. The future of the Arctic is in our hands.

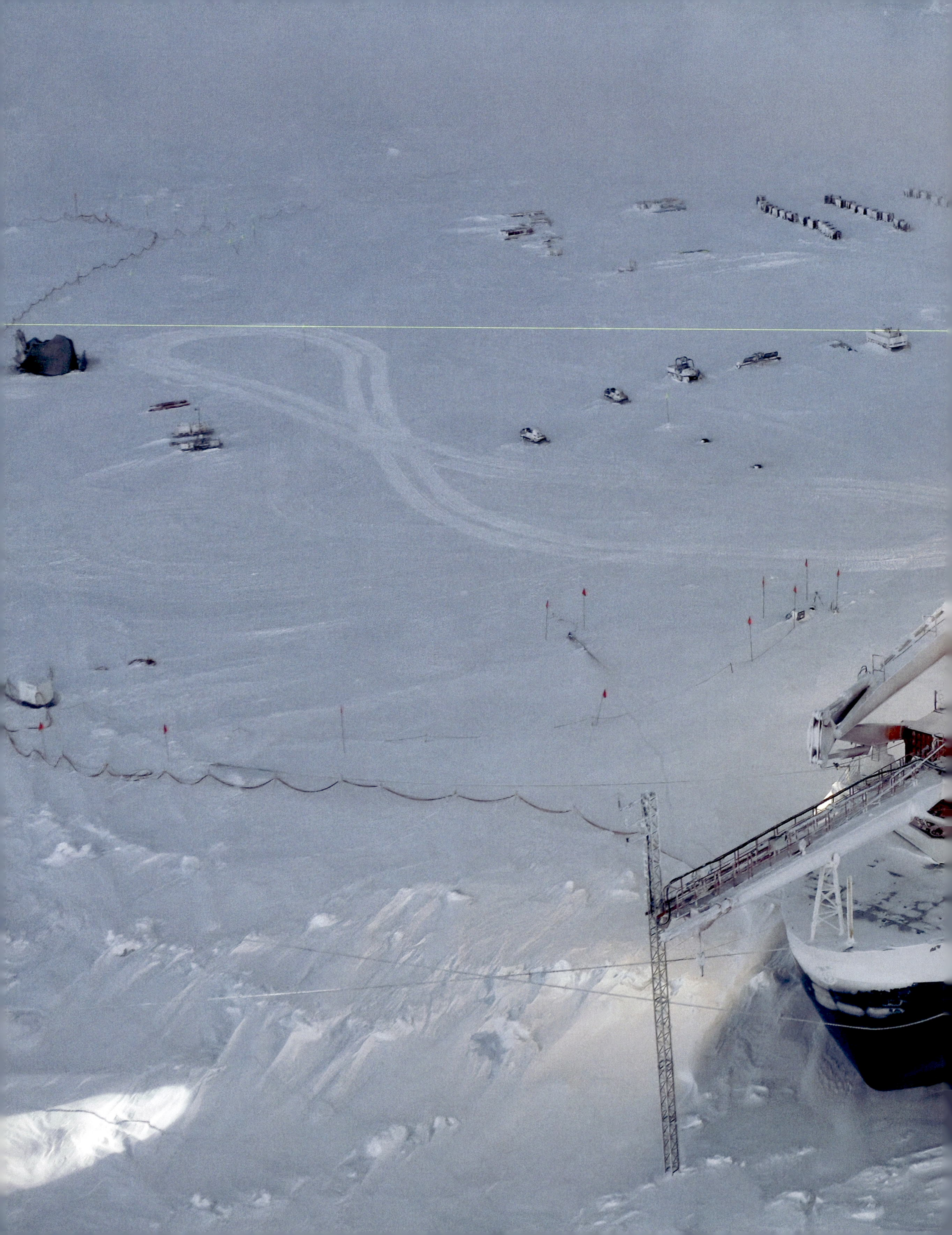

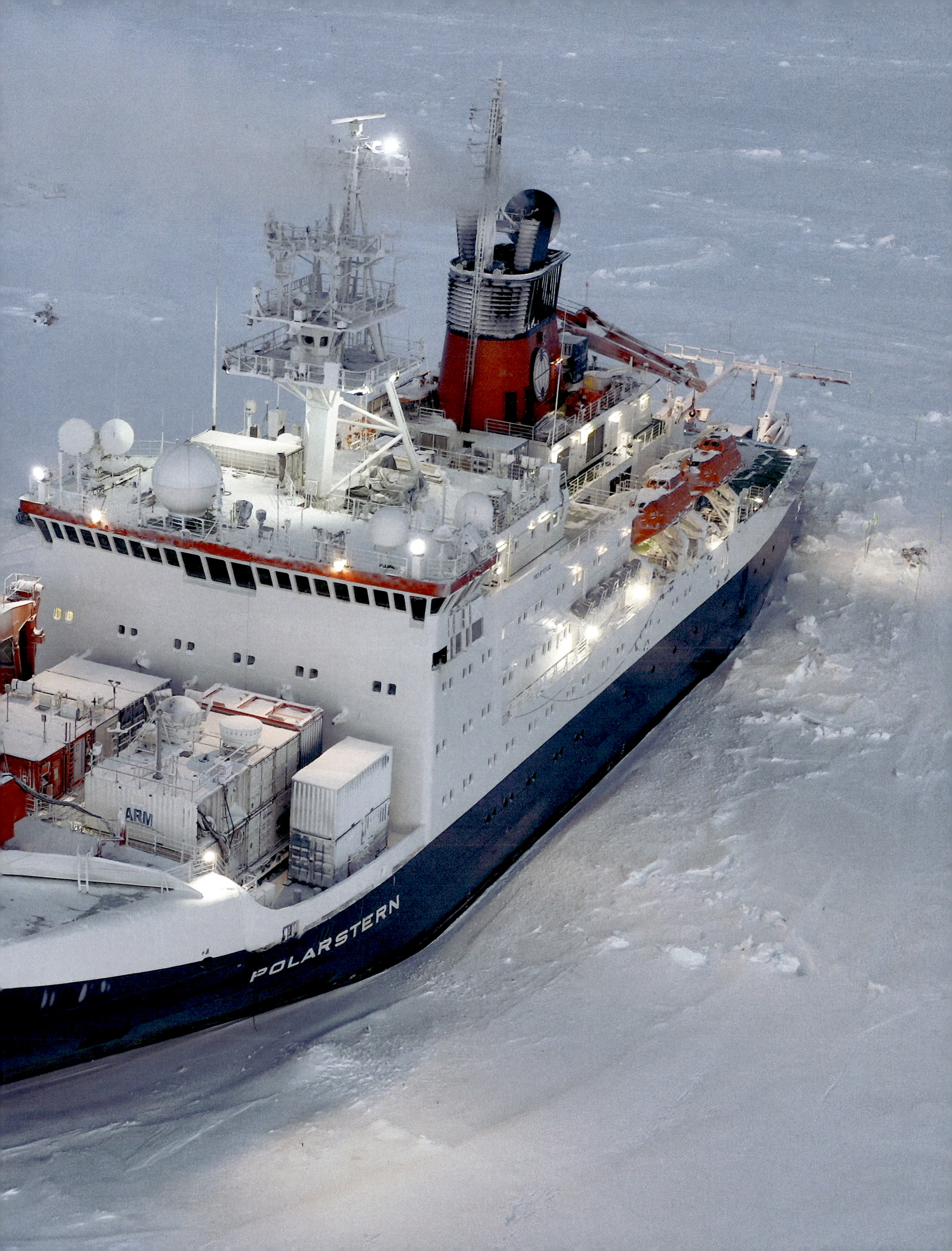
ARM
POLARSTERN

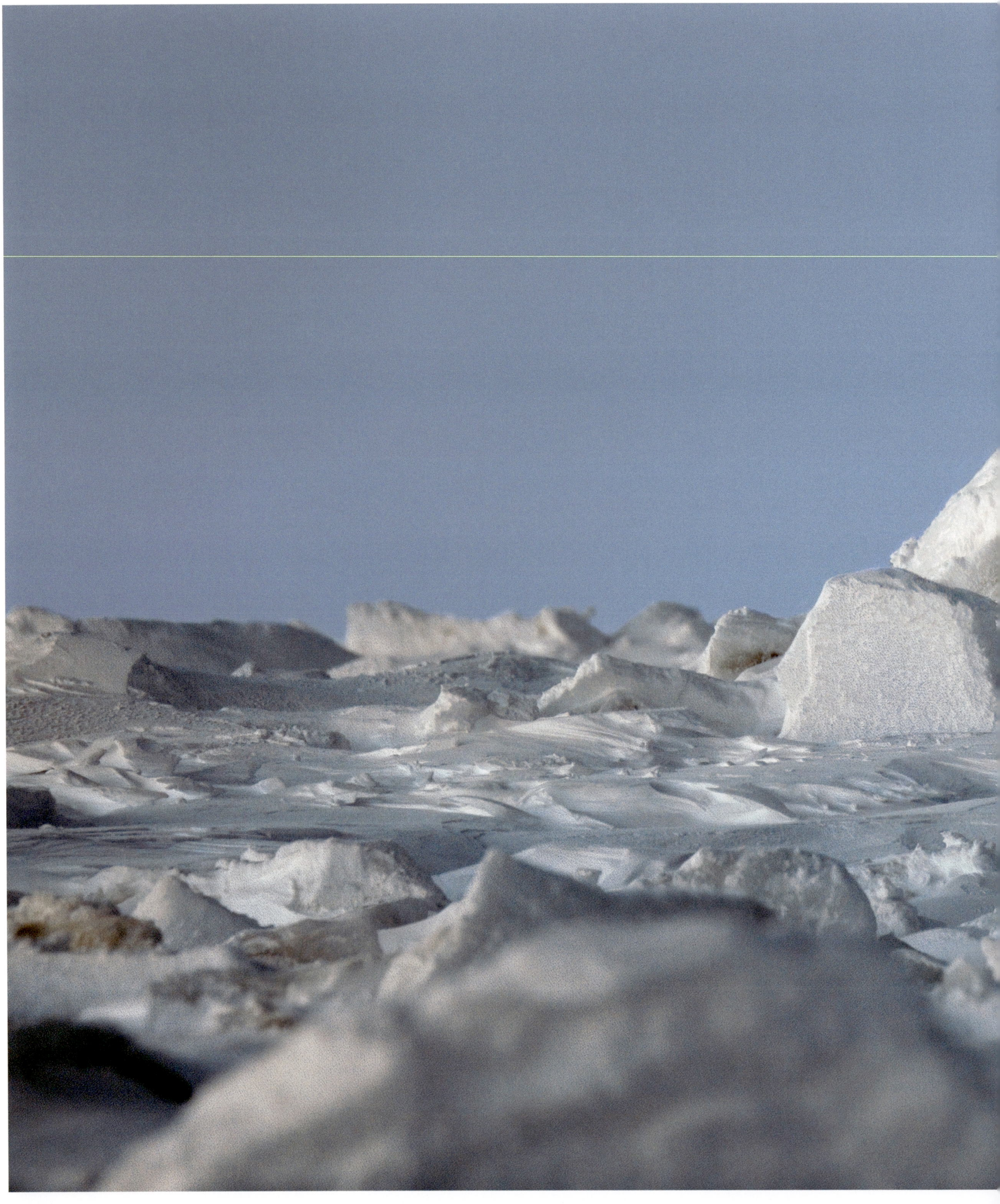

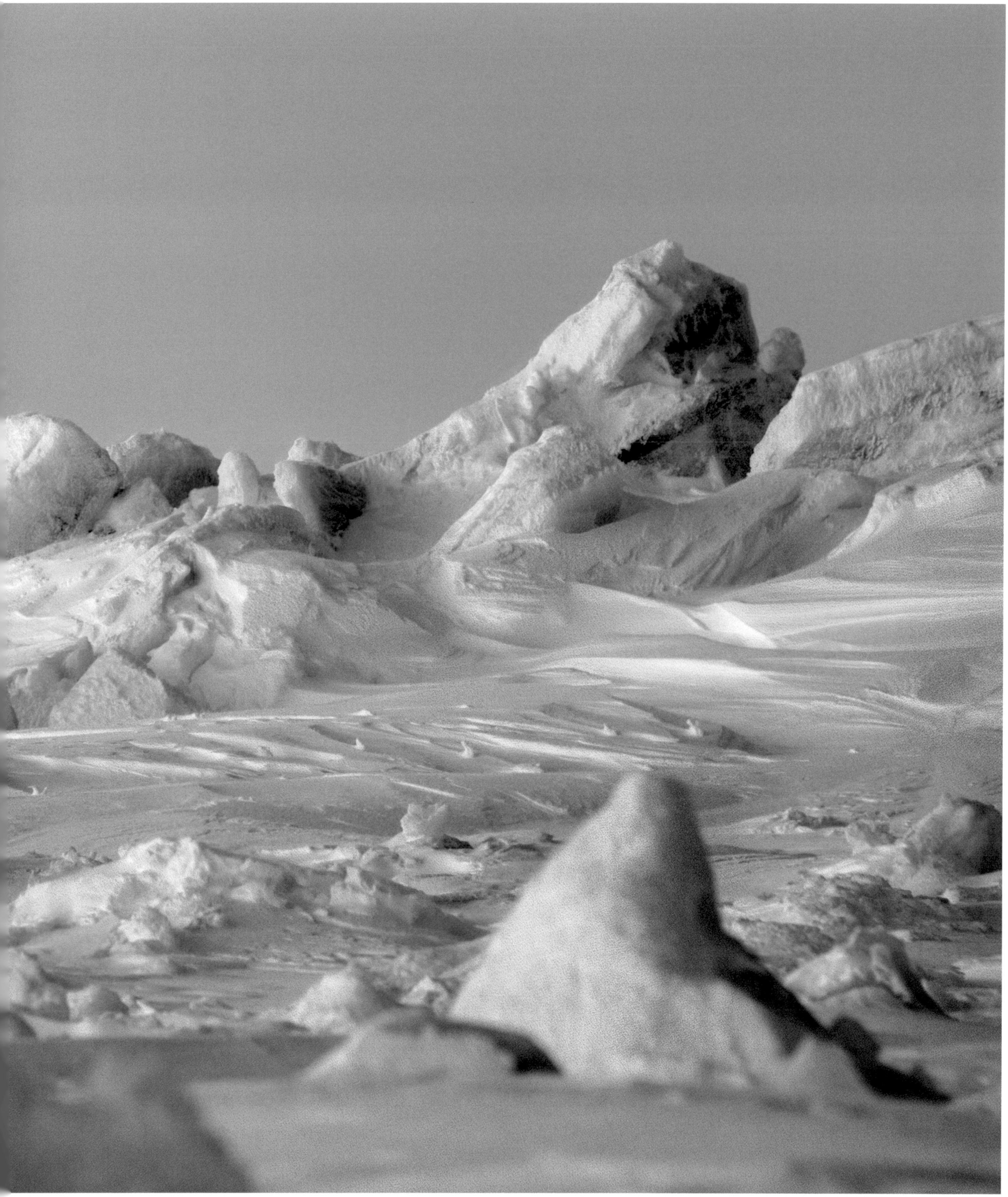

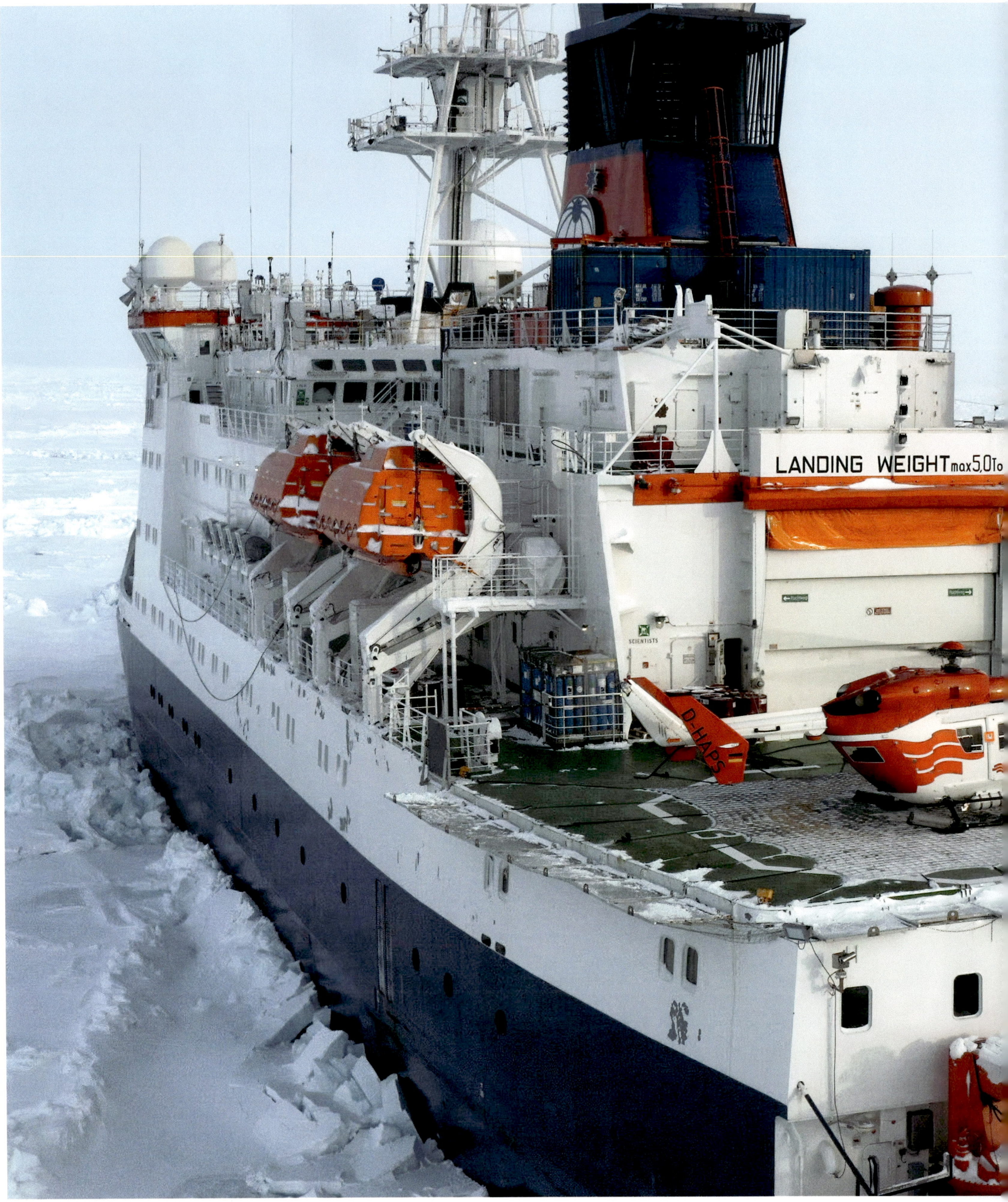
LANDING WEIGHT max 5,0 To
D-HAPS

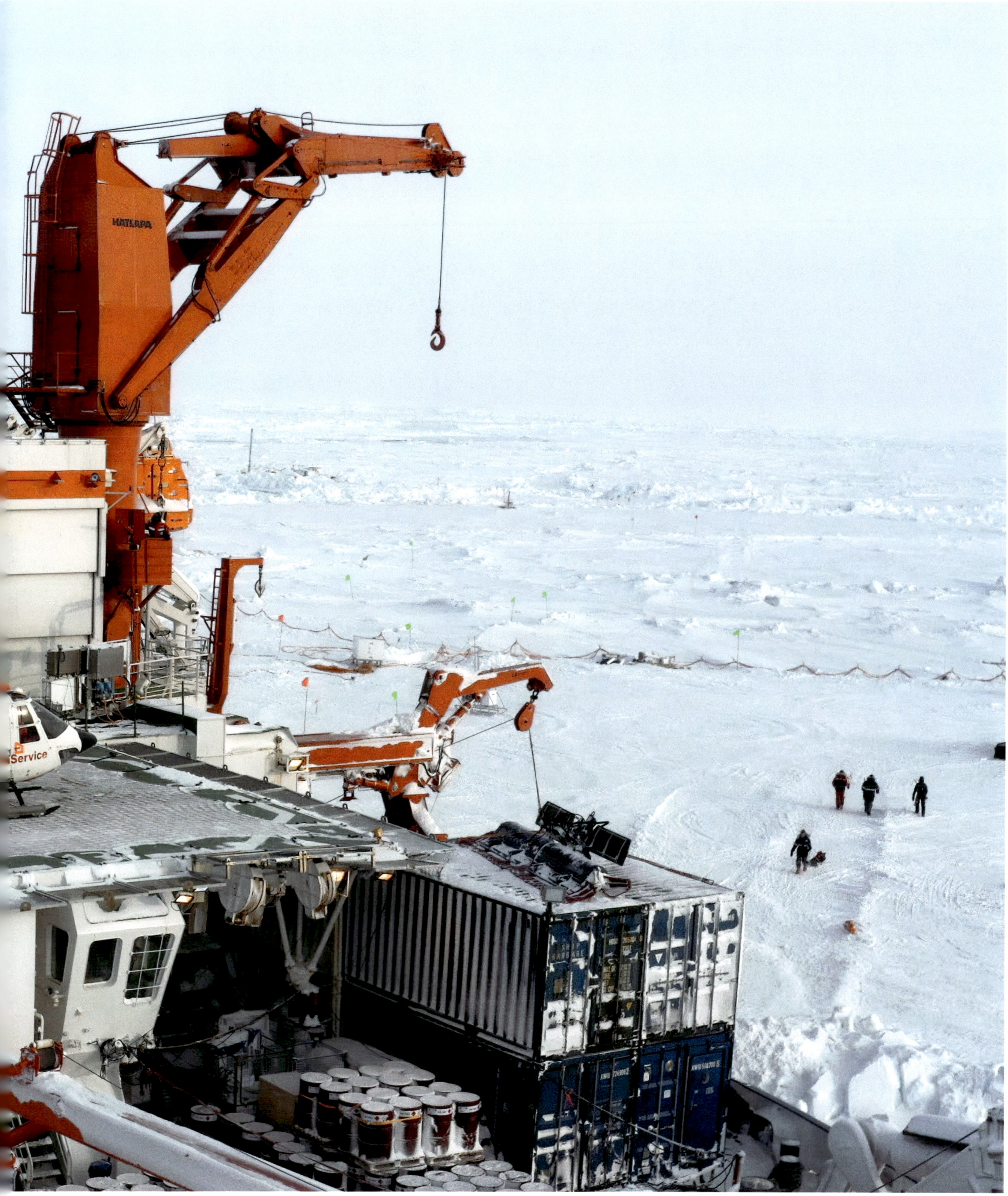
HATLAPA
Service

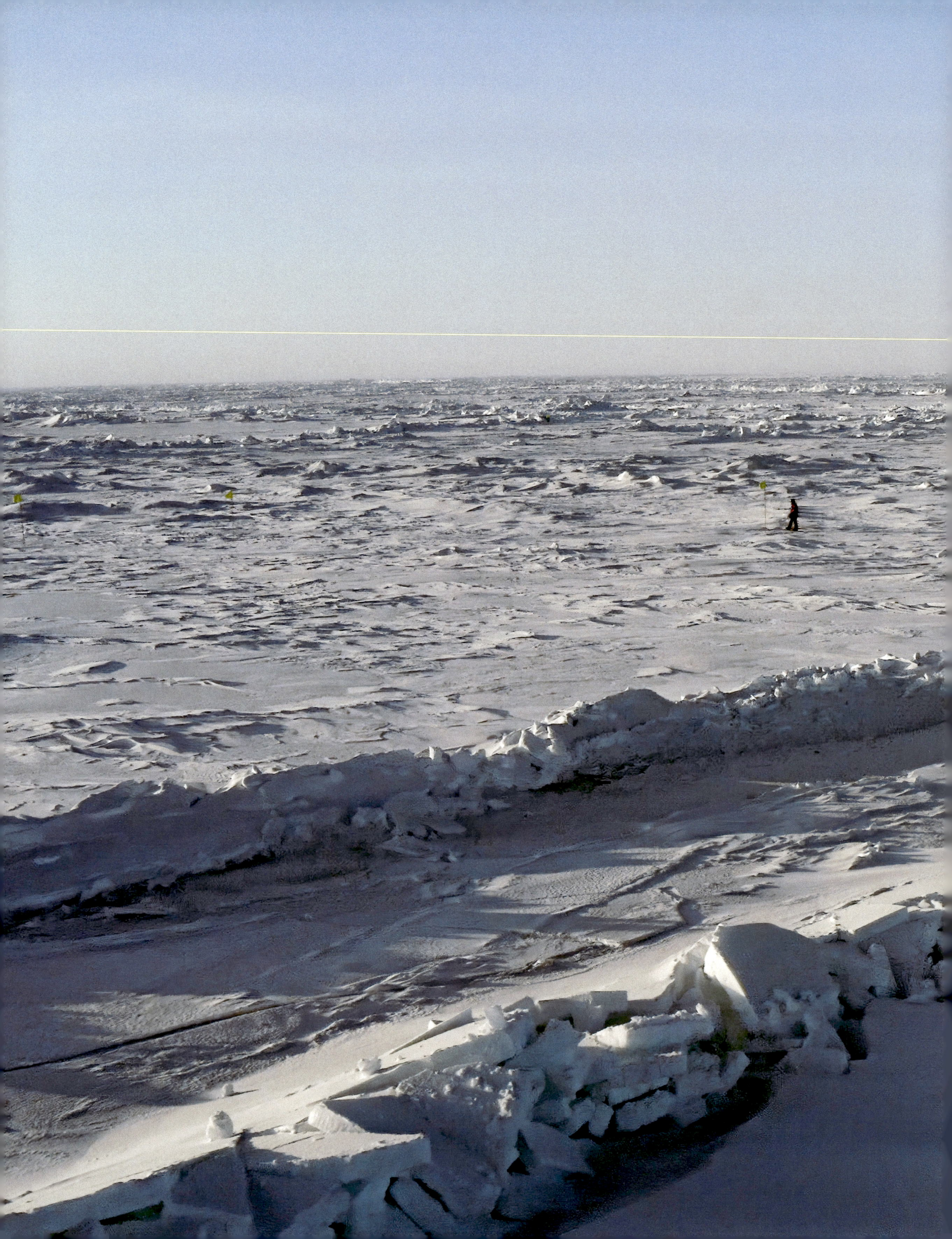

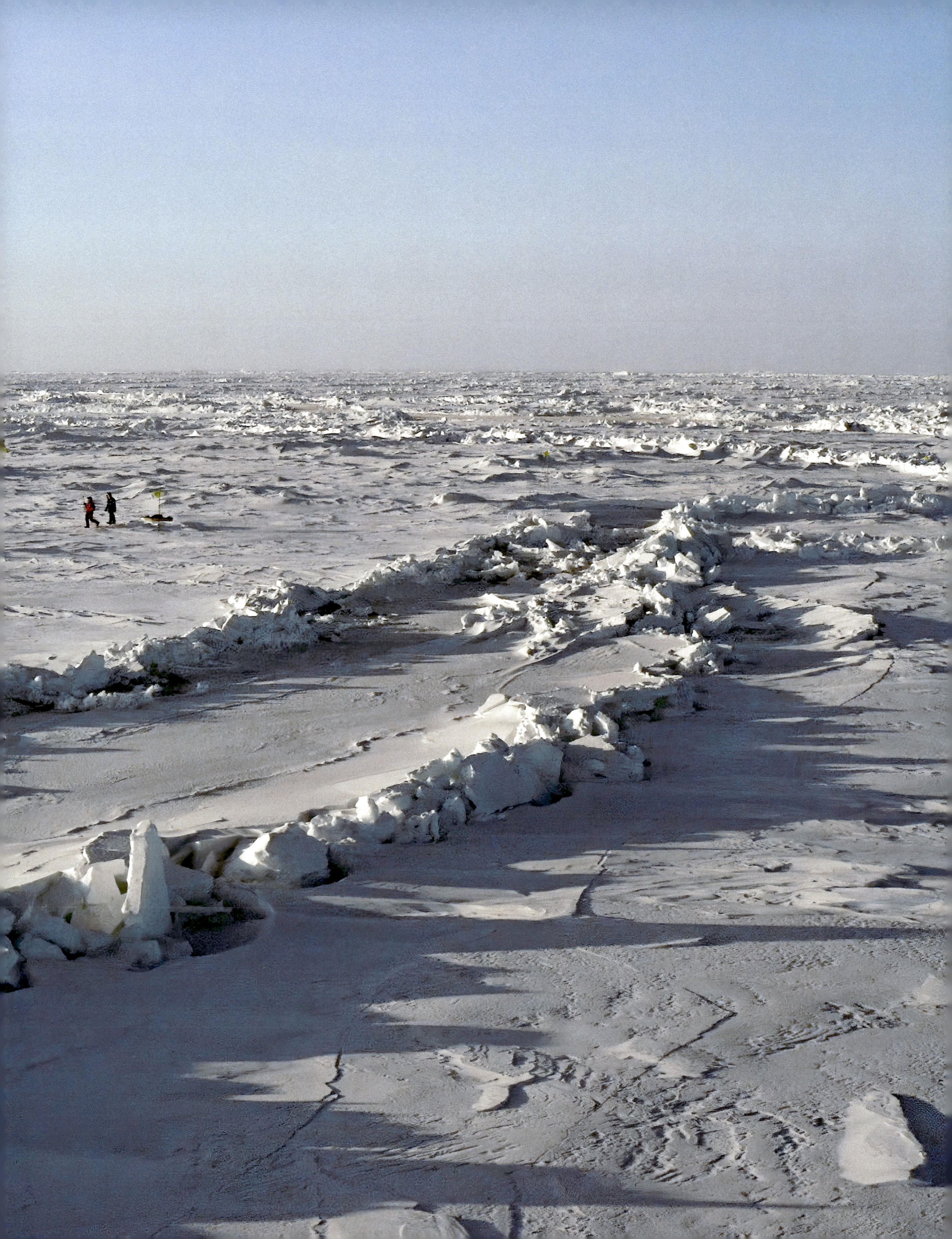

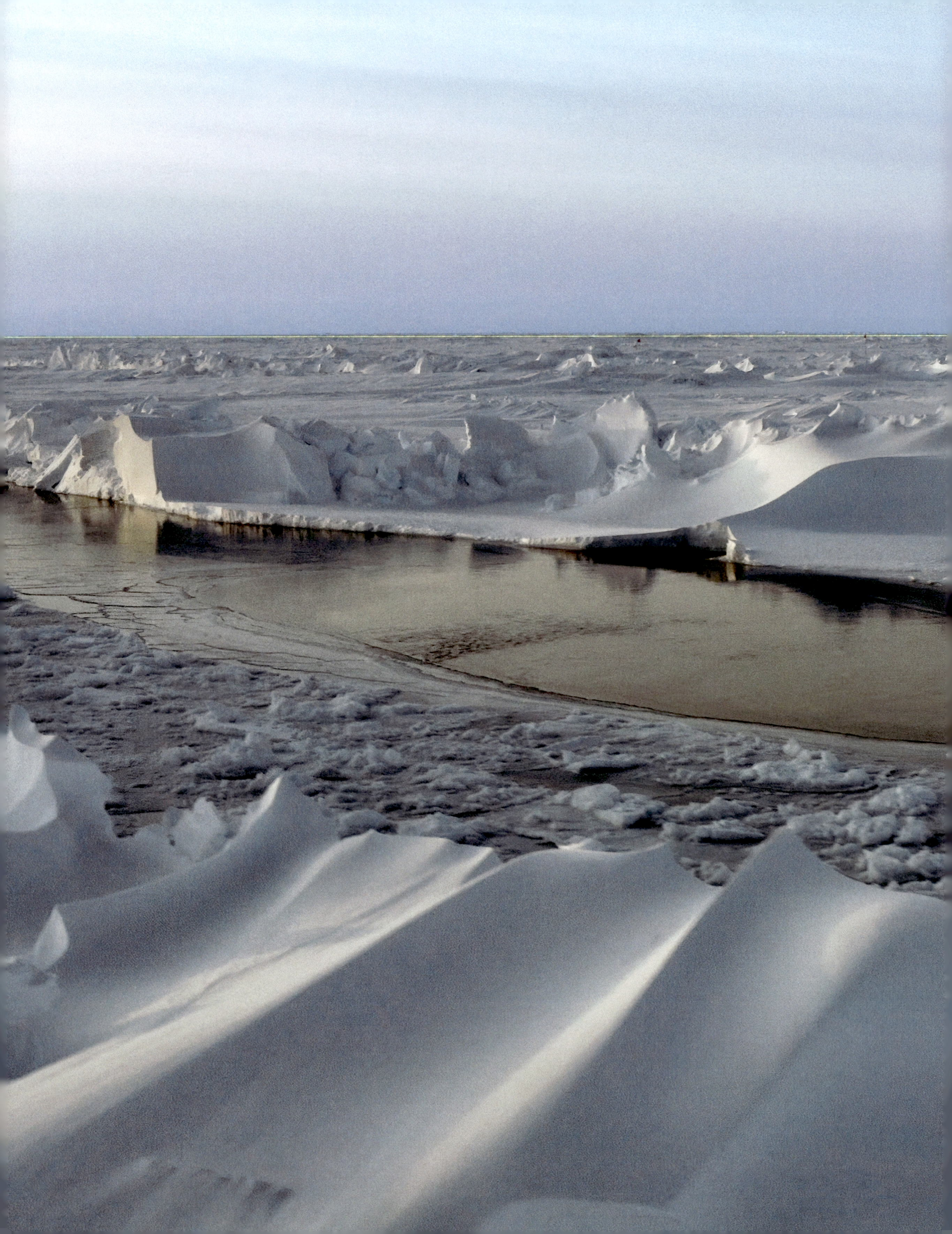

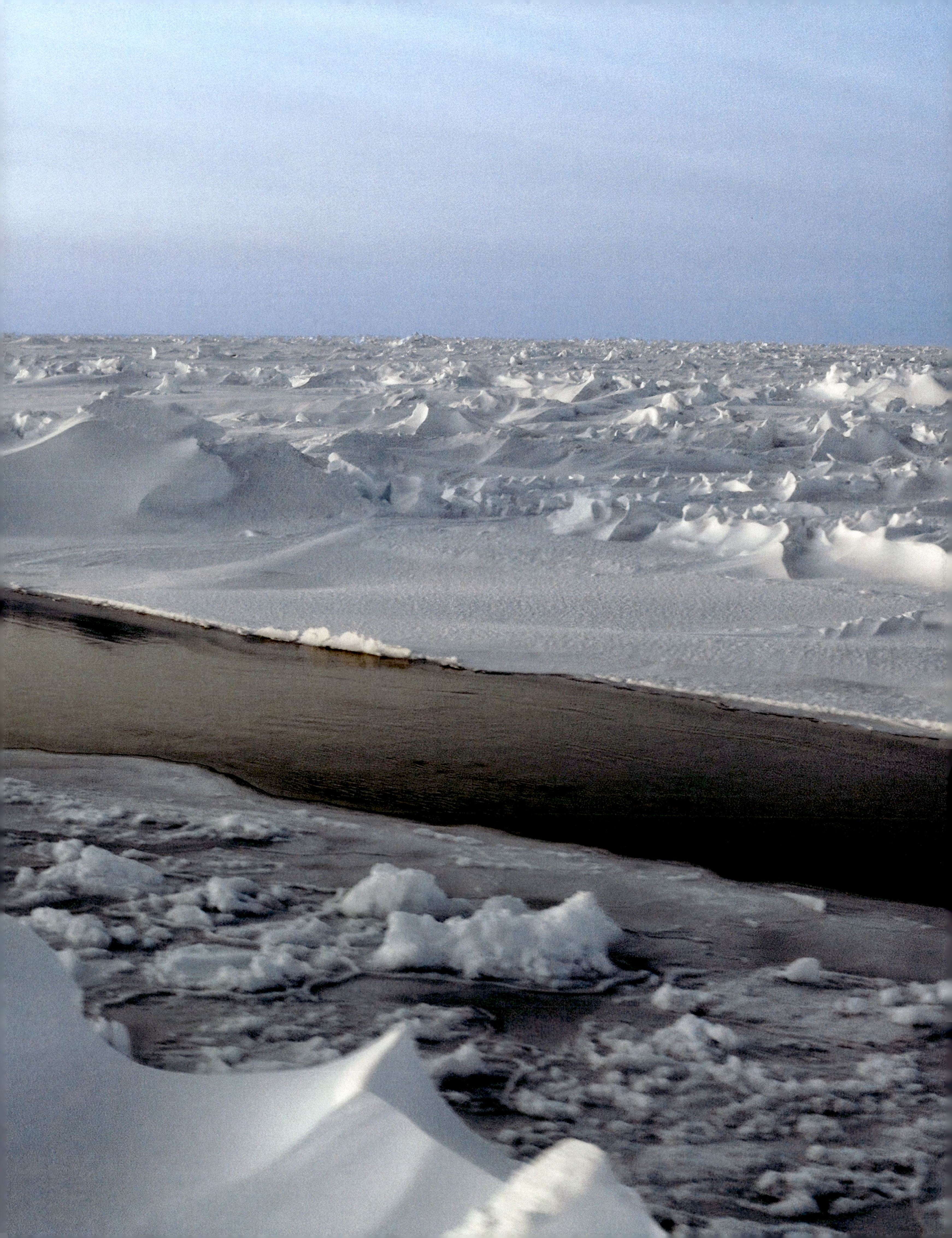

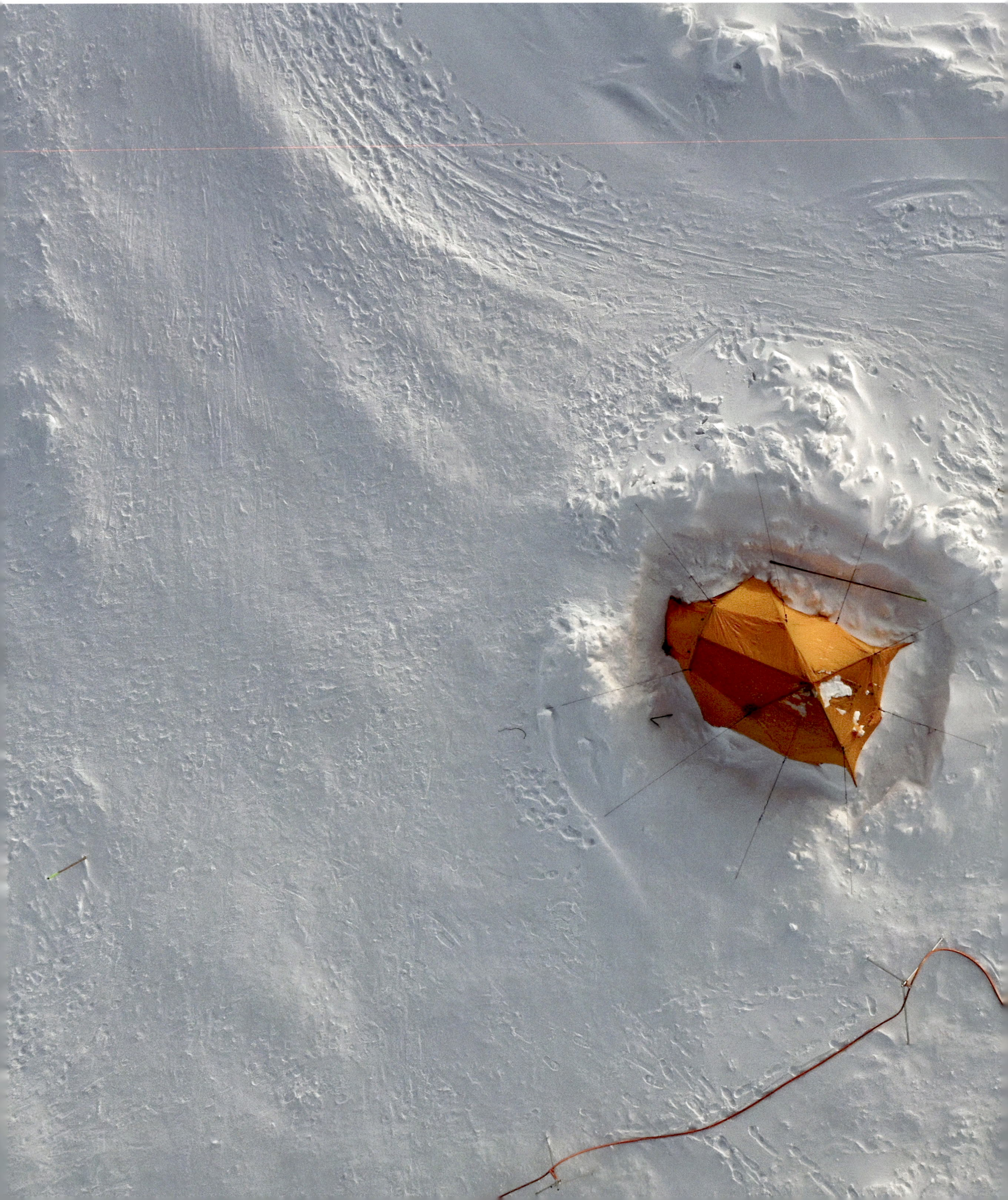

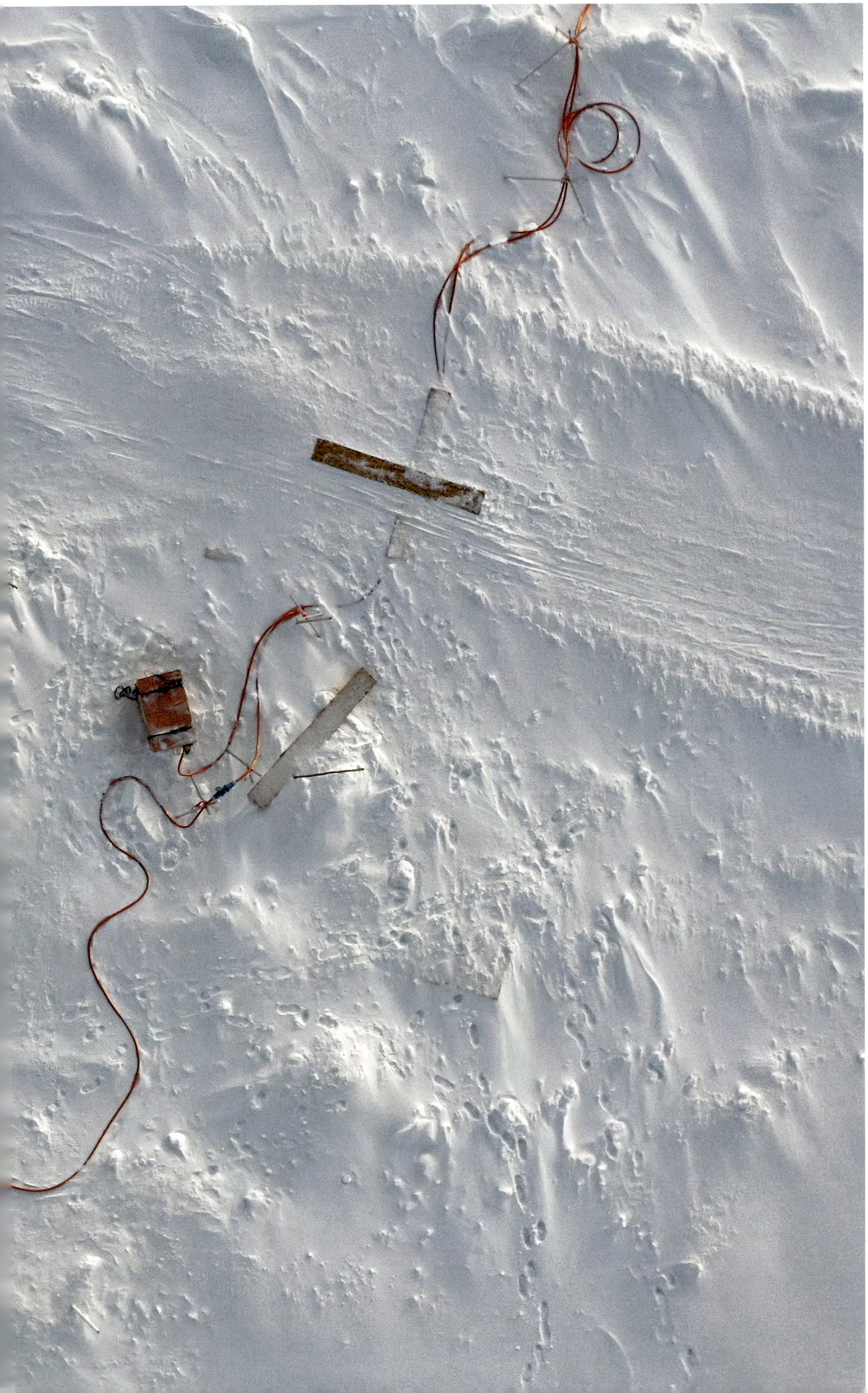

In winter there is unusually strong ozone depletion over the Arctic. The scientists of the MOSAiC expedition are the only ones able to quantify the full extent of this depletion taking place directly above their heads in the central Arctic.

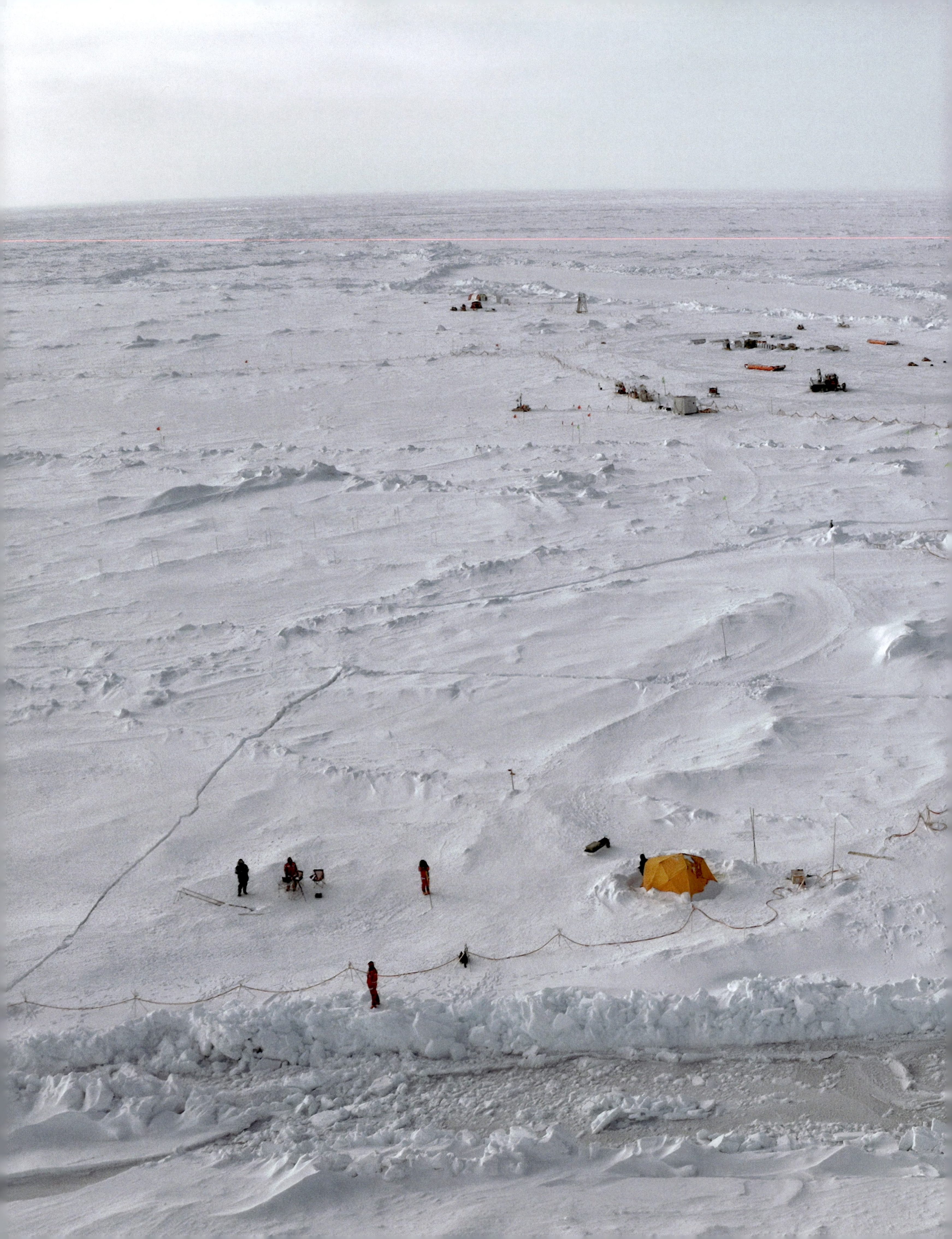

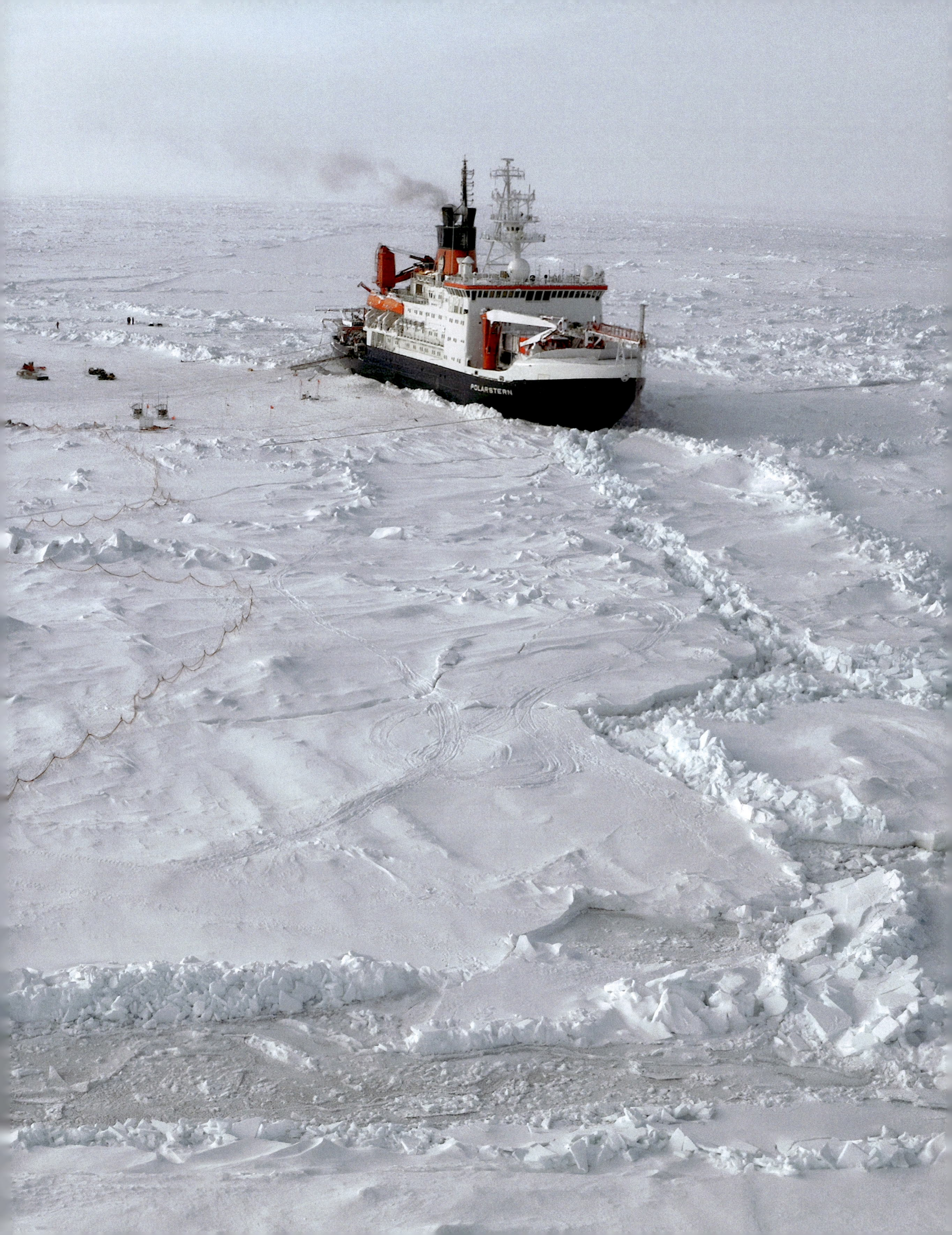
POLARSTERN

Every morning starts with a probing look out of the window: How did the floe change overnight? Over and over again leads both small and large form in the ice. At times, specific areas of the research camp are completely inaccessible. At the same time, every lead provides new opportunities for research, as these cracks in the ice play a major role in the exchange of energy and trace gases between the atmosphere, the sea ice and the ocean.

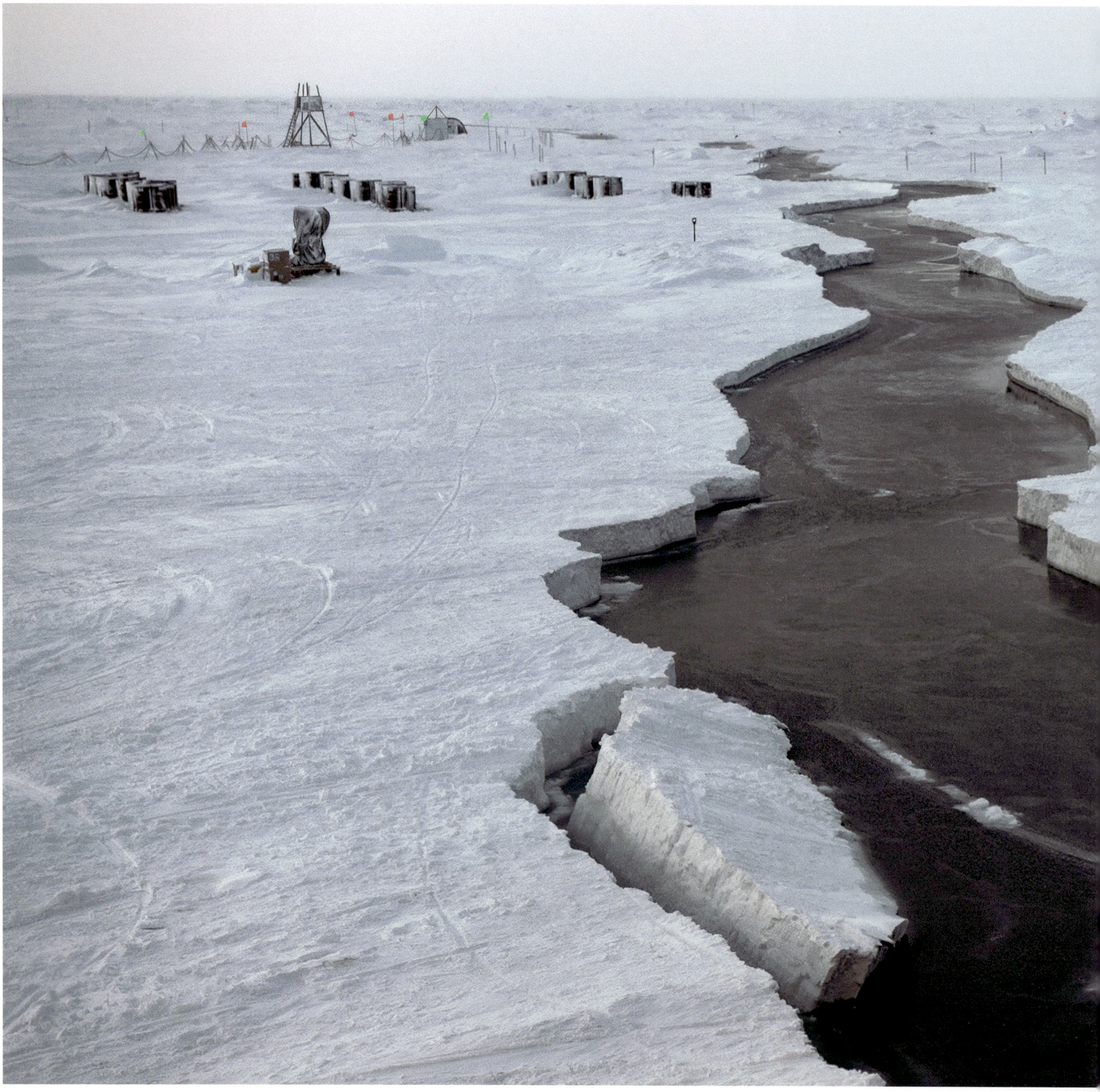

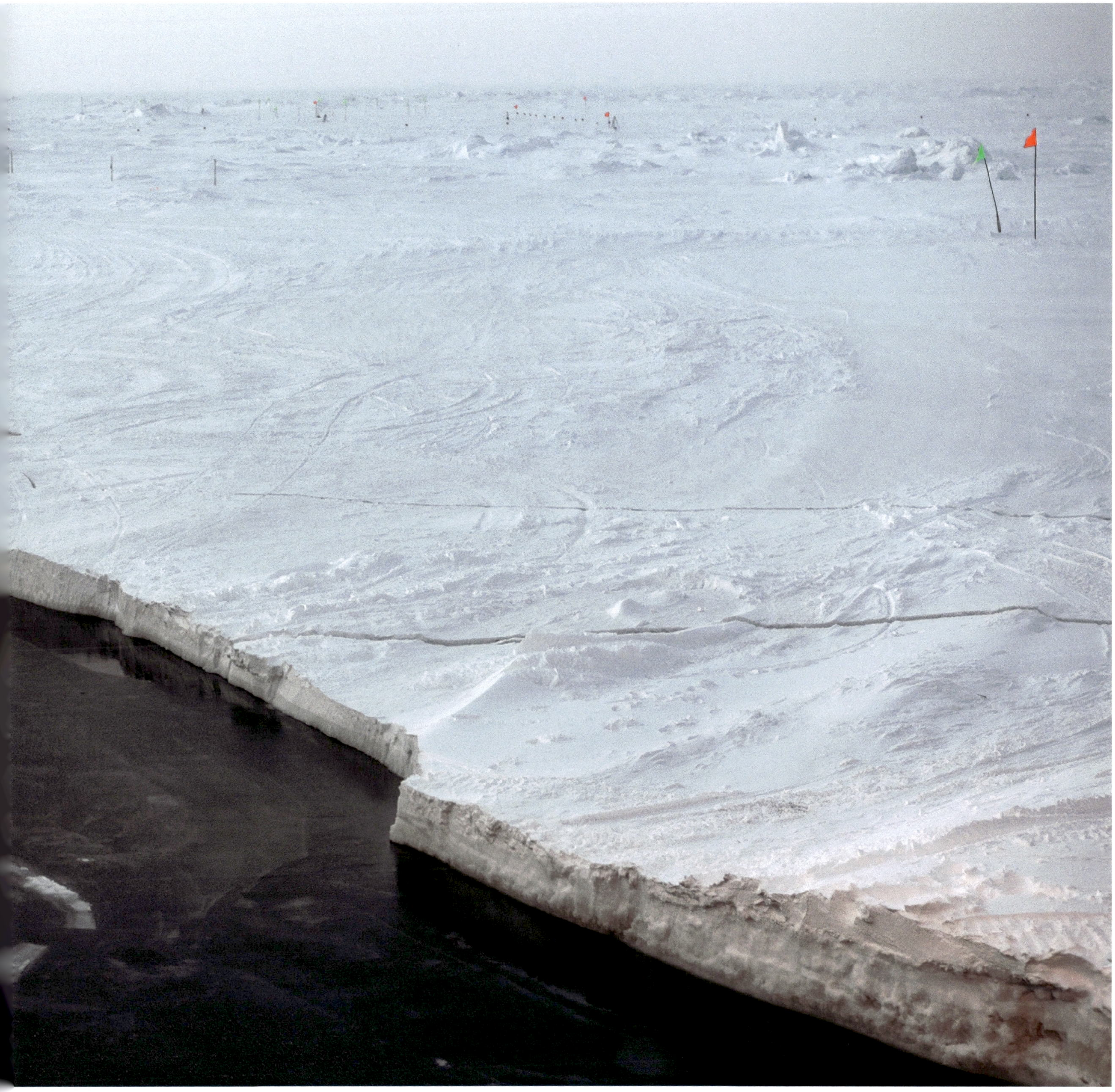

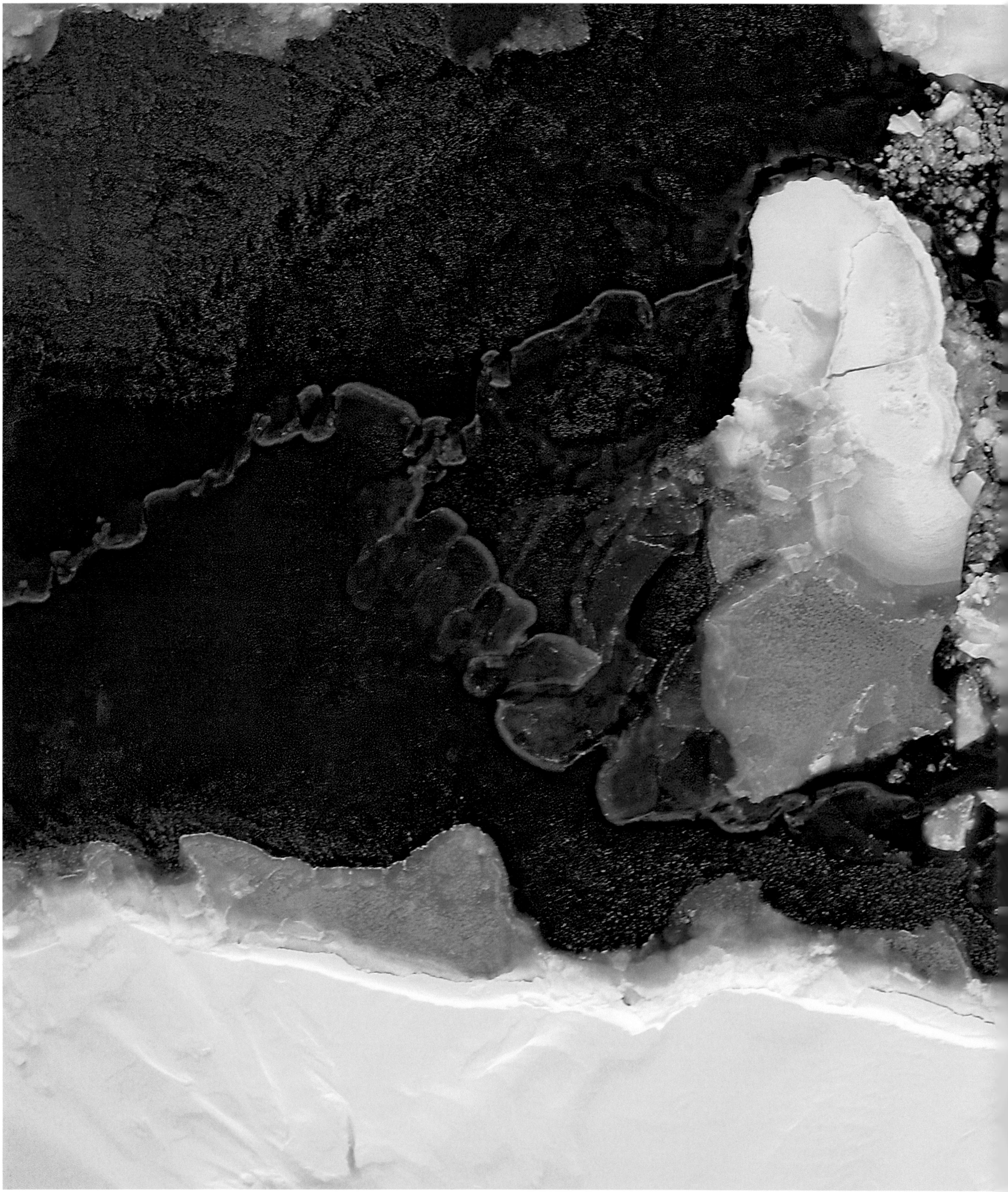

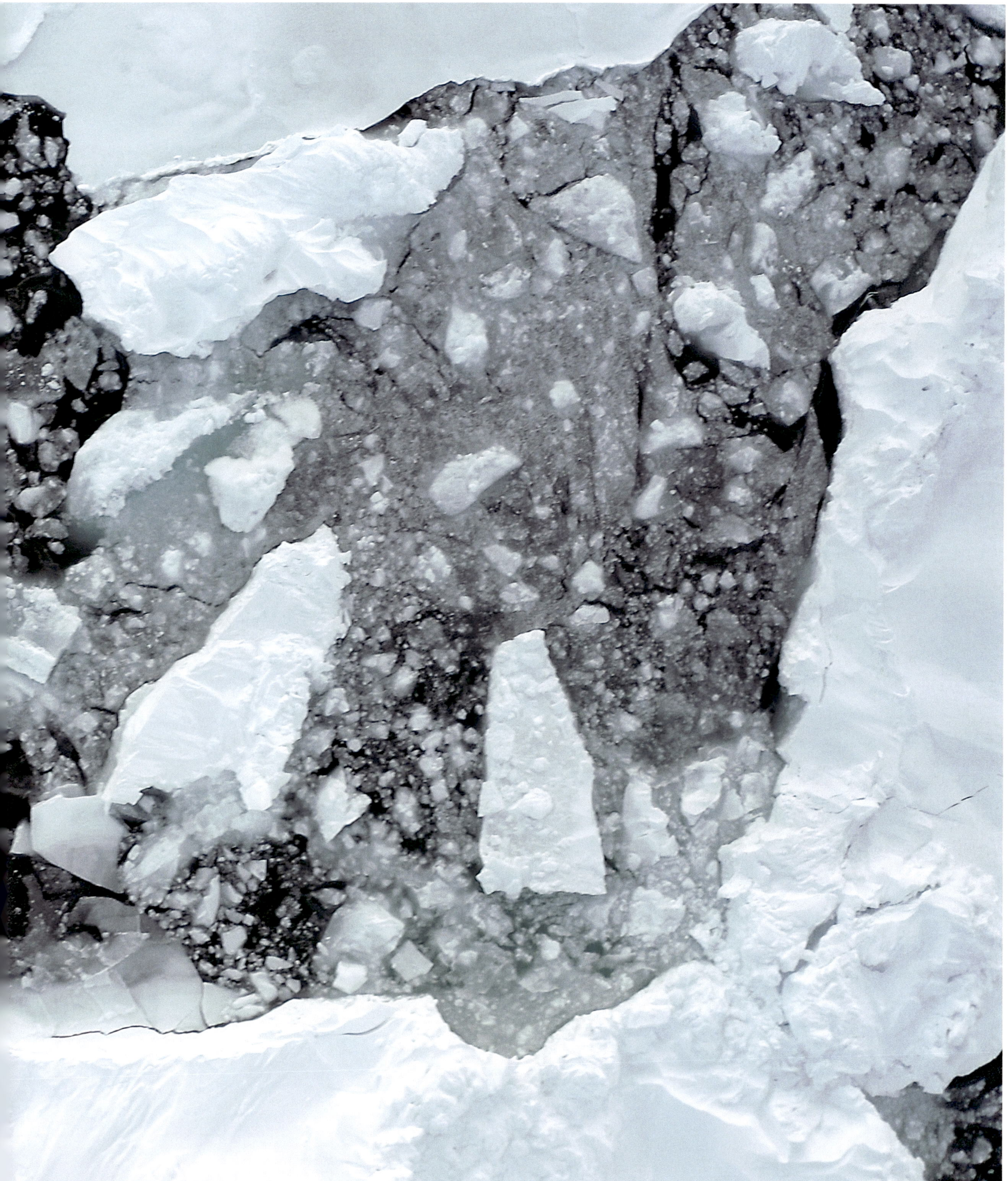

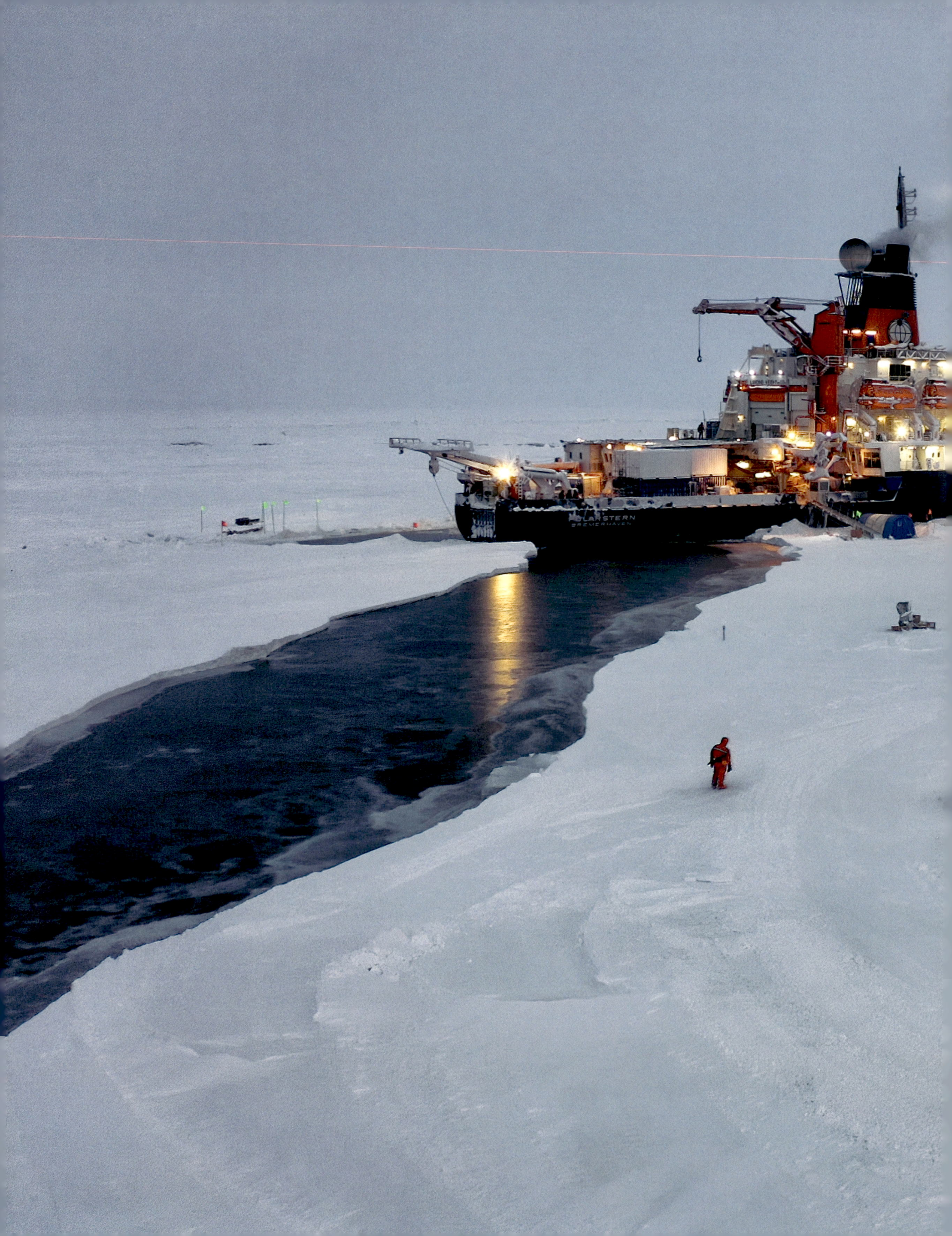

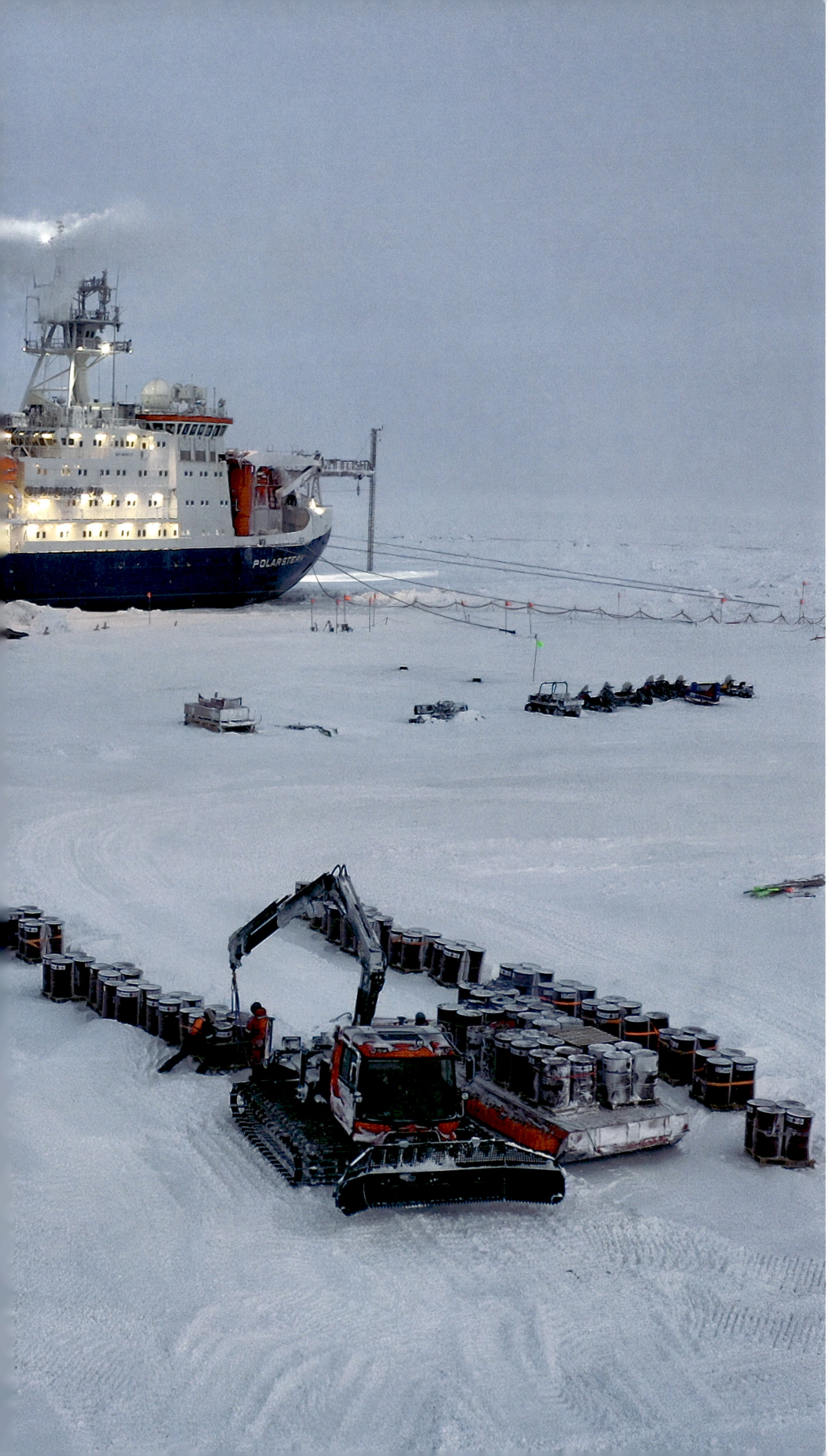

"For a long time, our floe was very stable, but for now that's over. Depending on the tides, we move in different directions. Many old fractures have opened again, and some parts of the camp are no longer accessible by the normal routes. This afternoon the ship moved so much as a result of ice pressure that for a while we could only go on to the floe with the crane. We are excited to see how things will progress. Tomorrow we are expecting a new storm."

Stefanie Arndt,
scientist

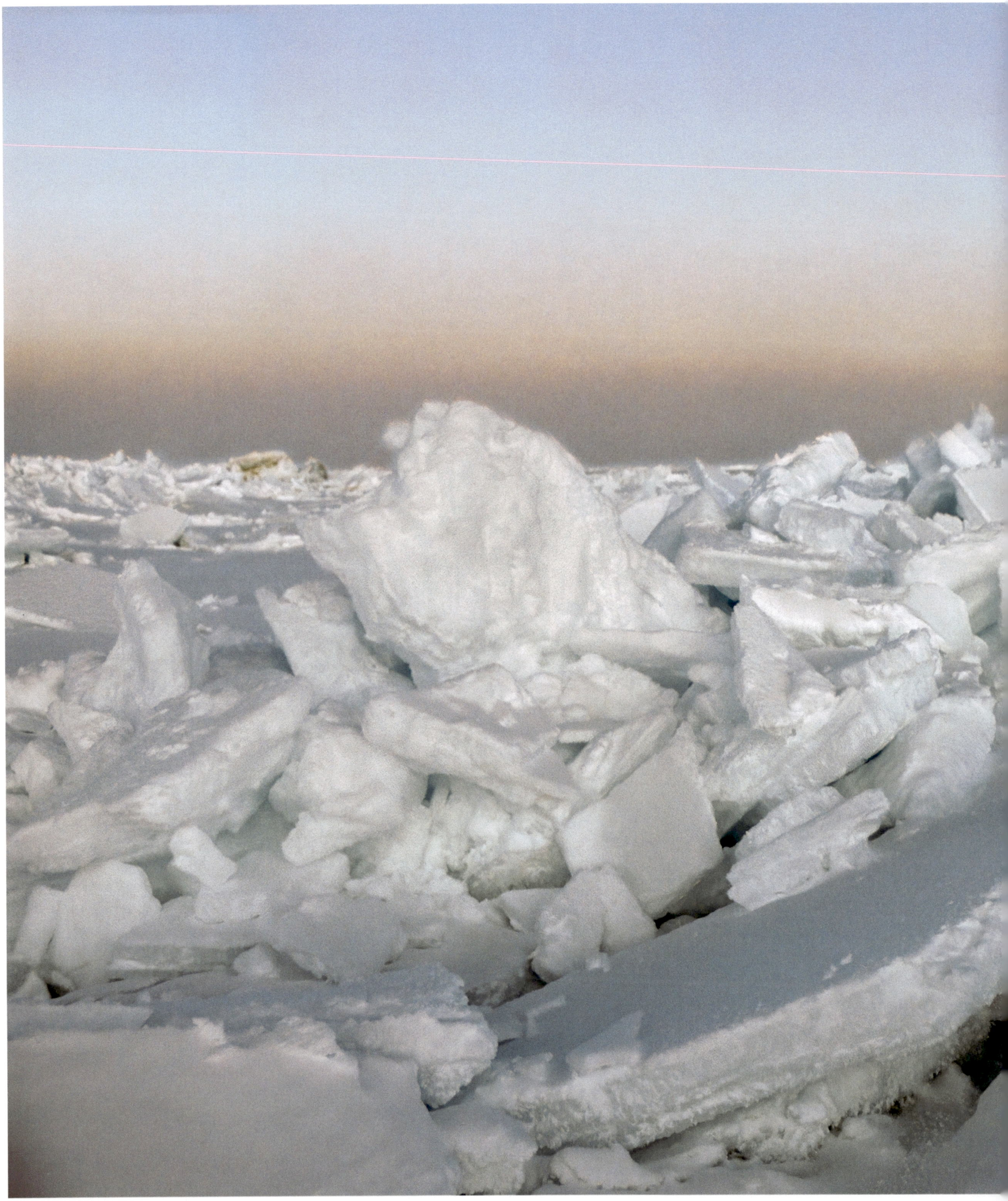

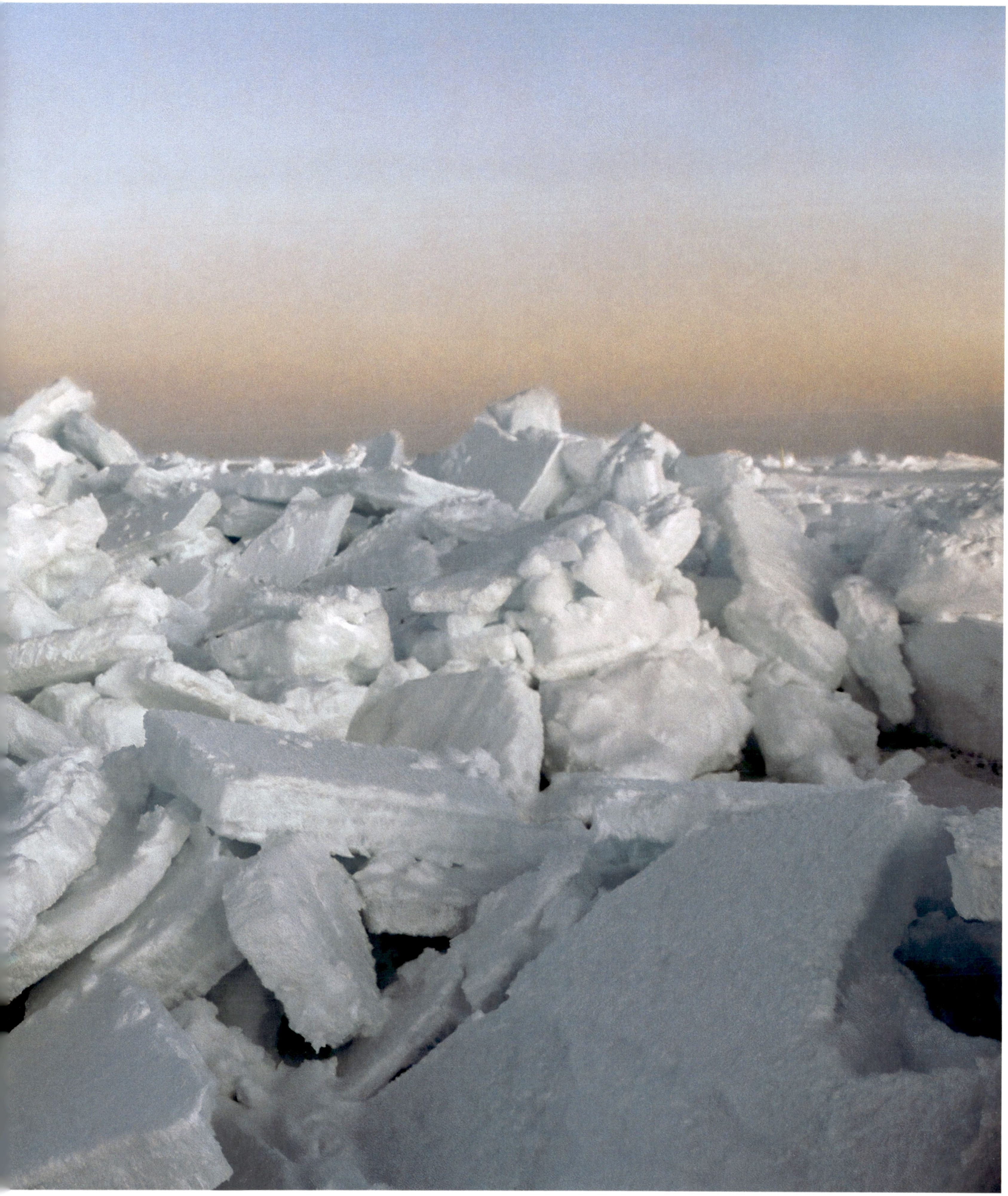

In a short time, the sea ice can pile up into metres-high ridges of compressed ice and swallow up the structures of the research camp. Instead of scientific work, everyone spends the day cleaning up. Over the course of the expedition year, time and again the sea ice demonstrates its spontaneous dynamism, making it clear that the scientists are only guests out here – and are travelling on their ship along with the forces of nature.

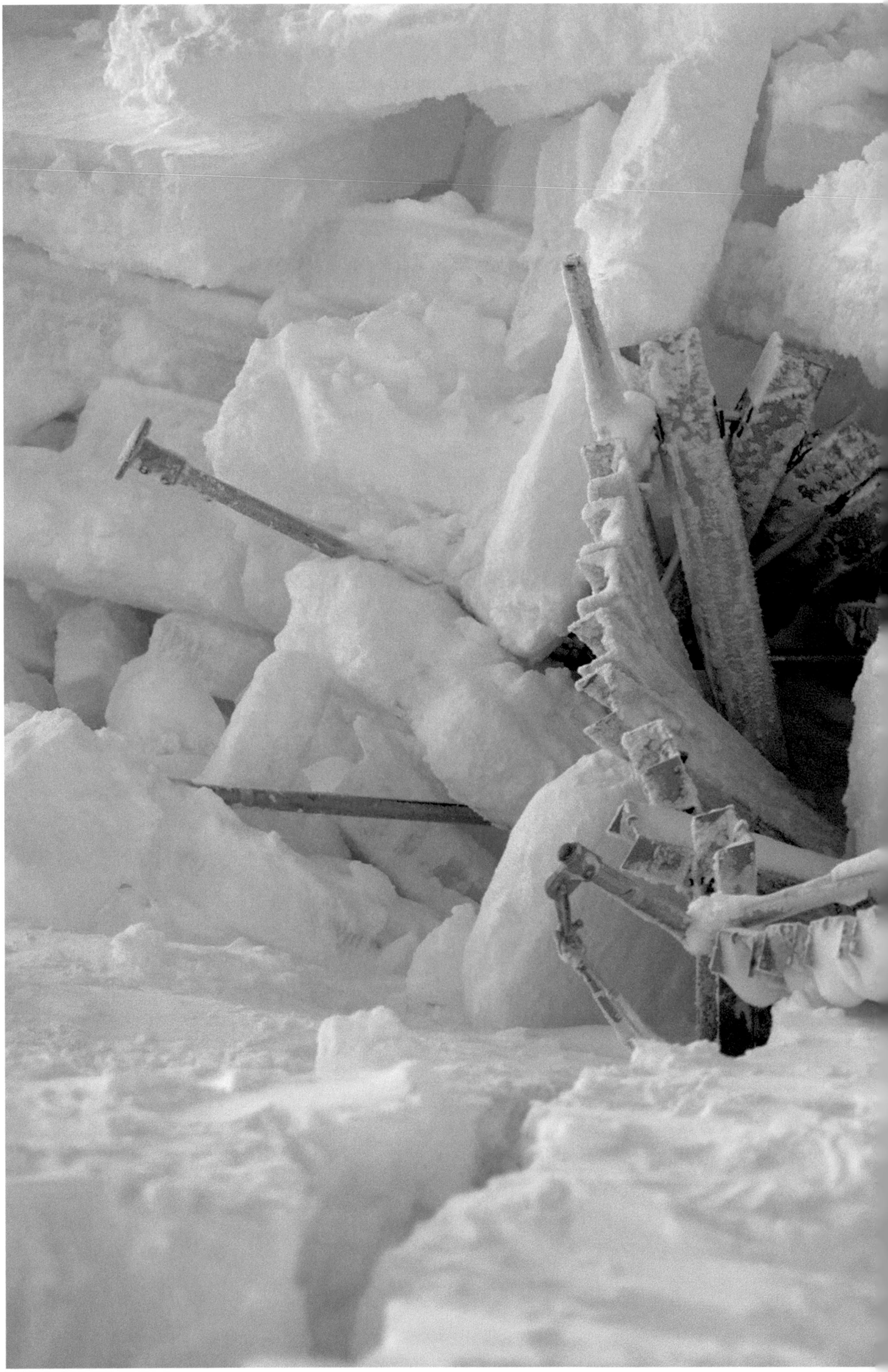

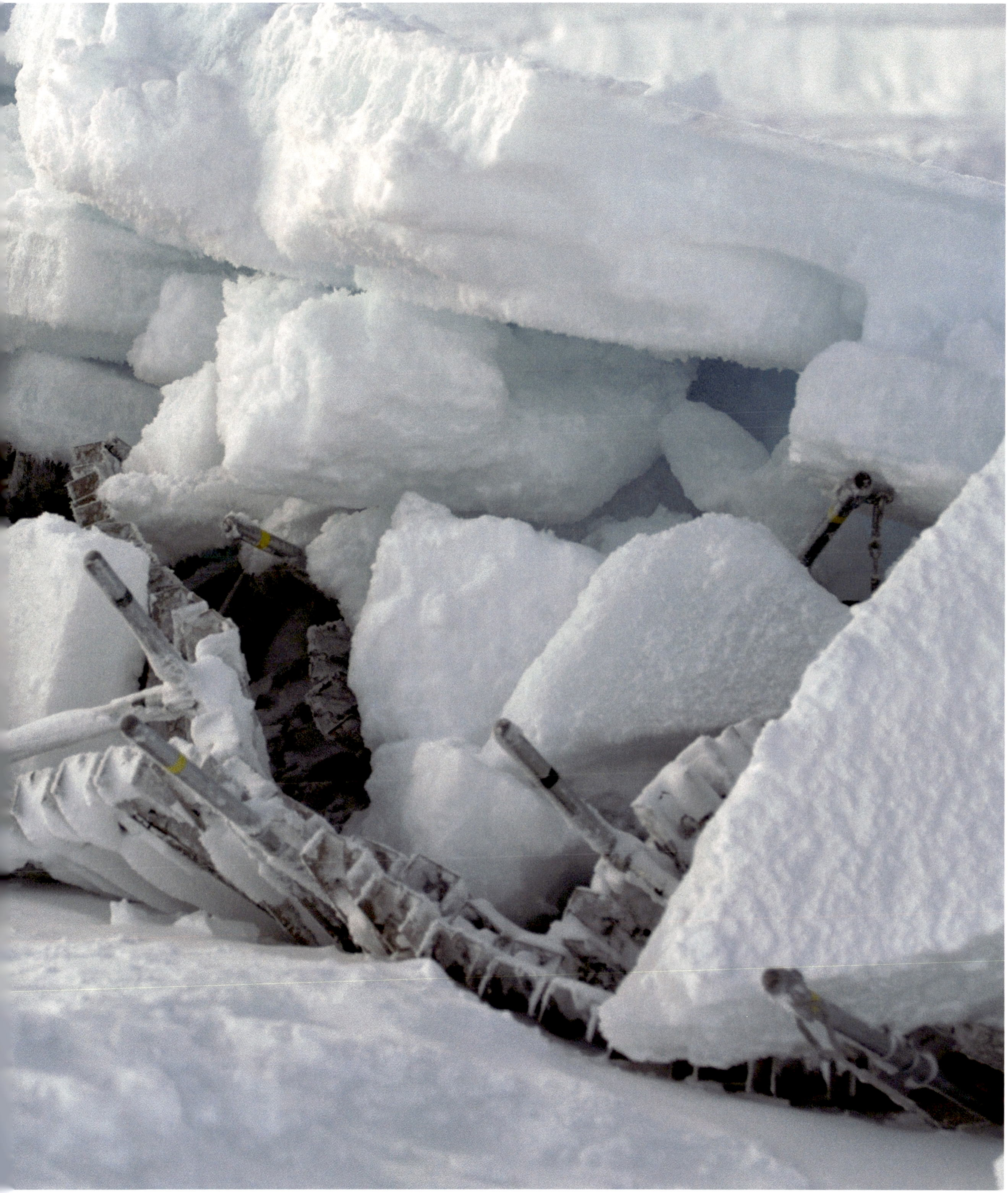

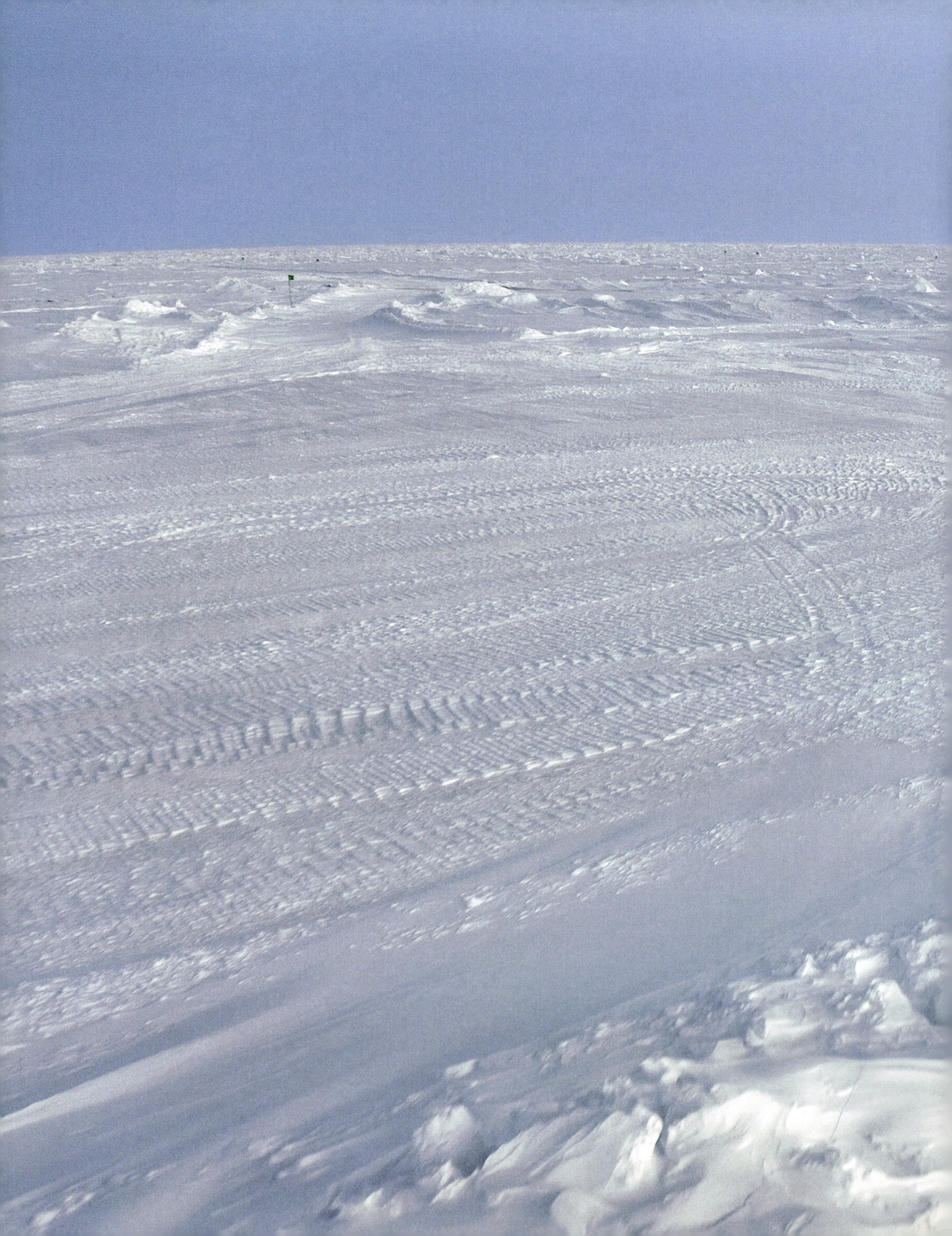

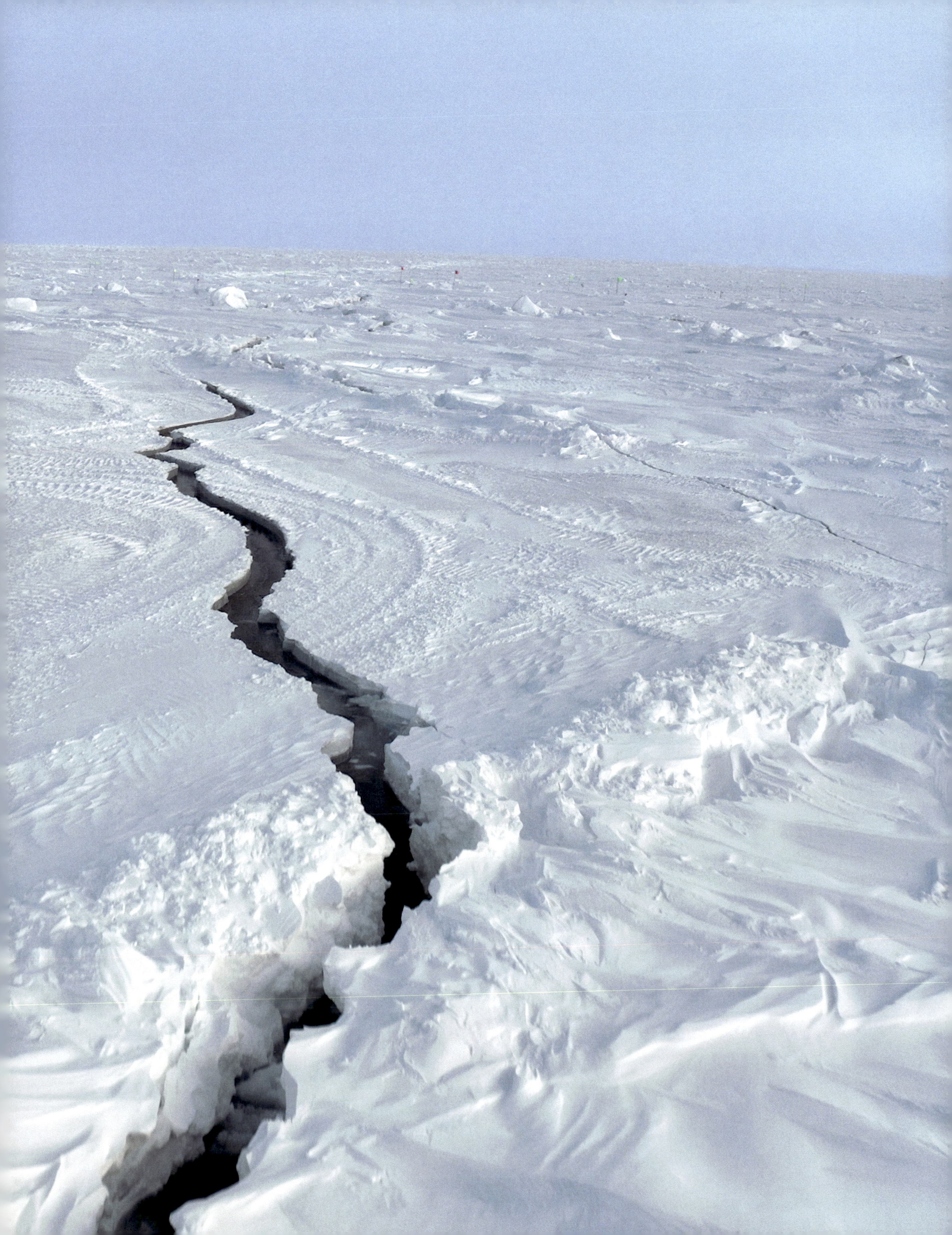

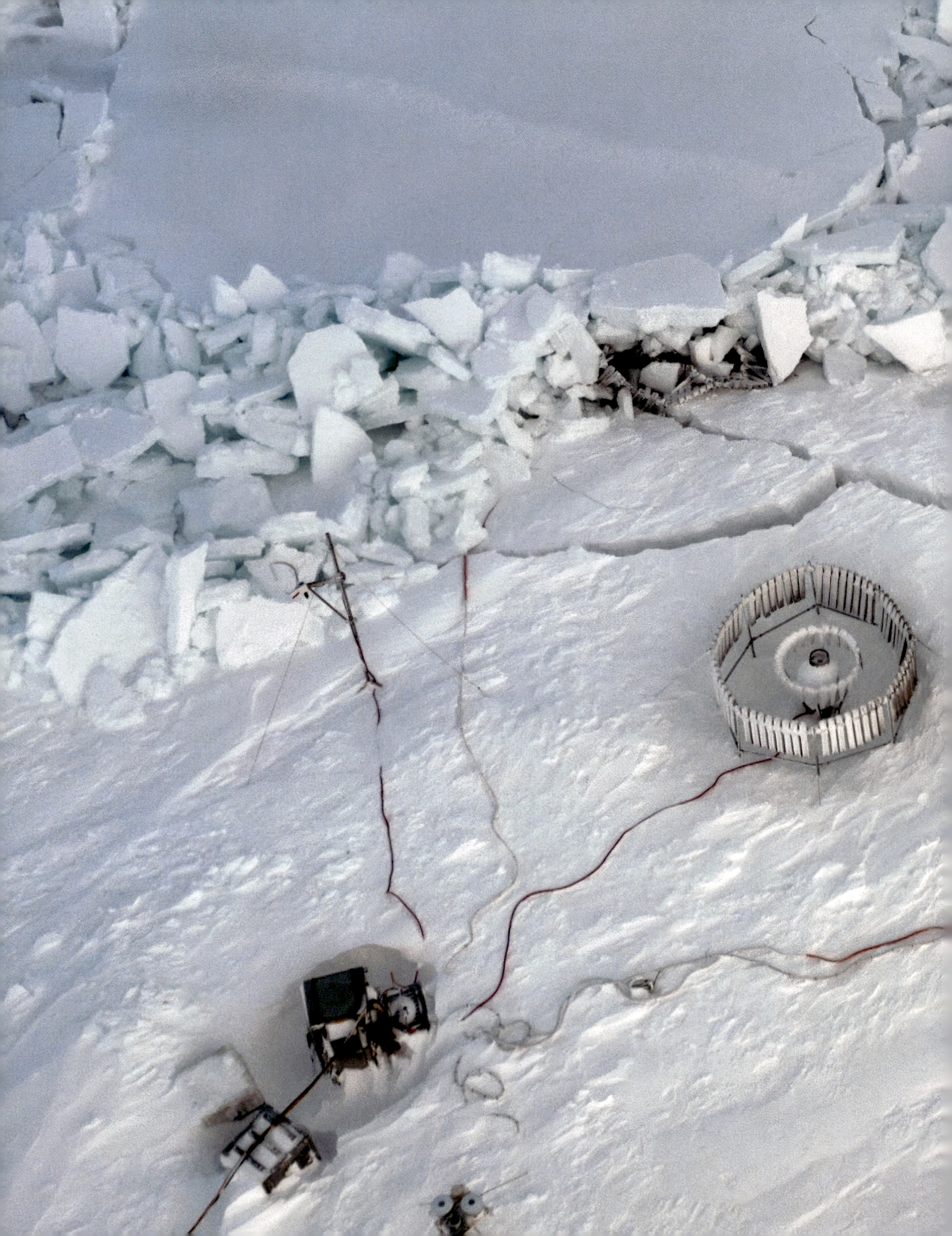

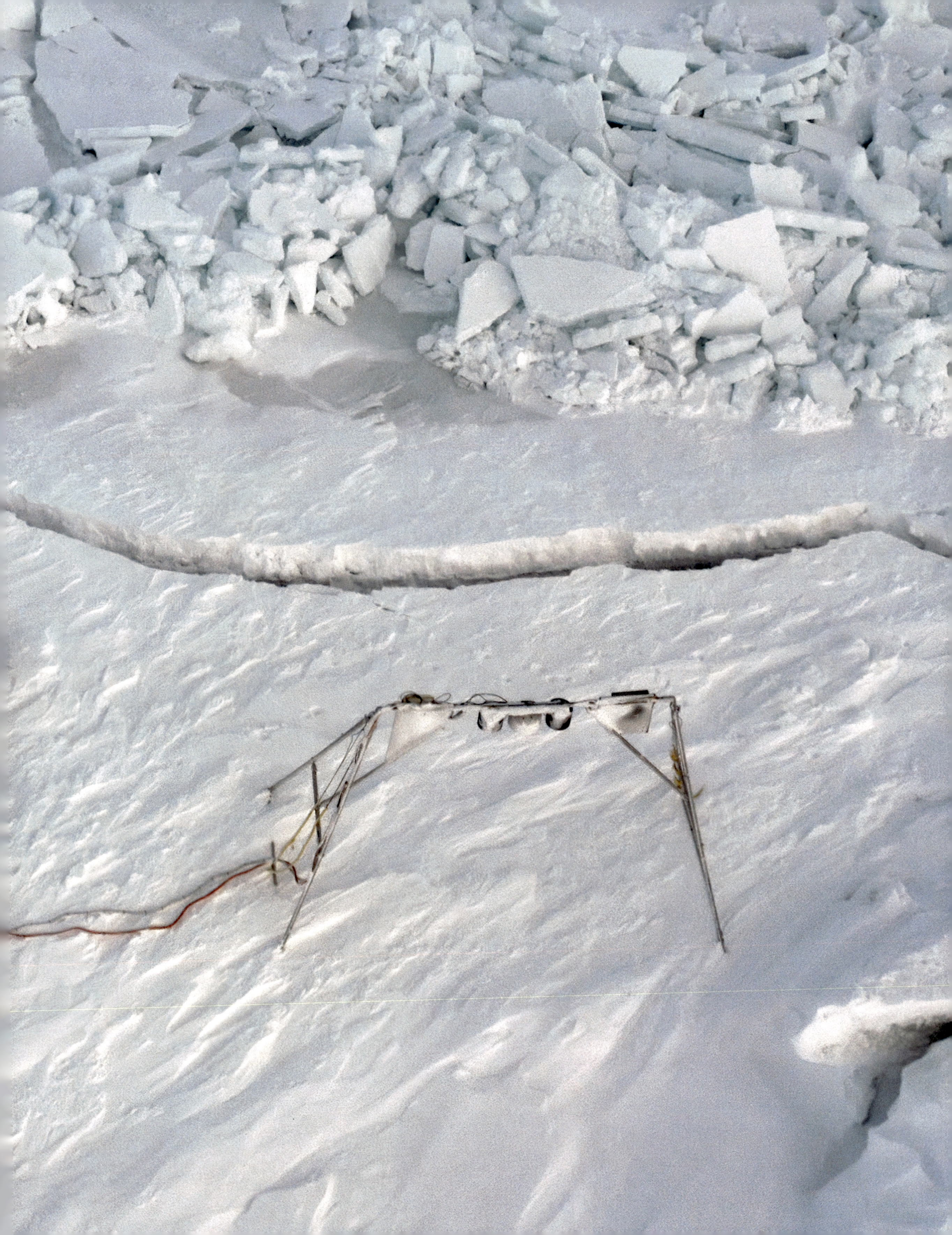

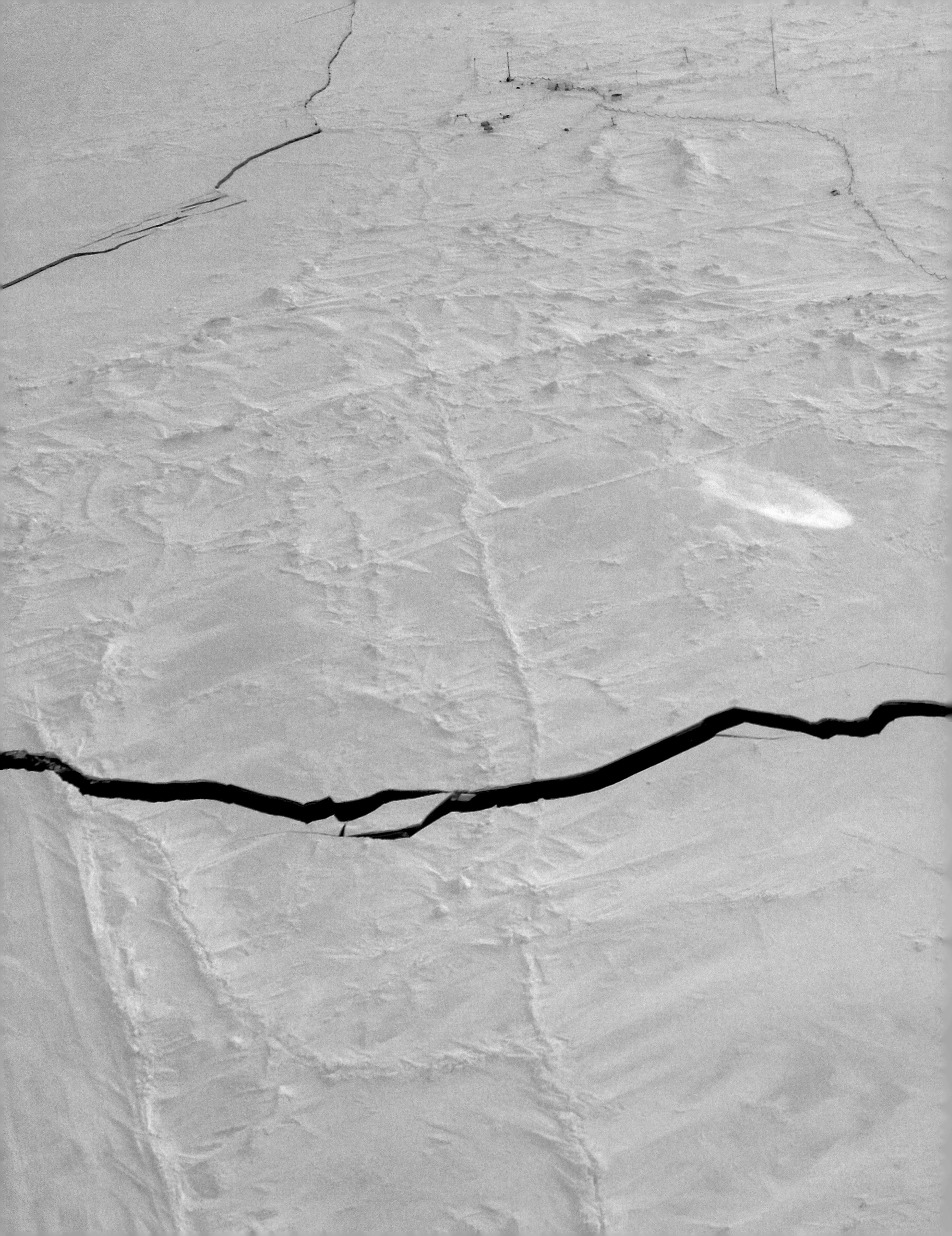

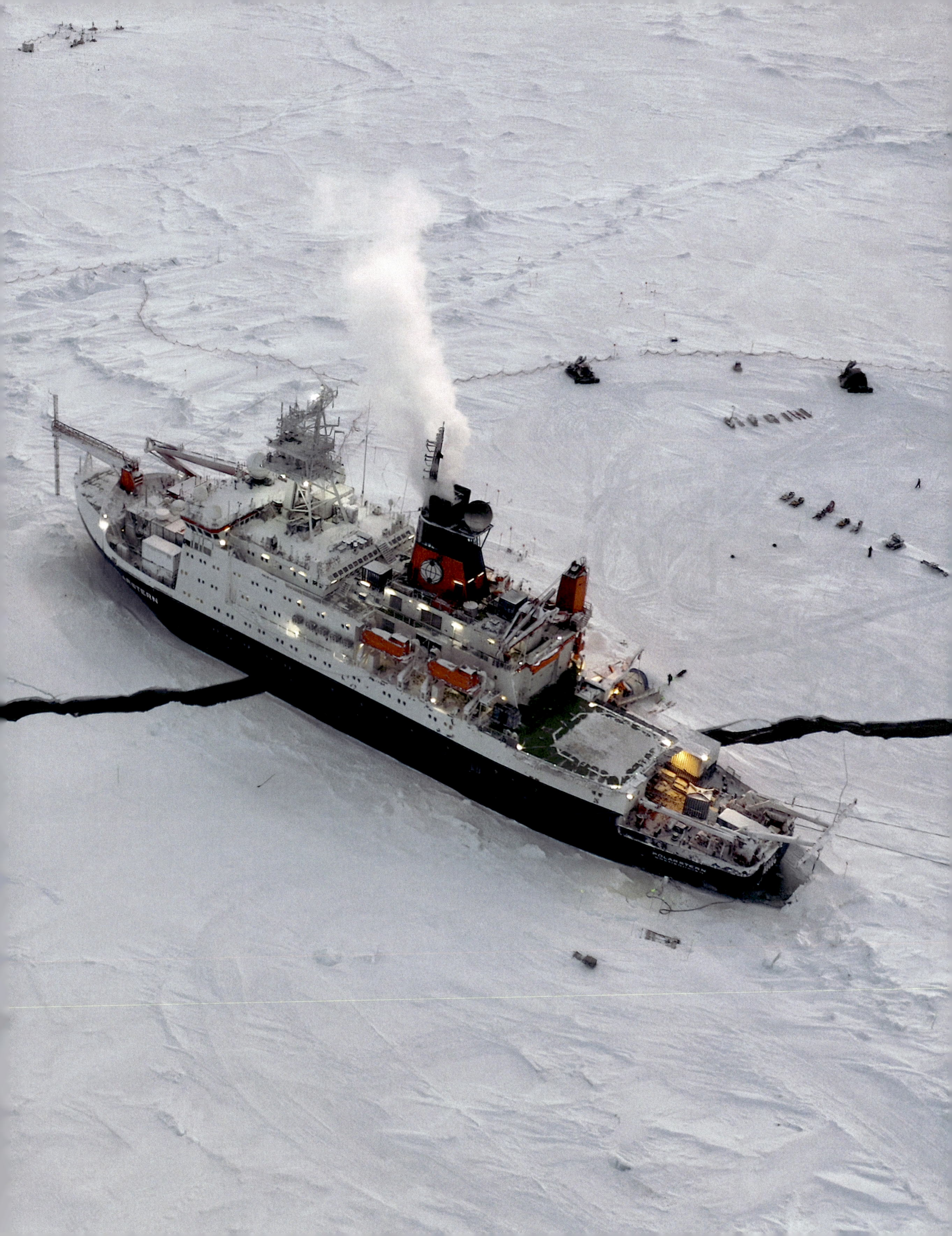
POLARSTERN

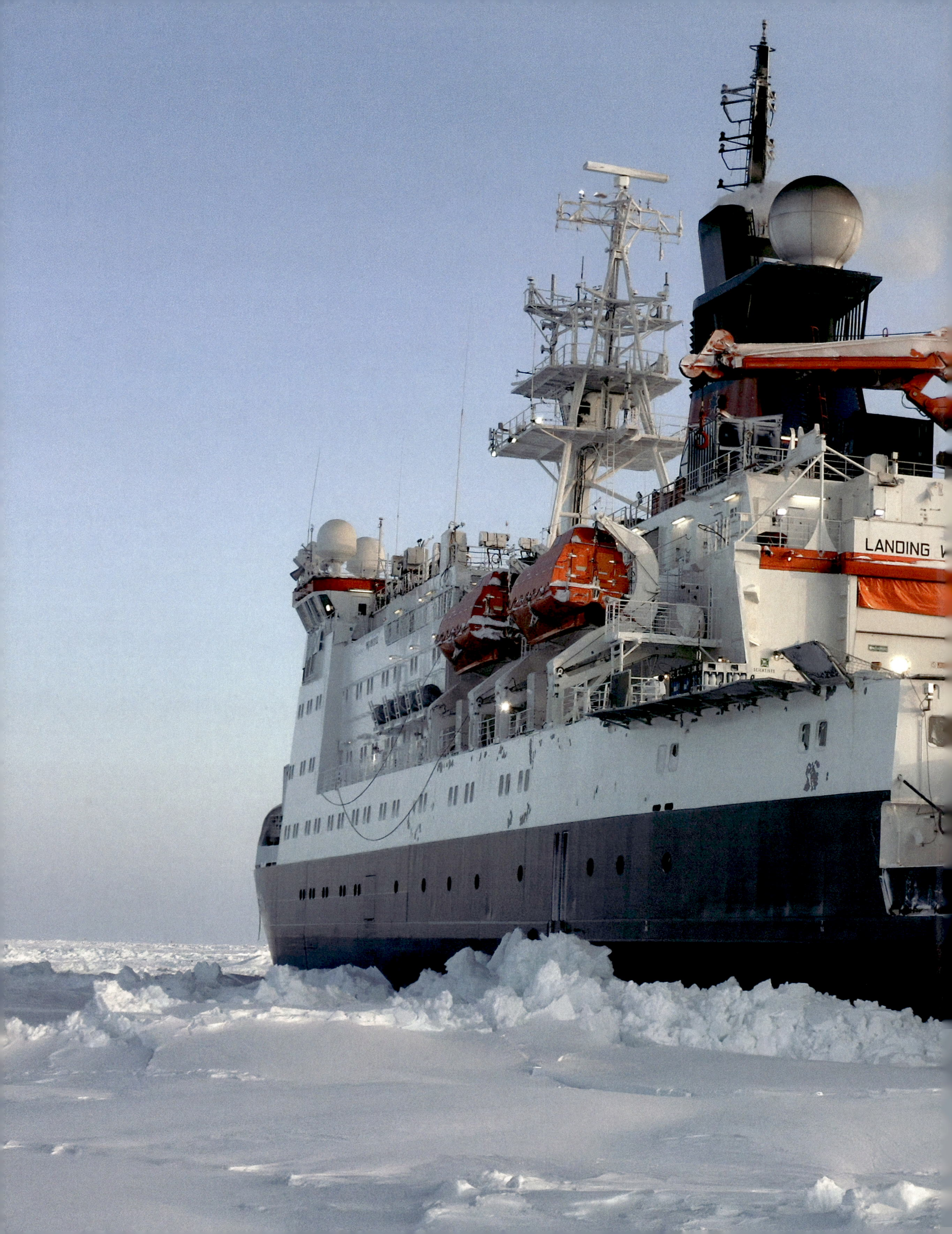
LANDING W

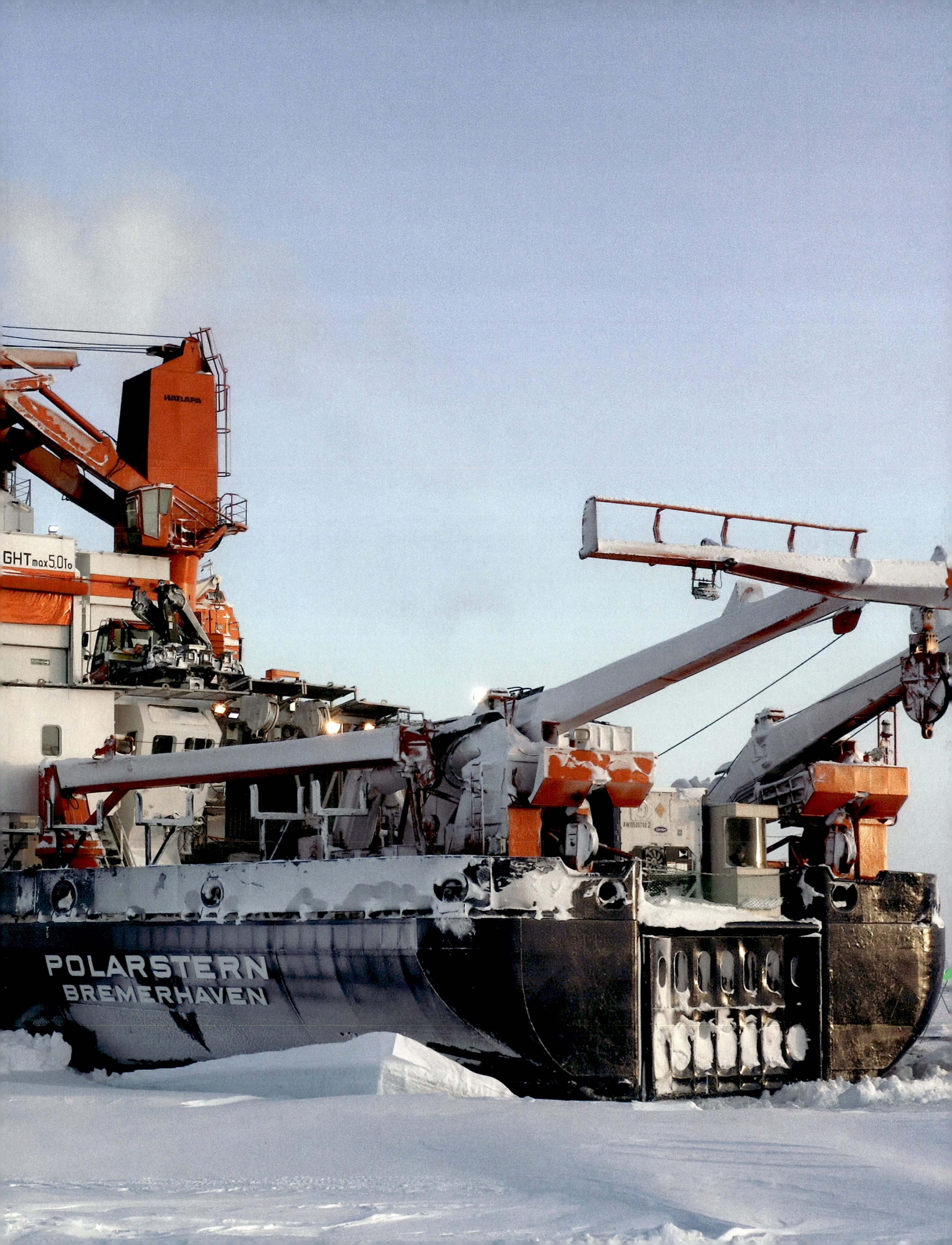
HATLAPA
GHT max 5,0 To
POLARSTERN
BREMERHAVEN

This book was only possible with the support of numerous individuals. Without their help, the images of and texts about this unique research project could not been realized. We would like to thank Prof. Dr. Antje Boetius, Dr. Karsten Wurr, Dr. Uwe Nixdorf and Prof. Dr. Markus Rex for giving us the opportunity to accompany a part of the MOSAiC expedition.

We thank Dr. Marcel Nicolaus for advice on the content and for his critical, sharp-eyed reading of the manuscript. We are grateful to Dr. Ellen Damm, Dr. Hauke Flores, Dr. Clara Hoppe, Dr. Benjamin Rabe, Dr. Christian Salewski, Dr. Anja Sommerfeld, Prof. Emeritus Timothy Stanton and Prof. Dr. Dieter Wolf-Gladrow for their advice on the content of the main themes. Dr. Folke Mehrtens and Dr. Stefanie Arndt not only provided us with invaluable advice but also kindly made available parts of the official on-board log. We are especially thankful to Isabelle Erler and Florian Frohnholzer for their meticulous and patient work on the editing and the design of the book. A special thanks to Sabine Winterstern for assisting the photographic project and its editing.

We thank our families and friends who remained close to us during our months-long journey to the end of the world. But above all we would like to thank all the members of the expedition, both crew and researchers, for letting us take part in their everyday life on the expedition. This book is dedicated to them.

Esther Horvath, Sebastian Grote and Katharina Weiss-Tuider, June 2020

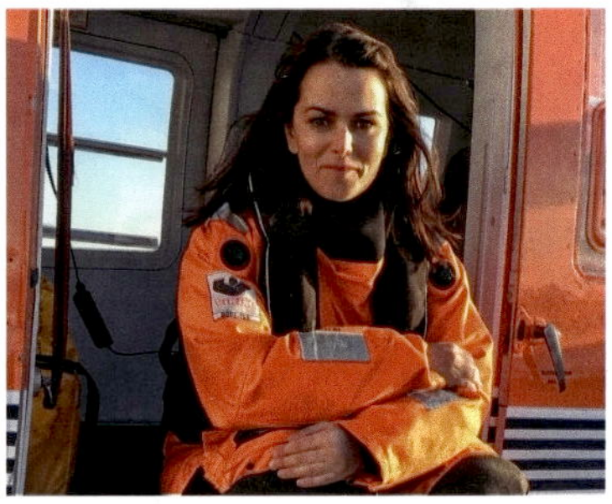

Esther Horvath is a documentary photographer of the Alfred Wegener Institute, a member of the International League of Conservation Photographers (iLCP), The Photo Society – a collective of National Geographic photographers – and The Explorers Club.

Horvath was born in Hungary, where she received her master's degree in economics at the University of West Hungary. In 2012 her passion for photography took her to New York City, where she studied at the International Center of Photography (ICP), graduating in photojournalism and documentary photography. She lived in New York City for six years before moving to the German city of Bremen in 2018. Since 2015, Horvath has dedicated her photography to the polar regions, especially the Arctic Ocean.

Horvath's photographs have appeared in, among other publications, *National Geographic*, the *New York Times*, the *Wall Street Journal*, *Time*, the *Audubon* and *GEO*.
In 2020 she won first prize in the Environment category of the World Press Photo Awards with an image from the MOSAiC expedition.

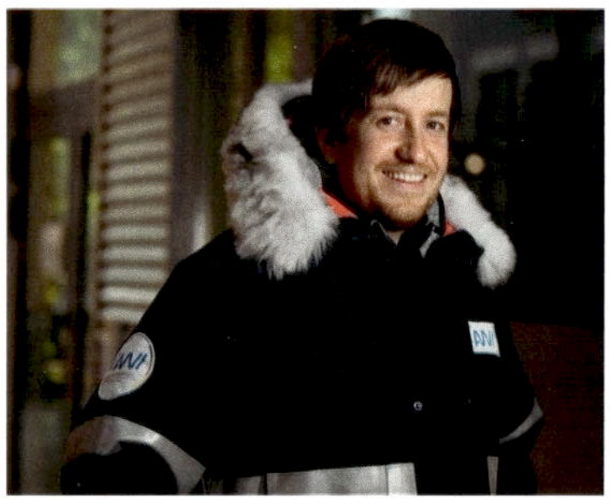

Sebastian Grote studied history and Scandinavian studies at the universities of Cologne, Helsinki and Uppsala. An internship in the press department at the University of Cologne led him to science journalism. After moving to the Communications and Media Relations department of the Alfred Wegener Institute, Helmholtz Centre for Polar and Marine Research, Grote has taken part in expeditions to both the Arctic and Antarctic regions. He reported from the *Polarstern* during the MOSAiC launch phase.

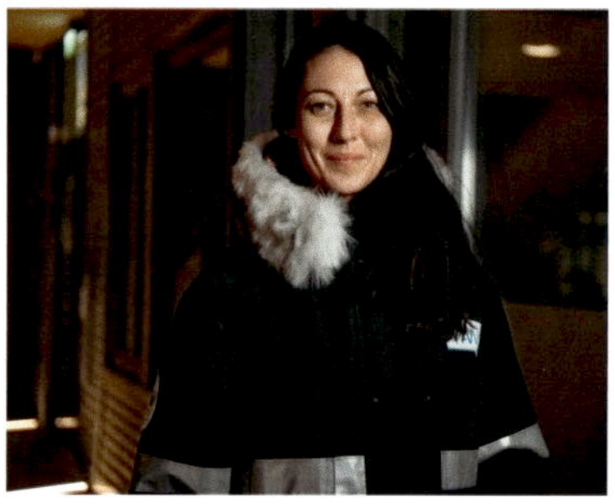

Katharina Weiss-Tuider works at the Alfred Wegener Institute, Helmholtz Centre for Polar and Marine Research in Potsdam as the communication and media manager for the MOSAiC expedition. During the expedition she coordinated the media work on board. She earned her PhD in Literature at the Ludwig Maximilian University of Munich and has several years of experience as a freelance author writing on topics such as environmental and climate protection as well as agriculture and nutrition. Katharina Weiss-Tuider lives in Berlin with her husband.

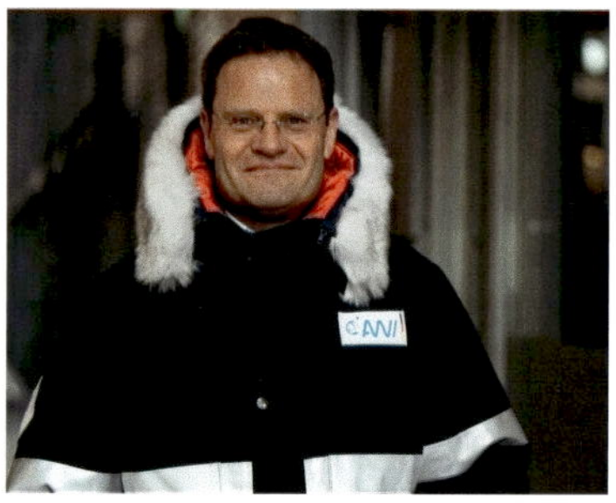

Markus Rex is the head of atmospheric research at the Alfred Wegener Institute, Helmholtz Centre for Polar and Marine Research, and he is professor for atmospheric physics at the University of Potsdam. He had already joined numerous expeditions to the Arctic, Antarctica and other remote regions of the world to research the complex climatic processes that lead to dramatic climate changes. He heads the MOSAiC project, a unique research collaboration of ninety institutions from twenty countries.

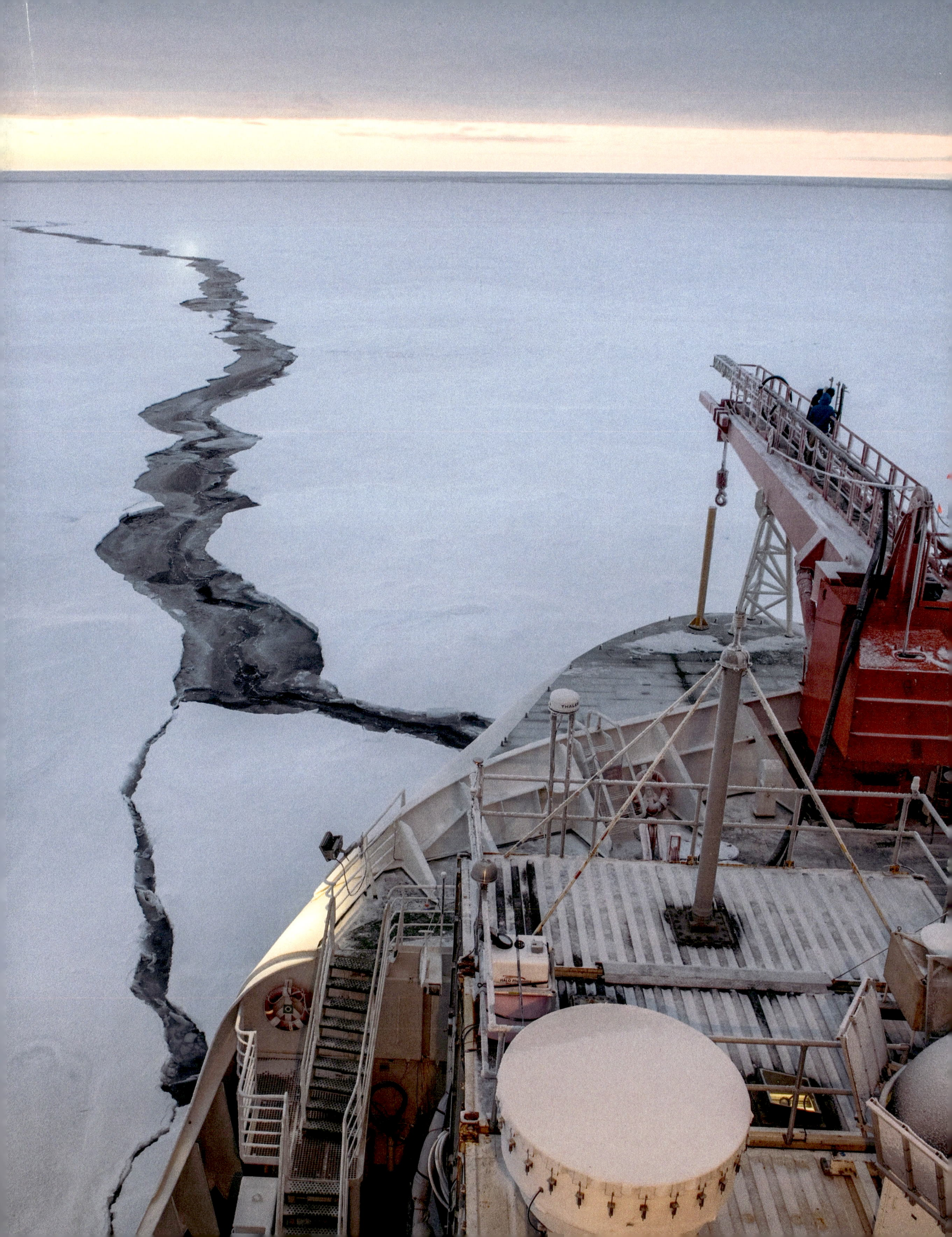

A member of Verlagsgruppe Random House GmbH
Neumarkter Straße 28 · 81673 Munich

The quotes on pages 74–77 are from Fridtjof Nansen's *Farthest North: Being the Record of a Voyage of Exploration of the Ship "Fram" 1893–96*, 2 vols., Westminster, 1897.

Library of Congress Control Number is available; British Library Cataloguing-in-Publication Data: a catalogue record for this book is available from the British Library.

Editorial direction: Curt Holtz
Editorial assistance: Josephine Fehrenz
Translation from German: John Sykes
Copy-editing: Jonathan Fox
Design, layout and typesetting: Sofarobotnik, Augsburg and Munich
Production management: Corinna Pickart
Separations: Reproline Mediateam, Unterföhring
Printing and binding: Grafisches Centrum Cuno, Calbe
Paper: Condat matt Périgord

FSC www.fsc.org MIX Paper from responsible sources FSC® C043106

Verlagsgruppe Random House FSC® N001967

Printed in Germany
ISBN 978-3-7913-8670-6
www.prestel.com

UFA SHOW & FACTUAL PRESTEL